REFRIGERATION
IN AMERICA

A HISTORY OF A NEW TECHNOLOGY
AND ITS IMPACT

43340

REFRIGERATION IN AMERICA

A HISTORY OF A NEW TECHNOLOGY

AND ITS IMPACT

BY OSCAR EDWARD ANDERSON, JR.

KENNIKAT PRESS
Port Washington, N. Y./London

REFRIGERATION IN AMERICA

TO MY FATHER AND MOTHER

PREFACE

THE study of refrigeration in the United States might be undertaken from any one of several possible approaches. A sociologist, for example, might make a survey in search of the basic principles governing invention and technological change. As a student of American history I have been concerned primarily with the task of showing the relation of refrigeration to our national development. I have sought to record the main trends in technological progress, to describe the uses of refrigeration, to explain resistance to its application, and to give some indication of its social effects. This is an introductory survey; many tasks remain. Of these, detailed analyses of the technical evolution of the domestic refrigerating machine, of air conditioning, and of quick freezing should have high priority. They should be supplemented by case studies of the role of the entrepreneur in the introduction of new devices, methods, and products.

This is an age of cooperative scholarship; no study of any complexity is possible without the aid of a great many individuals and institutions. I have tried to give full credit in my notes to those whose research and ideas have been incorporated in the text, but this does not indicate the full nature of my debt.

Most I owe to Frederick Merk of Harvard University. His keen criticism and advice he placed at my disposal at all stages in the preparation of this book. But more than that, it was his inspiration and encouragement that brought me back to historical studies after a five-year interruption occasioned by the second World War. Only those who have studied with Professor Merk can appreciate fully how generous he is with his time and his talents.

I am grateful to the Social Science Research Council for assistance in financing my research and to the Charles Phelps Taft Memorial Fund of the University of Cincinnati for the aid that made publication possible.

Unsung heroes (and heroines) in the annals of historical writing are the staffs of American libraries. Few groups mani-

Preface

fest a higher and more consistent level of professional competence and courtesy. Space does not permit the singling out of individuals, but I should like to pay tribute to the staffs of the libraries of Harvard University, the University of Cincinnati, and the Massachusetts Institute of Technology, of the U. S. Department of Agriculture Library, of the Library of Congress, and of the public libraries of Boston, Cincinnati, and Cleveland.

Many individuals have given of their time to answer patiently my questions. Among those who granted personal interviews were Paul C. Wilkins and Anne L. Gessner of the Farm Credit Administration, U. S. Department of Agriculture; Lee D. Sinclair and John A. Zelinski of the Production and Marketing Administration, U. S. Department of Agriculture; Donald R. Moore of the Office of Public Relations, Federal Trade Commission; Frank A. Taylor, Curator of the Department of Engineering and Industries at the National Museum; Charles B. Heinemann of the National Independent Meat Packers Association. I must mention specially the late Everett E. Edwards of the Bureau of Agricultural Economics, who not only answered questions, but served as a guide to the mysteries of the vast Department of Agriculture. Douglas Albert, an engineer of Berkeley, California, wrote a detailed letter in response to my questions, as did Margaret Ingels, Engineering Editor of the Carrier Corporation, Syracuse, New York.

The entire manuscript was read by Professor Merk, by Professor Reginald C. McGrane, Head of the Department of History at the University of Cincinnati, and by Professor Robert S. Fletcher of Oberlin College. Portions were read by Miss Ingels and by Dr. William Cross of Deep River, Canada.

I owe much to my colleagues in the Department of History at the University of Cincinnati, for the impulse to write is partly the product of stimulating associations. I must mention again Professor McGrane, for he encouraged the completion of this book.

Mr. A. C. Lemons of South Bend, Indiana, made a contribution when regularly he saved a summer job for me at the

Artificial Ice Company. This certainly helped finance my education; I hope it also freed this volume from some of the dust of the study.

Finally, I wish to thank my wife, Dorothy Sebelin Anderson, for her indispensable assistance. Scarcely a page does not bear the mark of her sense of style and her gift for analysis. But more than that, she has made ungrudgingly the very real sacrifices that are, but should not be, the lot of the historian's wife.

These people have done their best to save me from errors of fact and interpretation. If, despite their guidance, I have strayed, it is I, not they, who is to blame.

O.E.A., Jr.

CONTENTS

ILLUSTRATIONS

REFRIGERATION
IN AMERICA

A HISTORY OF A NEW TECHNOLOGY
AND ITS IMPACT

CHAPTER I

THE BEGINNINGS OF REFRIGERATION

ICE is an American institution—the use of it an American luxury—the abuse of it an American failing." So asserted a writer in *De Bow's Review* in 1855 as he compared the widespread employment of ice for domestic refrigeration in the United States to its use in Europe, where it was "confined to the wine cellars of the rich, and the cooling pantries of first class confectioneries."[1] This assertion was accurate.[2] Refrigeration for domestic purposes was and has continued to be an American institution. For various reasons, including climate, diet, custom, and average level of living,[3] the United States remains the only country where domestic refrigeration is extensively used.[4]

The United States has led not only in domestic, but in many other aspects of the broad field of refrigeration. The export of natural ice to tropic ports was a triumph of Yankee enterprise. Americans developed methods of harvesting, storing, and transporting natural ice on a commercial scale.[5] As

[1] "Ice—How Much of It is Used, and Where It Comes From," *De Bow's Review*, XIX (1855), 709.

[2] St. Petersburg, Russia, was unusual among European cities in that in the eighteen-sixties ice seems to have been used extensively for domestic purposes, not only in the homes of the aristocracy, but in those of the middle and lower classes. See "The Ice Trade," *The Leisure Hour*, XII (June 27, 1863), 414. For the small extent of domestic refrigeration in the British Isles in the eighteen-seventies see *Ice Trade Journal*, III (Nov. 1879), 2.

[3] J. C. Goosman, "History of Refrigeration," *Ice and Refrigeration*, LXVI (1924), 542.

[4] W. P. Hedden, "Refrigeration," *Encyclopaedia of the Social Sciences*, XIII, 196. A convenient summary of the status of domestic refrigeration outside the United States in the nineteen-twenties, based on information compiled by the Industrial Equipment Division of the U.S. Department of Commerce, can be found in *Ice and Refrigeration*, LXXI (1926), 104-105. For the situation in western Europe in 1949 see W. R. Woolrich, "Observations on Refrigeration in Great Britain and Western Europe," *Refrigerating Engineering*, LVII (1949), 658.

[5] Not long after Americans demonstrated the practicability of exporting ice, certain Norwegians visited the United States to learn American methods of harvesting, storing, and shipping. R. Maclay,

the inadequacies of dependence on nature's product became apparent, they made significant contributions to the development of machinery to make ice or to do its work. In the application of refrigeration to the problems of a great industrial and agricultural nation, Americans, spurred by the demands of climate, the vast distances between centers of production and consumption, and the requirements of industry, have held a position of world leadership. In no other country has refrigeration been used for purposes more varied or significant.

It is not, however, a phenomenon altogether American. Europeans not only played a leading part in the invention of refrigerating machinery, but they also supplied the scientific background upon which its development depended. Throughout the world refrigeration has been given a wide variety of applications;[6] in a few cases these have been more highly developed elsewhere than in the United States. Nor is the United States the only nation in whose economy the role of refrigeration bulks large. Ever since Great Britain became a great manufacturing nation, she has been unable to produce all the meat she needs. She has depended upon meat imports,

---

"The Ice Industry," C. M. Depew, ed., *One Hundred Years of American Commerce*, ii (New York, 1895), 468.

    [6] For discussions indicating the wide variety of uses of refrigeration in Europe see A. R. Leask, *Refrigerating Machinery. Its Principles and Management* (London, 1895), 169-217; R. Stetefeld, *Die Eis- und Kälteerzeugungs-Maschinen, Ihr Bau und ihre Verwendung in der Praxis* (Stuttgart, 1901), 328-411; E. Pacoret, *La Technique de la Production du Froid et de ses Applications Modernes* (Paris, 1920), 129-186, 189-222; G. Lehnert, *La Technique du Froid*, Deuxième édition, G. Dermine, tr. (Paris, 1920), 13-67; and M. Hirsch, *Die Kältemaschine; Grundlagen, Ausführung, Betrieb, Untersuchung und Berechnung von Kälteanlagen*, Zweite Auflage (Berlin, 1932), 275-333. For analyses of ice-making and cold-storage facilities outside the United States see U.S. Department of Commerce, Bureau of Foreign and Domestic Commerce, *Trade Information Bulletin Nos. 209, 229, 280, 330, 338*, and *388* (Washington, 1924-1926). For the status of quick freezing outside the United States see E. W. Williams, "Frozen Foods in Europe," *Quick Frozen Foods and The Locker Plant*, xiii (Oct. 1950), 39-43, 111-115; (Nov. 1950), 43-47, 110-115. In May 1949 a leading American technical journal, *Refrigerating Engineering*, recognized the importance of research and development outside the United States by establishing a department devoted to a review of progress abroad.

transported to her shores in refrigerated ships, for a large part of her requirements.[7] Great Britain also relies to a considerable extent upon fruits and dairy products imported under refrigeration.[8] Britain's food needs and the role of refrigeration in meeting them have affected vitally the development of producing countries, particularly Australia, New Zealand, and Argentina.[9]

The history of refrigeration cannot be written in terms of any one country, for its development is the result of the efforts of many workers of many nationalities, and its benefits have not been confined by national borders. But in order to understand its significance, it is useful to examine its development in a single country, to study there its evolution, its applications, its repercussions. There is no better national laboratory for such an investigation than the United States, for in the United States are to be found the most widely varied applications of refrigeration, the greatest development of most of them, and a society that depends upon them to an extent few Americans realize.

Methods of cooling were known to many ancient peoples. These methods, however, were used not for preserving foods, but for chilling beverages. Ice was harvested and stored in China earlier than the first millennium before Christ. The Jews used snow, as did the Greeks and Romans, who constructed

[7] A valuable source of information is J. T. Critchell and J. Raymond, *A History of the Frozen Meat Trade; An Account of the Development and Present Day Methods of Preparation, Transport, and Marketing of Frozen and Chilled Meats*, 2nd edn. (London, 1912).

[8] D. A. Willey, "The Diversity of Uses for Cold Storage," *Scientific American Supplement*, LV (Jan. 24, 1903), 22628. For recent observations on the importance of refrigerated shipping to Britain see "A Silent Service without which Britain Would Starve: Modern Methods in the Refrigerated Food Ships of To-Day," *The Illustrated London News*, CCXII (May 15, 1948), 540-541.

[9] John Cooke, in chapter contributed to Critchell and Raymond, *op.cit.*, 299-302; Sir Joseph Ward, in chapter contributed to Critchell and Raymond, 303-305; E. G. Jones, "The Argentine Refrigerated Meat Industry," *Economica*, IX (1929), 156-172; Y. F. Rennie, *The Argentine Republic* (New York, 1945), 146, 149-150; F. M. Surface and E. L. Thomas, Section on "Foreign and International Aspects" under "Meat Packing and Slaughtering," *Encyclopaedia of the Social Sciences*, X, 250.

large storage pits in which it was placed and covered with insulating material. To cool drinks, snow or water melting from it was added directly. According to Pliny, the emperor Nero discovered that liquids could be refrigerated by burying their containers in snow. To obtain cool water in Egypt, shallow earthen jars filled with boiled water were exposed to the night air on the roofs of houses. Throughout the dark hours slaves moistened the outside of the jars; the resulting evaporation cooled the water. This method, probably known to many peoples, was used in India to produce ice. It is fairly certain the ancients knew that some solutions produced cooling effects.[10]

It is likely that the practice of using ice or snow to cool liquors was unusual in Europe as late as the sixteenth century except in the southern countries, particularly Italy and Spain. But by 1600 the custom had its French devotees. In warm countries snow or ice was not always available; as a substitute wealthy Roman families about 1550 cooled liquors by rotating long-necked bottles in water in which saltpeter was dissolved. It was learned in the seventeenth century that a solution of snow and saltpeter or salt could be used to produce very low temperatures and to make ice. It was this discovery, the principle of the old-fashioned ice-cream freezer, that made frozen delicacies possible. By the end of the seventeenth century iced liquors and frozen juices were popular among wealthy classes in France.[11]

The use of low temperatures, or refrigeration, for food preservation did not become significant until the nineteenth century. Until then the main methods of keeping food were

[10] A. Neuberger, *The Technical Arts and Sciences of the Ancients*, H. L. Brose, tr. (New York, 1930), 122-123; J. Beckmann, *A History of Inventions and Discoveries*, 2nd edn., W. Johnston, tr. (London, 1814), III, 322-332. German edition published under title: *Beyträge zur Geschichte der Erfindungen*, 1783-1805. In 1828 ice still was being made in India by evaporating water from shallow vessels. See F. Parlby (Parks), *Wanderings of a Pilgrim, in Search of the Picturesque, During Four-and-Twenty Years in the East; with Revelations of Life in the Zeñana* (London, 1850), I, 78-81.

[11] Beckmann, *op.cit.*, III, 333-341, 344-354.

salting, spicing or pickling, smoking, and dehydration accomplished by drying in the sun. Some of the preservative effects of cold, nevertheless, had been known for centuries. Even primitive man must have learned that his meat kept better in winter than in summer and that a cool cave was a good place to store any surplus food he might have during the summer months. The medieval manor had its cellar where food stored received the benefits of a relatively low temperature. Fishermen must have observed that their catch kept fresh longer in the winter. Francis Bacon contracted a fatal illness while experimenting with the preservative effects of snow stuffed into a dressed chicken. But it was not until the nineteenth and twentieth centuries that scientists provided the explanation. They showed that the decay of foods was a complicated phenomenon, caused in part by the action of microorganisms—bacteria, yeasts, and molds—and in part by the self-digestion or ripening, accelerated by complex organic substances called enzymes, that took place in animal tissues after death and in plant tissues most rapidly after they were separated from the organism to which they belonged. They demonstrated that low temperatures, by providing an unfavorable environment for these forces causing decay, retarded their activity and that temperatures sufficiently low, although they did not kill micro-organisms, slowed their growth.

Ice was not used to a significant extent for refrigerating foods in the United States before 1830. Although there was an occasional icehouse, in general what little refrigeration was used in colonial days was provided by spring houses, wells, or cellars.[12] In the early years of the nineteenth century for temporary storage of meats and milk some city-dwellers used "safes," boxes with sides and ends of woven wire which permitted air but not insects to come in contact with the food, but

[12] J. T. Adams, *Provincial Society, 1690-1763* (A. M. Schlesinger and D. R. Fox, eds., *A History of American Life*, III) (New York, 1927), 90. See J. B. McMaster, *A History of the People of the United States from the Revolution to the Civil War* (New York, 1890-1913), I, 17, for a discussion of the primitive substitutes for refrigeration used in Boston, 1784.

the main means of preserving foods continued to be those used for centuries: salting, spicing, smoking, and drying.[13]

One reason for the limited use of refrigeration before 1830 was that the need for it was little felt. Foods whose preservation required low temperatures—fresh meats, fish, milk, and fruits and vegetables of inferior keeping qualities—did not play as important a part as they do now, for American diet was based on foods not so perishable; it was dominated by bread and salted meats. This diet was largely the result of long-established habits, and although it was short on many of the foods we now value so highly, it was fairly adequate for people who led active lives in the open. The fare of the wealthy, of course, was more varied than that of the poor. Not only were foods that required coolness less used than at present, but refrigeration was not as necessary for transporting them as it was to become, for no great distances separated grower from consumer.[14]

Although refrigeration was not a principal method of food preservation before 1830, its use was by no means unknown. In the first decade of the nineteenth century Thomas Moore, a Maryland farmer, had a refrigerator of his own invention to carry butter to market and to keep it hard until sold. Early in the century fish packed in ice were carried in light carts from the New Jersey coast to supply the Philadelphia market, while after 1825 lots of ice-packed fish were occasionally shipped on the Erie Canal. Refrigeration, however, was not yet practical for general transporting of perishables. Shipment was much too slow when any considerable distance was involved, and packing the product to be preserved in the ice itself was not a method that could be applied to all types of foods. Farmers continued to carry their products to market during the night

[13] R. O. Cummings, *The American Ice Industry and the Development of Refrigeration, 1790-1860* (Unpublished doctoral dissertation, Harvard University), 21-22.

[14] *ibid.*, 180-181; R. O. Cummings, *The American and His Food: A History of Food Habits in the United States* (Chicago, 1940), 10-42; J. A. Krout and D. R. Fox, *The Completion of Independence, 1790-1830* (A. M. Schlesinger and D. R. Fox, eds., *A History of American Life*, v) (New York, 1944), 101-103.

to receive the benefit of the comparatively low temperatures, and the coastwise trade in perishables was confined to the cool seasons of the year. There was little use of refrigeration in the market place. In northern cities a few butchers used ice for holding their fresh meat, and in southern ports some products were stored in the icehouses built by Yankee traders. But ice, in both the North and the South, was difficult and expensive to obtain before 1830, and in its absence most marketing was done in the early morning hours. Most butchers slaughtered for only a day's fresh-meat trade, while dealers in fresh fish tried to keep their product alive until sale.[15]

Ice refrigeration was used to a very limited extent on the consumer level for food preservation by individuals and by such institutions as hotels, taverns, and hospitals. Butter, fish, and meat were probably the products refrigerated most frequently. When icehouses were available, foods were stored there. To supply those who did not have them and to provide a more convenient means of holding foods, iceboxes or refrigerators slowly came into use.[16] The refrigerator that Thomas Moore developed and patented was an oval cedar tub with an inner sheet metal butter-container which could be surrounded on four sides by ice. A cloth lined with rabbit skin was provided to serve as an insulated cover for the tub. Moore also developed an insulated box which featured an ice-container attached to the lid and a six-cubic-foot storage space below. He recommended that the refrigerator be used not only in transportation of perishables but also in the market place and the home. "Every housekeeper," he declared in a pamphlet describing his device, "may have one in his cellar, in which, by the daily use of a few pounds of ice, fresh provisions may be preserved, butter hardened, milk, or any other liquid preserved at any temperature. . . ."[17] Insulated boxes

---

[15] Cummings, *The American Ice Industry*, 48, 50-51, 55-58, 60-61.

[16] *ibid.*, 23-24, 61-63; "Of Ice and Ice Houses," *The American Museum, or, Universal Magazine*, XII (1792), 175; Krout and Fox, *op.cit.*, 103-104.

[17] T. Moore, *An Essay on the Most Eligible Construction of Ice-Houses. Also, A Description of the Newly Invented Machine Called the Refrigerator* (Baltimore [1803?]), 16-21.

to contain ice were introduced in southern ports that received ice shipments from the North. Although it was advocated that the boxes be used for the preservation of butter, fruit, fish, and meat in addition to the cooling of wine and water,[18] this suggestion was not generally adopted, for ice was expensive, costing about five to six cents a pound as late as 1827, and it melted more quickly when the boxes were filled with provisions. Because of the cooler climate iceboxes were used less in the North than in the South. Their more general adoption depended on a greater demand for refrigerated preservation of foods and upon methods of making ice available at lower prices.[19]

The principal use for ice in the United States before 1830 was, as it had been for many years in Europe, for cooling drinks and for making frozen delicacies. After 1790 the practice, derived from the French, of using it in wines, punches, and lemonade became popular in the United States. An accompanying trend was the increased consumption of ice cream among the wealthy. In some quarters ice was desired for medicinal purposes. It was taken internally by cholera patients, and some recommended it in summer as a tonic to counteract the "putrescent tendency of the fluids."[20] The demand for ice for these purposes alone could never have been large, but these early uses were significant in that to provide for them the first steps were taken in the development of methods of harvesting and storing natural ice.

Icehouses began to appear in increasing numbers in the seventeen-eighties and seventeen-nineties. Although Andrew Craigie found one in bad condition on the Cambridge estate he acquired in 1791,[21] most of the early icehouses were in Virginia, Maryland, and Pennsylvania. In the autumn of 1785

[18] Charleston *Courier*, March 8, 1817, quoted in *Ice and Refrigeration*, LII (1917), 273.

[19] Cummings, *The American Ice Industry*, 23-25, 55, 99-100.

[20] *ibid.*, 11-13, 16; Krout and Fox, *op.cit.*, 104; "Of Ice and Ice Houses," *loc.cit.*, 175, 177-178. (In the footnotes of this book *loc.cit.* is used throughout to refer to previously cited periodical articles and to contributions to composite publications; *op.cit.* refers to previously cited publications in book form.)

[21] Cummings, *The American Ice Industry*, 18-19.

Washington supervised the construction of one at Mount Vernon, and in January 1786 harvesting and storing operations began on the Potomac.[22] Jefferson had an icehouse at Monticello, and the writer of a 1792 treatise on icehouses knew of four on the Chesapeake peninsula and of one in Philadelphia. Later they became common on plantations of the border states.[23] These houses, as well as those in other states, were square pits or vaults, either lined or unlined, built as much as possible below ground, and roofed with boards, sod, or thatch. Some had no means of draining off the meltwater. Harvesting methods were primitive. It was difficult to cut square blocks with the main tools, the ax and the saw. Since cutting was slow with such implements, it was often not safe to wait for ice of proper thickness because of the danger of sudden thaws capable of melting the entire crop. As a result, ice was harvested in a great many irregularly shaped blocks. Because in this form it melted very rapidly in storage, it was pounded and sometimes sprinkled with water to consolidate it into a solid mass.[24]

These methods of harvesting and storing may have served fairly well the needs of the gentleman farmer who cut ice for his own use, but they proved unsatisfactory for operations on the commercial scale necessary to supply the trade that had developed in shipping it, principally from Boston, to tropic ports and to the ports of the southern United States.

The export trade began in 1806 when Frederic Tudor shipped a cargo from Boston to Martinique. Tudor's early experiments in this trade were cut short by the War of 1812, but he resumed operations afterwards, sending cargoes to Havana and other Caribbean ports. Although he had competitors, Tudor made most of the shipments from Boston before

[22] J. C. Fitzpatrick, ed., *The Diaries of George Washington, 1748-1799* (Boston, 1925), II, 428, 434, 445, 452; III, 7-9.

[23] Cummings, *The American Ice Industry*, 67-68; "Of Ice and Ice Houses," *loc.cit.*, 177, 179; "Ice: and the Ice Trade," *Hunt's Merchants' Magazine and Commercial Review*, XXXIII (1855), 170.

[24] Cummings, *The American Ice Industry*, 81, 83; Moore, *op.cit.*, 11-13; Goosman, *loc.cit.*, LXVII (1924), 110.

about 1827.[25] The coastwise trade may date from as early as 1799, when a New York-to-Charleston consignment is said to have been made on the initiative of a Charleston gentleman.[26] Other ventures were made in the early years of the nineteenth century, and in 1817 Tudor entered the coastwise ice trade, establishing a business at Charleston. Tudor widened his operations to Savannah and New Orleans, maintaining until 1828 a virtual monopoly in all three cities. In 1826 Tudor's competitors extended their business to Wilmington and Norfolk and in 1827 to Mobile. The trade to the South, however, grew slowly until after 1827.[27]

Tudor, faced with the problem of storing ice under the adverse conditions of southern latitudes, was forced to devise efficient methods. Influenced by Thomas Moore's treatise on icehouses and refrigerators, which emphasized the importance of proper insulation and drainage, and by a report of an above-

[25] Cummings, *The American Ice Harvests; A Historical Study in Technology, 1800-1918* (Berkeley, 1949), 7-11; Cummings, *The American Ice Industry*, 90n-91n. Cummings' work, based on the Tudor papers in Baker Library of the Harvard Graduate School of Business Administration, is the best secondary source of information on Tudor and the ice trade. Descriptions of the Tudor papers can be found in "Frederic Tudor—Ice King," *Bulletin of the Business Historical Society*, VI (1932), 1-8, and in "Supplementary Material on Frederick Tudor Ice Project," *Bulletin of the Business Historical Society*, IX (1935), 1-6. Interesting material on Tudor's personality can be found in H. G. Pearson, "Frederick Tudor, Ice King," Massachusetts Historical Society, *Proceedings*, LXV (1932-1936), 169-215, while a brief account of the export ice trade appears in S. E. Morison, *The Maritime History of Massachusetts, 1783-1860* (Boston, 1921), 280-283. A popular account is included in S. H. Holbrook, *Lost Men in American History* (New York, 1946), 114-123.

[26] L. Weatherell, "The Ice Trade," U.S. Department of Agriculture, *Report of the Commissioner . . . for the Year 1863* (Washington, 1863), 439n.

[27] Cummings, *The American Ice Industry*, 75-81, 92. According to Tudor, he shipped 1,506 tons to ports of the southern United States in 1827 and his competitors 375. Although he sometimes had to resort to price cutting to maintain his position, Tudor was able to keep prices high through 1827. Before 1827 he usually bought ice wholesale in the Boston area and engaged ships to carry it to his southern depots, but on occasion he bought ice, as his competitors did, from vessels in the coastwise trade which had carried it as ballast. Although Tudor had been involved in ice production before 1827, that year marked his real entrance into that phase of the business. He entered in an effort to protect his monopoly in the South.

ground icehouse in Philadelphia, he erected in 1816 at Havana an above-ground storage which had been prefabricated in Massachusetts. This building, twenty-five feet square on the outside, had insulation three feet thick and a capacity of 150 tons. After this and other experiments Tudor constructed in 1817 a well-insulated and ventilated house at Charleston, where it gave him a great advantage over his competitors. In 1824 and 1825 Tudor built icehouses of this improved type at Fresh Pond in Cambridge, Massachusetts, the chief source of the ice for the export and coastwise trade. After 1827 many like his were built there. These reduced estimated wastage during a season from more than sixty-six per cent to less than eight per cent.[28]

The methods of cutting ice were revolutionized in 1825, when Nathaniel J. Wyeth, who soon took charge of harvesting operations for Tudor, invented a horse-drawn ice-cutter.[29] This instrument, which was patented in 1829, consisted of a quadrangular iron frame about three feet long and twenty inches wide. Projecting down at right angles from each of the long sides of the frame were four steel or steel-pointed teeth which increased in length from front to back. When this was drawn plow-like over a field of ice, parallel grooves were cut. They were deepened to four inches by running the frame back and forth in them. When similar grooves were cut at right angles to the first, the ice was divided into squares cut partially through its thickness.[30] Square blocks could then be obtained by sawing or splitting with iron bars along the grooves.[31] This invention, improved and supplemented by other devices, was of prime importance in the development of the ice industry and in the use of refrigeration.

[28] *ibid.*, 81-83; Pearson, *loc.cit.*, 182-183.

[29] Cummings, *The American Ice Harvests*, 19-21; J. Schafer, "Nathaniel Jarvis Wyeth," *Dictionary of American Biography*, xx, 576-577. Wyeth is best known as a pioneer trader in the Oregon country.

[30] "American Patents," *Journal of the Franklin Institute of the State of Pennsylvania*, iii n. s. (1829), 417.

[31] Wyeth also invented a horse-drawn circular saw to be used to cut through the ice after the grooves had been made. This device, however, does not seem to have been practical, and circular saws were not used for cutting ice until many years later. For a description of Wyeth's circular saw see *ibid.*, 417-418.

CHAPTER II

THE APPLICATION OF NATURAL-ICE REFRIGERATION
TO PROBLEMS OF FOOD SUPPLY, 1830-1860

AMERICAN urban diet was improved between 1830 and the Civil War by an increasing use of fresh foods. Although this improvement did not do away with the deficiencies in the diet of American city-dwellers and although prejudice still lingered against fresh fruits and vegetables, demand for them as well as for fresh meats and milk was increased in eastern cities.[1] Fashion was a factor in this development, for French influence, which had stimulated the enjoyment of iced drinks and ice cream, had also made fashionable the consumption of green vegetables, which were in demand by the middle of the eighteen-thirties. Another factor, the movement for diet reform, part of the wave of reform movements that was sweeping the country, increased the desire for fresh fruits and vegetables and created a prejudice against certain items in the traditional fare, such as salted meats. Perhaps the most important factor influencing the new demand was the growth along the eastern seaboard of urban centers, great cities filled with people whose sedentary lives made unsuitable the food habits of their fathers. For the most part the economic status of city-dwellers improved during these years, enabling them to afford the more expensive fresh foods they required. The increased use of fresh fruits, vegetables, meats, and milk made refrigeration more necessary for preservation than it had been before. At the same time the growth of cities helped to increase the demand for these foods, it made them more difficult to obtain by removing great numbers of people from the area where they were produced. Between 1830 and 1860 the first

[1] R. O. Cummings, *The American and His Food* (Chicago, 1940), 43-44, 51; R. O. Cummings, *The American Ice Industry* (Unpublished doctoral dissertation, Harvard University), 151-152; C. R. Fish, *The Rise of the Common Man, 1830-1850* (A. M. Schlesinger and D. R. Fox, eds., *A History of American Life*, VI) (New York, 1927), 140.

attempts were made to use refrigeration for spanning the gap between producer and consumer.[2]

It was Nathaniel J. Wyeth's method of cutting quickly and cheaply blocks of like size that made the bountiful natural-ice resources of the United States available for food preservation. His invention resulted in a complete transformation of the ice industry.[3] Now the process of removing the ice from the body of water on which it had formed could be mechanized. For elevating it to the storage house from the surface of the water, a steam-powered endless chain was developed with which by 1855 six hundred tons could be housed in a single hour.[4] Blocks of uniform size made possible speedier and less wasteful handling techniques in storage, transportation, and retail distribution. Ice now could be stacked in compact, regular piles. To keep the blocks from freezing together, Tudor patented a process devised by Wyeth which involved separating them from each other by layers of insulating material, usually sawdust.[5] Machinery was invented for discharging ice from storage and lowering it into the holds of ships.[6] Special insulated compartments no longer had to be constructed on vessels used in the trade, for the square blocks could be stacked in the holds with sawdust spread over the tops of the piles and between them and the hull of the ship.

[2] Cummings, *The American and His Food*, 31-36; Cummings, *The American Ice Industry*, 130-132, 134, 181, 184.

[3] A full discussion of the significance of this method can be found in Cummings, *The American Ice Industry*, 85-89.

[4] *ibid.*, 84-86; "Ice: and the Ice Trade," *Hunt's Merchants' Magazine*, xxxiii (1855), 178; J. Schafer, "Nathaniel Jarvis Wyeth," *Dictionary of American Biography*, xx, 577. For descriptions of some of Wyeth's inventions see "List of American Patents which Issued in December, 1841," *Journal of the Franklin Institute*, iii 3rd s. (1842), 125-127, 129; "American Patents," *Journal of the Franklin Institute*, x 3rd s. (1845), 24-25; and "American Patents," *Journal of the Franklin Institute*, xviii 3rd s. (1849), 502.

[5] Cummings, *The American Ice Industry*, 88-89; "Specifications of American Patents," *Journal of the Franklin Institute*, xxiii n.s. (1839), 244-245. Icehouses increased in size in these years, becoming great structures one to two hundred feet in length constructed mostly of wood, but sometimes of brick. "Ice: and the Ice Trade," *loc.cit.*, 175.

[6] L. Weatherell, "The Ice Trade," U.S. Department of Agriculture, *Report of the Commissioner, 1863* (Washington, 1863), 440.

Uniform blocks resulted in reduced waste during distribution at the retail level and in delivery to the consumer in a convenient and attractive form. Wyeth's method of cutting reduced the great risk involved in ice production, making it an enterprise more attractive to capital. Now it was fairly certain that any ice that formed could be harvested. With this assurance an increasing number of firms entered the business.[7] All of these effects of Wyeth's work—quicker and cheaper cutting, mechanical conveyance to storage, more efficient handling techniques, reduced waste and risk, increased competition, and larger-scale operations—lowered production and distribution costs and contributed to lower prices both North and South.[8] Lower prices gave more people an opportunity to receive the benefits of refrigeration.

Ice harvesting, under the influence of Wyeth, had become a very efficient process by the eighteen-fifties.[9] If snow had fallen on the field, preparation began by clearing it away with scoop-like wooden scrapers pulled by horses. If water had mingled with the snow and subsequently been frozen, the resulting "snow ice" was removed with an ice plane, a horse-drawn frame which mounted a steel cutting edge capable of removing the snow ice to a depth of two inches.[10] The scraps then were cleared away with the snow-scraper, and the field was ready to be marked for cutting. To begin the harvesting operations proper, a single straight groove was made with a hand tool. Then the surface was marked into squares of grooves twenty-two inches on a side[11] by means of the swing-guide-marker, a descendant of Wyeth's primary invention.

[7] Cummings, *The American Ice Industry*, 87-89.

[8] *ibid.*, 100-102; "Ice: and the Ice Trade," *loc.cit.*, 174n.

[9] R. Maclay, "The Ice Industry," C. M. Depew, ed., *One Hundred Years of American Commerce*, II (New York, 1895), 468, gives a John Barker part of the credit for the invention of ice tools.

[10] The removal of snow ice was a necessary, yet risky and expensive operation. If it were not done, the snow ice acted as insulation and prevented the ice from becoming thicker should the weather continue sufficiently cold before the ice was cut. If a thaw intervened between removal of the snow ice and cutting, the cost of removal was a total loss. Weatherell, *loc.cit.*, 422, 447.

[11] The dimensions of the squares varied in some cases.

Instead of having two lines of cutting teeth, the swing-guide-marker had only one line of teeth and an iron guide which fitted into the groove already made.[12] The next step was to draw ice-cutters through every other groove in each direction to cut two-thirds through the ice. These cutters, also descendent from Wyeth's basic invention, were plow-like devices with a single line of cutting teeth that increased in length from front to back. After the cutters had done their work, large strips of ice were sawed off and floated through channels to a basin near the elevator, where they were split with iron bars into cakes usually forty-four inches square. The elevator was an inclined plane on which moved a steam-powered endless chain. As the cakes moved up, they were drawn under an adjustable blade which shavèd them to the thickness desired, leaving two raised ribs of ice which kept the blocks from fusing when stacked. When the blocks reached a' suitable elevation above the level of the top of the ice in the storage house, they were removed from the elevator and allowed to slide down a ramp into the storage, where workers guided them into position.[13] An indispensable aid to the harvesting was whiskey, which steeled the workers against the rigors of winter.[14]

Boston, favored with a cheap supply and low freight rates, continued to dominate the export and coastwise trade.[15] Ice,

[12] W. E. Wood, "Twenty Five Years' Development in Harvesting and Housing Ice," *Ice and Refrigeration*, LI (1916), 191.

[13] This description is based on Weatherell, *loc.cit.*, 442, 447. Although it appeared in the *Report of the Commissioner of Agriculture for the Year 1863*, it is descriptive of practices that prevailed in the late eighteen-fifties. There were, of course, variations in the methods of harvesting. Techniques less highly developed are described in "Ice: and the Ice Trade," *loc.cit.*, 174. Figure 1, although it pictures operations in the eighties, is illustrative of the methods that have been described. An account of harvesting about 1848 is N. J. Wyeth, "The Ice-Trade of the United States," *The American Almanac and Repository of Useful Knowledge, for the year 1849* (Boston, 1848), 178-180.

[14] Communication from "The Ice Lifter," *Ice Trade Journal*, XVII (Oct. 1893), 1.

[15] F. Tudor and T. T. Sawyer, "Ice Trade," Boston Board of Trade, *Third Annual Report of the Government, Presented to the Board at*

however, was shipped from Maine as early as 1819, and in years when the crop was short on Fresh, Spy, Newham, and other ponds, Massachusetts producers turned to Maine. For example, Wyeth harvested Kennebec River ice for Tudor in 1828. But operations there were inefficient, resulting in high costs. These, combined with high freight rates, hampered the development of the Maine industry, which grew slowly until after 1860.[16]

The export trade expanded, both in volume and in geographical scope, after 1827. The number of participants in the traffic increased, and Tudor, the pioneer, found himself faced with competition from several rival firms.[17] In 1833 a cargo of 180 tons was sent to Calcutta, while in 1834 a shipment was made to Brazil. These ventures marked a new era in the trade. Previously, ice had been exported only to the Caribbean. By 1850 shipments had been extended to the large ports of South America and the Far East.[18] During the eighteen-forties Massachusetts firms shipped to England and for a while controlled the London market. Not many years passed, however, before Americans were compelled to withdraw from the English trade by Norwegian competition.[19] By the year ending June 1860 the nation's export ice trade amounted to 49,153 tons valued at $183,134. Yet despite this expan-

---

the Annual Meeting on the 21st of January, 1857 (Boston, 1857), 80. In 1841 a railroad was built to connect Fresh Pond with the ice wharves at Charlestown. This strengthened the advantages the port of Boston enjoyed in the ice trade. R. O. Cummings, *The American Ice Harvests* (Berkeley, 1949), 43-44. By 1863, 550 vessels were employed in the Boston ice trade. Weatherell, *loc.cit.*, 448.

[16] Cummings, *The American Ice Industry*, 94-95.

[17] Tudor and Sawyer, *loc.cit.*, 79-80; "Ice: and the Ice Trade," *loc.cit.*, 173.

[18] Cummings, *The American Ice Industry*, 91n; S. E. Morison, *The Maritime History of Massachusetts, 1783-1860* (Boston, 1921), 282-283; "Ice: and the Ice Trade," *loc.cit.*, 170; Letter, F. Tudor, Boston, Mass., to Robert Hooper, Esq., Jan. 22, 1849, in *Proceedings of the Massachusetts Historical Society*, III (1855-1858), 56.

[19] H. Hall, "The Ice Industry of the United States, with a Brief Sketch of Its History and Estimates of Production in the Different States," U. S. Department of the Interior, Census Office, *Tenth Census, 1880*, XXII (Washington, 1888), 3.

sion, it always was overshadowed in volume by the coastwise traffic.[20]

The growth of the coastwise trade was reflected by the statistics of the port of Boston. Only 1,911 tons were shipped in 1827, but the total reached 43,125 tons in 1848 and 97,211 in 1860. Some years, when the local harvests of Philadelphia, Baltimore, and other cities failed, compelling them to buy from northern firms, the tonnage was greater.[21] In 1856 shipments were made from Boston to Philadelphia and Baltimore, to the District of Columbia, and to every southern state bordering the sea from Virginia to Texas, except, it seems, Mississippi.[22] After his storage methods were copied by competitors, Tudor found it impossible to retain his monopoly in ports of the southern states. The increased competition—along with the reduced costs of harvesting and distributing and the economies permitted by operations on a large scale—was a factor in the decline of ice prices in the South after 1828 to between one-half and three cents a pound. Although prices occasionally rose under the pressure of abnormal circumstances, the high levels of earlier years were never again maintained.[23]

The benefits of ice brought to the South in the coastwise trade were restricted for the most part to port cities and their vicinities, although the development of railroads made possible shipments to inland cities. Towns of the lower Mississippi not reached by New England vessels could obtain ice from the northern tributaries of the river. Just when the practice of carrying it down the Mississippi started and how extensive was the traffic is uncertain. Peru, a small town on the Illinois

[20] U. S. Treasury Department, *Report of the Secretary of the Treasury, Transmitting a Report . . . of the Commerce and Navigation of the United States for the Year Ending June 30, 1860* (Washington, 1860), 24-25. The $183,134 figure presumably represents valuation at the port of export and does not include costs of ocean transportation and retail distribution. Boston ice shipments for 1860 are summarized by Weatherell, *loc.cit.*, 449, as follows: to the East Indies and Ceylon, 16,210 tons; to Cuba, South America, and other southern ports, 29,042 tons; to U. S. ports, 97,211 tons.

[21] Cummings, *The American Ice Industry*, 92-93; Weatherell, *loc.cit.*, 449.

[22] Tudor and Sawyer, *loc.cit.*, 80.

[23] Cummings, *The American Ice Industry*, 80-81, 100-102.

River about sixty miles above Peoria, was a center of the trade. Here it was the custom to allow flatboats to become icebound during the winter. Ice, as soon as harvested, was loaded into the boats. When spring arrived, all that remained to be done was to float the loaded craft down the river to market.[24]

The northern states were fortunate in having natural ice close at hand. In most places north of the Potomac and the Ohio winter brought the possibility of a local harvest, although its dependability varied in direct proportion to the severity of the winter weather. If there were ice-crop failures in Maryland and Pennsylvania, comparatively short coastwise shipments helped alleviate the shortage, while the availability of canals and later of railroads made it possible to transport ice to other areas where warm winters had blighted the harvest.[25]

Between 1830 and 1860 the business of supplying the cities of the North became localized at various lakes, ponds, and rivers where the factors of weather, distance from the city, and transportation facilities were present in favorable combination. Boston was supplied by the same ponds that provided ice for the export and coasting trades, while New York City by 1855 depended upon Highland and Rockland lakes and the Hudson in the vicinities of Tarrytown, Kingston, Catskill, and Athens. Philadelphia, Baltimore, and Washington were able in favorable seasons to obtain a large part of their supply from local rivers and lakes. In unfavorable years, which occurred about once in every three, ice was imported from Boston. On at least one occasion Philadelphia bought from Halifax and other Canadian ports. Cincinnati drew upon local sources, which it came to supplement in bad years with ice from Lake Erie, and Chicago cut from local rivers and lakes. Smaller towns throughout the North depended on nearby sources.[26]

The business of harvesting and distributing in northern

[24] *ibid.*, 95-97; "Ice: and the Ice Trade," *loc.cit.*, 172.

[25] Cummings, *The American Ice Industry*, 96-97.

[26] "Ice: and the Ice Trade," *loc.cit.*, 171-172; Hall, *loc.cit.*, 4, 30-32; Tudor and Sawyer, *loc.cit.*, 82.

cities started slowly, the first dealers usually being business-men or farmers who sold the surplus of the ice they used for their trade in perishables. In some cases dealers were dairy farmers or market gardeners; in others they were fishmongers, butchers, hotel-keepers, or confectioners, while in Philadelphia the Pennsylvania Hospital was an early distributor. With the increase in demand others entered the business, not handling ice as a by-product, but specializing in its sale. As the business evolved, large firms came to displace the small dealers, especially in cities. These companies, easy to finance after the improvements in cutting methods, were necessary if cities were to have dependable supplies. This was particularly true in a city like Philadelphia, where the local resources were uncertain. A large company could afford the equipment necessary to make the most of the ice formed during a mild winter and could afford to put capital into the great storage houses necessary to carry the surplus over from a good year. When all other expedients to procure a supply locally failed, a large company could arrange for shipments from New England producers on more favorable terms than could a small firm.[27] Ice companies in Boston delivered to families at a flat charge for a specified amount daily, either on a monthly or seasonal basis. For example, about 1855 fifteen pounds delivered daily cost two dollars a month or eight dollars for the season May through October. Commercial users who needed one hundred pounds daily paid by the hundredweight at reduced rates, while those who used five hundred pounds received a further reduction.[28]

The cities of the northern states were able to obtain ice with relative ease, but on the Pacific coast San Francisco, suddenly made a great city by the discovery of gold, faced a serious

[27] Cummings, *The American Ice Industry*, 51, 71-75; Hall, *loc.cit.*, 4. For a description of how an ice-dealer started business in 1844 or 1845 in Syracuse, New York, see letter from Joseph Savage, Esq., Syracuse, N.Y., Jan. 15, 1855, quoted in "Ice: and the Ice Trade," *loc.cit.*, 172.

[28] "Ice," Boston Board of Trade, *Second Annual Report of the Government, Presented to the Board at the Annual Meeting, on the 16th of January, 1856* (Boston, 1856), 57.

problem of supply. An abundance of ice was formed on lakes in the Sierra Nevadas, but lack of suitable means of transportation prevented its use. In 1850 ice was being imported around Cape Horn from New England, but the quantities shipped were not large enough to satisfy the demand. A more satisfactory source was opened by San Francisco capitalists who formed the American Russian Commercial Company, which first imported ice from Alaska in February 1852. The ice, furnished under contract by the Russian-American Company, seems first to have been obtained from Sitka, but Sitka's climate proved unpredictable, and the harvesting operations were shifted to Woody Island in Kodiak Harbor. Prices for the product obtained were high, for at the retail level five to six cents a pound was asked. At such rates the trade never attained a volume comparable to that of other sections of the United States. The larger part of San Francisco's ice was brought from Alaska, although some was supplied by Boston as late as 1855.[29]

The increased consumption of fresh foods between 1830 and 1860 made preservation by low temperatures more necessary than it had been before. As cellars and wells proved increasingly unsatisfactory, there was a growing interest in ice-cooled refrigerators. Between 1830 and 1860 they improved slowly. Both their use and their design were hampered by a confusion of their function with that of icehouses and by a general lack of understanding that ice produced refrigeration by melting.[30] The delusion was widespread—it is not yet extinct—that the important thing was to save the ice. Ice

[29] E. L. Keithahn, "Alaska Ice, Inc.," *Pacific Northwest Quarterly*, xxxvi (1945), 121-122; Hall, *loc.cit.*, 37; "Ice: and the Ice Trade," *loc.cit.*, 178. Ulysses S. Grant, while on duty at Fort Vancouver in 1852, sought with two others to exploit the high ice prices in San Francisco. They cut one hundred tons, but lost their investment when the vessel carrying the harvest was delayed by adverse winds and the arrival of ice ships from Alaska drove down the price. L. Lewis, *Captain Sam Grant* (Boston, 1950), 312.

[30] Cummings, *The American Ice Industry*, 134-138, 140; "The New Refrigerator," *New-York Mirror*, xvi (July 14, 1838), 23. In changing from a solid to a liquid, ice absorbs heat which is known as its latent heat of fusion. The heat is absorbed from surrounding substances.

frequently was wrapped in a blanket. This prevented it from melting so rapidly, but it also prevented the refrigerator from functioning with anything like proper efficiency.[31] Although Thomas Moore had devised a box with the ice compartment at the top, most early manufacturers did not follow his example, probably because ice melted more quickly when placed there.[32] Refrigerators in 1840 were nothing more complicated than insulated wooden boxes, tin- or zinc-lined, with shelves or sometimes compartments to hold the food. The ice section was located at the bottom to one side and was equipped with a drain to carry off meltwater. Heat transfer, or cooling, was accomplished mainly by conduction.[33] These refrigerators were inadequate, for in the absence of any effective circulation of air, they were inefficient and had no means for the removal of odors or excess moisture. Proper temperatures were not maintained, food flavors were mixed, and the food cabinet became moldy, musty, smelly.

For many years it was not realized that air circulation was necessary for proper functioning of refrigerators. Later it was discovered that the ice should be elevated so that the heavier, cooled air could drop freely to the bottom of the chamber and displace warmer and lighter air, which would rise to the top. As the warm air passed over the surface of the ice, moisture from it would be condensed and deposited along with odor- and taste-bearing particles. The air—dried, purified, and cooled—would return to the food compartment, while the impurities deposited on the ice would be removed from the refrigerator in the meltwater through the drainpipe.

Gradually the deficiencies of early refrigerators were diagnosed and some makers began to see the value of circulating cool, dry air through the food space.[34] John C. Schooley of

---

[31] The use of blankets is recommended by [E.] Leslie, *The House Book; or, A Manual of Domestic Economy. For Town and Country*, 3rd edn. (Philadelphia, 1840), 243-244.

[32] Cummings, *The American Ice Industry*, 138-139.

[33] Leslie, *op.cit.*, 243-244; "The New Refrigerator," *loc.cit.*, 23.

[34] Several systems for accomplishing this are outlined in E. W. Blake, "Pure Air and Pure Food—Their Connection and Relationship as an Item in Domestic Economy and Hygiene," *The Boston Medical and Surgical Journal*, LX (June 23, 1859), 411-413.

Cincinnati, in connection with experiments in cooling a pork house so hams could be cured in summer, devised a system for "obtaining a dry cold current of air from ice" which he patented in 1855. His system, applied to domestic and commercial refrigerators, involved admitting outside air to the ice chamber through a register. As the air passed over the ice, it was cooled and dried as part of its moisture condensed on the ice surface. The cooler, drier, and heavier air then flowed into the food compartment, where it absorbed moisture and became warmer, eventually passing from the refrigerator through a second register without returning to the ice compartment.[35] A similar system was patented in 1856 by a Massachusetts manufacturer named Winship.[36] Although refrigerators constructed on this principle provided a current of relatively dry, cool air, it was inefficient to chill air from outside temperatures, pass it over the food once, and then expel it. The ice consumption must have been prohibitive. At least as early as 1859 Schooley had modified his design by providing valves for closing the inlet and outlet registers and by arranging two openings between the ice chamber and the food compartment so that when the registers were closed, circulation would take place inside the refrigerator as it does in a modern unit. It was recommended that the valves be kept open until fresh or warm provisions placed inside had cooled, but that the valves be kept closed when articles were stored that would not contaminate the air. An advertising leaflet published about 1861 in the interests of Schooley's improved refrigerator observed that "by repeated contact with the Ice, a low temperature is maintained, the air is dried, and to a considerable extent purified."[37] The correct principle of the refrigerator had been discovered and was understood.

[35] J. C. Schooley, *A Process of Obtaining a Dry Cold Current of Air From Ice, and Its Different Applications* (Cincinnati, 1855), 1-5, 11-15. Soon after Schooley got his patent, he introduced his refrigerator into Boston. In 1858 prices for various domestic sizes of his invention, which was produced in chest and upright form, ranged from twelve to thirty dollars. [S. Harnden], *Description of Schooley's Patent Refrigerator and Fruit Preserver* . . . (Boston, 1858), 4, 10.

[36] Cummings, *The American Ice Industry*, 139.

[37] [S. Harnden], *Description of Schooley's Improved Patent Refrigerator and Fruit Preserver* . . . (Reading, Mass. [1861?]), 3-6.

The extent to which improved units were used by 1860 is not clear.[38] It is likely, however, that the older, less satisfactory types had not been displaced to any considerable degree. Many years after 1860 refrigerators were being constructed which did not differ materially from those of 1830 and 1840.

There was less progress between 1830 and 1860 in the development of ice-cooled storages for the preservation of food products on a large scale than there was in refrigerators for domestic and commercial use. Indeed, only a start was made in the direction of what has come to be known as the cold-storage or refrigerated warehouse. One of the earliest ice-cooled storages was developed by John C. Schooley, who was interested in evolving a method of cooling a building in which pork could be cured in summer as well as in winter. To cure pork in summer, cooling was necessary to prevent spoilage before the preservative mixture could penetrate to all parts of the meat. Schooley examined various processes for summer curing with ice, concluding that they did not work well because they provided no means of removing excessive moisture. In an effort to combine low temperatures with a dry environment, he contacted a Cincinnati friend in the ice business who planned to build adjoining ice and pork-curing houses. He persuaded him to construct them with the pork-house cellar and the bottom of the icehouse on the same level and connected by an opening. This done, Schooley made the cellar airtight, except for a ventilator which connected the cellar with the outside air and was equipped with valves so it could be opened and closed. His theory was that dry, cold air from the icehouse would rush through the opening, absorb moisture from the meats as it cooled them, and then pass off through the ventilator as it became warmer and lighter. According to Schooley's report of his experiment, the arrangement worked successfully throughout the summer of 1854, five thousand dollars' worth of hams and shoulders being cured with only a comparatively small amount of spoilage.[39]

[38] Cummings, *The American Ice Harvests*, 59-60, indicates that at least two improved makes were marketed widely.

[39] Schooley, *op.cit.*, 1-5. Schooley's system was used for fruit storage in 1855. See letter, William Sims, Dayton, O., to Schooley, Feb. 21, 1856, in [Harnden], *Description of Schooley's Patent Re-*

This device was clumsy, but it showed an understanding of the basic problems involved in cooling large storage rooms.

Another experimenter in ice-cooled storage before the Civil War was Benjamin M. Nyce,.of Decatur County, Indiana, who began work in 1856. He seems to have been interested principally in fruit preservation. After building several experimental houses, he developed a system which he patented in 1858. Nyce's storage house was constructed carefully and was exceptionally well-insulated. The ice compartment was placed above the space to be cooled and was completely separated from it by a floor of sheet iron on which the ice rested. Air in the storage space, cooled by contact with the iron, became heavy and fell to the floor, displacing the warmer and lighter air and setting up a natural circulation. Nyce, recognizing the importance of removing excess moisture, accomplished this by exposing an absorbent, calcium chloride, in broad trays on the floor of the storage space. His system worked well enough to find commercial application for fruit after 1860, and as late as 1907 a few houses built on the Nyce design were still in use.[40]

The consumption of ice, stimulated by the greater need of refrigeration and by the decline in prices after 1827, increased greatly. In southern cities the increase came rapidly in the ten years after 1827. Although reliable statistics do not exist, it has been estimated that New Orleans consumed 375 tons for all purposes in 1827, while in 1838 she used 3,600. By 1847 consumption there had risen to 8,026 tons and by 1860 to 24,000. The greatest expansion in northern cities came between 1840 and 1850. During the five years 1843 through 1847 the annual consumption for all purposes of

---

frigerator, 9-10. Schooley believed his process could be used for preserving fruit, producing lard oil during summer, and ventilating buildings. Schooley, *op.cit.*, 15-17.

[40] W. A. Taylor, "The Influence of Refrigeration on the Fruit Industry," U. S. Department of Agriculture, *Yearbook* . . . , *1900* (Washington, 1901), 564; T. L. Rankin, "Prof. Benj. M. Nyce," *Ice and Refrigeration*, VI (1894), 405; M. Cooper, "Cold Storage By Means of Ice," *Transactions of the American Society of Refrigerating Engineers*, III (1907), 45; *Massachusetts Fruit Preserving Company* (Boston, 1866), 11-12, 15-16, 18.

New York City jumped from 12,000 to 65,000 tons, and the ice required by Boston moved from 6,000 to 27,000 tons. By 1856 New York needed 100,000 tons, while by 1860 Boston was using 85,000.[41] Although these figures show consumption for commerical as well as domestic purposes, there is no doubt that they reflect a growing use of ice in the home. Ice continued to be popular for cooling beverages; in New York some families chilled their drinking water with it even during the winter. It still was administered both internally and externally for medicinal ends.[42] More significant was its increasing application to home food preservation. A growing number of city-dwellers between 1830 and 1860 acquired refrigerators, which they employed not merely for cooling beverages, but also for preserving fresh meat and fish, for hardening butter, and "as a general cooler of most articles of food and drink." It is clear that the number of refrigerators in the cities increased, for references to them in a wide variety of publications multiplied with the years. They were viewed more often as necessities, not merely as luxuries to be enjoyed by the wealthy.[43] It is impossible to determine exactly how general was their use, but only a small percentage of city families must have had them, for as late as 1926 about one-half of the homes in Rochester, New York, relied for food preservation throughout the entire year not on the refrigerator, but on the pantry or cellar.[44]

Farmers made little use of ice for domestic food preserva-

[41] Cummings, *The American Ice Industry*, 102, 188. Cummings stresses the absence of reliable statistics on the ice industry and claims no more for these estimates than that they show the general trend of consumption. It required 93 wagons and 150 horses to distribute the 60,000 tons of ice consumed by Boston and its vicinity in 1856. Tudor and Sawyer, *loc.cit.*, 81.

[42] "Ice—How Much of It Is Used and Where It Comes From," *De Bow's Review*, XIX (1855), 709; Cummings, *The American Ice Industry*, 137n. Temperance societies viewed ice "as a promoter of the use of the healthful beverage of cold water, and thus of the cause of temperance." "Ice: and the Ice Trade," *loc.cit.*, 175.

[43] Cummings, *The American Ice Industry*, 134-136; "Ice: and the Ice Trade," *loc.cit.*, 175.

[44] Advertisement of Rochester Ice and Cold Storage Utilities, Inc. reproduced in *Ice and Refrigeration*, LXXI (1926), 302.

tion. Wells and cellars remained the chief means of keeping fresh foods in rural areas, although if there were an icehouse on the farm, it could be utilized, as Daniel Webster did his at Marshfield, for preserving meat and fish and for hardening butter during the summer months. Rural icehouses, however, generally were not constructed to hold food for the dinner table, but to help in the business of producing and distributing dairy products.[45]

Ice was adapted to the market place as well as the home. Dealers handling fresh meat, fresh fish, and butter found it essential to their business in warm weather. As refrigerators were developed large enough for market purposes, they were employed increasingly, but it is difficult to estimate the extent to which they were adopted. Although an observer of 1855 commented on the widespread use of ice in the markets of Boston and other American cities, the deficient refrigeration in many of these markets today raises doubt if any but the most progressive merchants used it before 1860.[46]

Ice found a place in all varieties of public eating establishments. Hotels, saloons, and confectioneries used it in greater quantities. Refrigerators were added to the equipment of passenger vessels, making it possible to have fresh provisions during a lengthy voyage. No longer was it necessary to carry livestock for slaughter aboard ship if fresh meat were to be enjoyed at sea.[47]

The introduction of lager beer into the United States from Germany during the early eighteen-forties made refrigeration a necessity in the brewing industry. American beers until then had been of the top-fermentation type, which, except for

[45] Cummings, *The American Ice Industry*, 68-69; "Ice: and the Ice Trade," *loc.cit.*, 176. Rural icehouses continued to be built as they had been at the beginning of the nineteenth century. Not until 1846 was information on the proper construction of icehouses made generally available to farmers. Cummings, *The American Ice Industry*, 69-70.

[46] Weatherell, *loc.cit.*, 449; "Ice: and the Ice Trade," *loc.cit.*, 176; [Harnden], *Description of Schooley's Improved Patent Refrigerator*, front cover.

[47] Tudor and Sawyer, *loc.cit.*, 81; "Ice—How Much of It is Used and Where It Comes From," *loc.cit.*, 710.

strong ales, required only a few days of fermentation and did not necessitate exacting temperature control. But lager beer was worked by yeast of the bottom-fermentation type, so-called because when it had completed its function, it settled to the bottom of the tank. Afterwards the beer had to be stored at low temperatures. During storage the beer matured, becoming clear, mellow, and charged with carbon dioxide. Temperatures during fermentation had to be kept approximately forty-seven to fifty-five degrees Fahrenheit, while during storage temperatures near the freezing point of water had to be maintained without fluctuation. Lager beer, sparkling, effervescent, and more palatable, was soon in great demand in the United States, for it appealed to beer-drinkers, whose numbers were increasing rapidly with the arrival of German immigrants. To attain the low temperatures so essential to the brewing of lager beer, operations were confined mostly to the winter months, and to provide a place where proper storage temperatures could be maintained throughout the summer, great subterranean cellars were constructed. Sites where the cellars could be extended into natural rock formations were especially desirable, for they provided greater coolness. The beer produced under such conditions varied in quality with the seasons, the climate, and the temperatures prevailing in the cellars. In an effort to make a uniform product throughout the year, brewers turned to ice. It is difficult to determine when the various techniques of applying it to brewing were introduced, but before 1860 they probably were limited to adding it to the water used to cool the hot wort and to piling it around the fermentation and storage tubs. Although these techniques required improvement, by 1860 they had made it possible to brew a uniform product throughout the year.[48]

[48] J. P. Arnold and F. Penman, *History of the Brewing Industry and Brewing Science in America* . . . (Chicago, 1933), 54, 56, 61, 93; V. S. Clark, *History of Manufactures in the United States, 1607-1860* (Washington, 1916), 481; E. H. Vogel, Jr., et al., *The Practical Brewer, A Manual for the Brewing Industry* (St. Louis, 1946), 102, 186, 190; G. J. Patitz, "Progress in Brewery Refrigeration," *Ice and Refrigeration*, LI (1916), 166-167; J. E. Siebel, "Refrigeration and

Problems of Food Supply, 1830-1860

The problem of supplying highly perishable foods to cities located many miles away from producing areas brought forth pioneer efforts at refrigerated transportation. As railroads extended during the eighteen-forties into the hinterlands of eastern cities, they were used to bring milk from areas that began to specialize in its production.[49] The Erie Railroad brought more than three million quarts into New York in 1842-1843 and over nine million in 1848-1849. As early as 1843 the farmers of Orange County had learned the value of refrigeration in marketing their milk. In the vicinity of Goshen milk was taken from cows early in the morning, poured into cans, and cooled by stirring it with a tin tube filled with ice. Many dairies, of course, depended on cold spring water for cooling. The milk was then sent to the railroad for the haul of four and one-half hours into New York City, where the tin tube again was inserted to keep the milk sweet until sale.[50] In 1851 refrigerator cars were used to bring butter from as far as Ogdensburg, New York, to Boston; in 1853 and 1854 refrigerator-car service for butter shipments was extended to Rutland and Franklin counties, Vermont.[51]

In the eighteen-forties ice was used to convey fresh sea-

---

the Fermenting (Brewing) Industry in the United States," *Premier Congrès International du Froid* . . . , III (Paris [1908?]), 72; F. P. Siebel, Jr., "Beer," *Encyclopedia Britannica* (Chicago, 1946), III, 314; J. C. Goosman, "History of Refrigeration," *Ice and Refrigeration*, LXIX (1925), 149. Lager beer was named after the storage phase of the brewing process; *lager* is the German word for storehouse.

[49] Before the railroad New York City depended for most of its milk on mash-fed cows kept under unsanitary conditions in the heart of the city. As late as 1853 more than half of the milk supply of New York came from this source. Cummings, *The American and His Food*, 53, 55.

[50] *ibid.*, 53-55; "Milk and Railroads," *The Cultivator*, x (June 1843), 1; J. Mullaly, *The Milk Trade in New York City* (New York, 1853), 84-85.

[51] Letter from J. Wilder, Bristol, Tenn., to the Editor of *The Railroad Gazette*, May 1, 1886, in *The Railroad Gazette*, XVIII (May 14, 1886), 325; E. Wiest, "The Butter Industry in the United States: An Economic Study of Butter and Oleomargarine" (Faculty of Political Science of Columbia University, eds., *Studies in History, Economics and Public Law*, LXIX, Number 2) (New York, 1916), 140-141.

foods to inland cities. Before that time there had been occasional shipments of fish packed in ice, and it was not unusual for fish frozen by winter weather to be transported two or three hundred miles inland from Boston and New York.[52] As with all early applications of refrigeration it is difficult to estimate the extent of the traffic. Albany-to-Buffalo shipments of oysters by express started in the early forties. By 1844 salmon from Maine and cod and other fish from Boston were being packed in ice and sent by rail to western New England, with some lots going as far west as Buffalo. Boston was selling cod and haddock to Chicago by 1852. Oysters were carried inland from Baltimore and New Haven.[53] About the same time that ice was applied to transporting seafood to the interior, fishing vessels began to carry it for keeping their catch fresh until unloaded. Ice was used on boats supplying the Philadelphia market in 1816, and its presence on a Gloucester halibut smack in 1838 has been recorded. By 1845 it was commonly carried. At first the ice was placed in a corner of the hold and kept separate from the fish, but when it was discovered that it did no harm for the fish to come in contact with it, the pack was made directly in crushed ice. As a result of these trials there was a decline in the trade in live fish north of Cape Cod.[54]

Refrigerated transportation before 1860 was limited for the most part to seafoods and dairy products.[55] Foods generally

[52] C. H. Stevenson, "The Preservation of Fishery Products for Food," U. S. Fish Commission, *Bulletin*, xviii (1898), 368.

[53] A. H. Clark, "Notes on the History of Preparing Fish for Market by Freezing," U. S. Commission of Fish and Fisheries, *Bulletin of the United States Fish Commission*, vi (1886), 467; "The Ice Trade," *Hunt's Merchants' Magazine*, xi (1844), 378; Cummings, *The American and His Food*, 59-60; Cummings, *The American Ice Industry*, 165-166.

[54] Cummings, *The American Ice Industry*, 55; Clark, *loc.cit.*, 467; Stevenson, *loc.cit.*, 359; H. F. Taylor, *Refrigeration of Fish* (U. S. Department of Commerce, Bureau of Fisheries, *Document No. 1016*) Appendix viii to the Report of the U. S. Commissioner of Fisheries for 1926 (Washington, 1927), 622; Tudor and Sawyer, *loc.cit.*, 81.

[55] Game probably was carried under refrigeration by express companies before 1860. Cummings, *The American Ice Industry*, 172-173; H. Wells, *Sketch of the Rise, Progress, and Present Condition of the Express System* (Albany, 1864), 21. In 1857 fresh meat and poultry

were sent in small containers, for although the fifties saw experiments with refrigerator cars for transporting fresh meats, poultry, and dairy products,[56] a practical car was not developed. Long-distance refrigerated shipments were first made in package lots, and to provide the careful supervision required, express companies specialized in them.[57] Fast-freight lines, which appeared in the fifties, aided the transportation of perishable food by providing through-routing over the many independent railroads.[58]

The application of refrigeration to the problems of supplying cities with perishable foods was extremely limited in comparison with that of today. Ice was not used to bring in fruits and vegetables. Although there were experiments in fruit storage before 1860, there were no great cold-storage houses to prolong the life of fresh fruits for the benefit of city-dwellers. Railroad and steamship lines by 1860 had extended the areas from which cities could obtain this fresh produce,[59] but there were no refrigerator cars or ships in the traffic.[60] Their absence limited the distance from which shipments could be made and added to the risk of those that were attempted. Since there was no significant use of refrigeration for eggs and poultry, supplies of these foods alternated between seasons of

---

shipments from the West to New York were made by means of a "rail car, fitted up on the principle of a refrigerator." "Transporting Produce," *Scientific American*, XIII (Nov. 7, 1857), 70.

[56] "Transporting Produce," *loc.cit.*, 70; L. D. H. Weld, "Private Freight Cars and American Railways" (Faculty of Political Science of Columbia University, eds., *Studies in History, Economics and Public Law*, XXXI, Number 1) (New York, 1908), 12; Letter from J. Wilder, Bristol, Tenn., to the Editor of *The Railroad Gazette*, *loc.cit.*, 325.

[57] Cummings, *The American Ice Industry*, 172-173; Wells, *op.cit.*, 20-21.

[58] Cummings, *The American Ice Industry*, 173-174.

[59] Cummings, *The American and His Food*, 57-59; Weld, *loc.cit.*, 58-59; U. S. Department of the Interior, Census Office, *Twelfth Census of the United States, Taken in the Year 1900* (Washington, 1901-1902), VI, 304.

[60] Tudor and others shipped perishable foods south with coastwise ice cargoes, but there was no northbound traffic, although Tudor did attempt to bring tropical fruit from Cuba in 1816. Occasional attempts were made to send foods to Europe under refrigeration. Cummings, *The American Ice Industry*, 49, 171-172.

glut and scarcity.[61] There was some refrigeration of dairy products during transportation, but there were no large cold-storage warehouses for holding butter over from the season of surplus production. Low temperatures for processing dairy products were attained by deep cellars or, preferably, spring water to which ice sometimes was added when available.[62]

Little refrigeration was to be found in slaughtering and distributing meat before the Civil War. Although even before 1830 some butchers in northern cities had icehouses,[63] and after 1830 there was a growing use of refrigerators in marketing meat, low temperatures were sought only for short-term preservation. Refrigeration was not applied, save for a few tentative attempts, to the problem of providing meat for the cities of the Atlantic coast.

The growth of eastern cities created a serious problem of meat supply. Slaughtering became a specialized industry, but slaughterhouses, because of the lack of refrigeration, had to be located in the centers of cities. If they were situated too far out, there was danger that meat would spoil before it could be carried to market. A slaughterhouse in the heart of the city was a problem with which municipalities had to contend for years.[64]

As cities grew, the livestock resources of the Atlantic seaboard proved inadequate to satisfy the demand. The earliest method of augmenting the supplies of eastern cities was to drive animals on the hoof from the West, eventually from as far as the valley of the Ohio River. As early as 1805 cattle from Kentucky and Ohio were driven to Philadelphia and Baltimore, and in 1817 the first cattle from the West reached New York. The traffic in cattle and sheep was greater than in hogs, for hogs lost more weight and suffered greater deterioration in meat quality on the road. Besides, pork was a

[61] For a sketch of the egg and poultry industry to 1880 see M. A. Jull et al., "The Poultry Industry," U. S. Department of Agriculture, *Agriculture Yearbook, 1924* (Washington, 1925), 383-396.

[62] H. E. Alvord, "Dairy Development in the United States," U. S. Department of Agriculture, *Yearbook . . . , 1899* (Washington, 1900), 387-388; Wiest, *loc.cit.*, 14.

[63] Cummings, *The American Ice Industry*, 60.

[64] *ibid.*, 163-164.

better product when cured or pickled than beef or mutton. When railroads to the East became available, livestock droving declined, and an enormous trade developed in freighting live animals to eastern markets in stock cars.[65] Since hogs still suffered more than cattle and sheep, fewer were shipped.[66]

The second way of augmenting the meat supplies of eastern cities was the development of slaughtering in the Ohio valley, close to the source of raw materials. Since there were no means of shipping fresh meat to the East, it was preserved by salting and smoking and then packed in barrels.[67] The location of the first big packing center in the West, Cincinnati, was determined partially by the fact that refrigeration had not been applied to packing, for nearby were large supplies of salt for preserving the meat, and the city was blessed with ideal winter temperatures for processing. Throughout the winter it was cold enough to prevent spoilage while curing was underway, but it was not often sufficiently cold to freeze the meat and interfere with cutting.[68] More pork was packed than beef

[65] L. D. H. Weld, "The Packing Industry: Its History and General Economics," *The Packing Industry: A Series of Lectures Given under the Joint Auspices of the School of Commerce and Administration of the University of Chicago and the American Institute of Meat Packers* (Chicago, 1924), 69-70; R. A. Clemen, *The American Livestock and Meat Industry* (New York, 1923), 72-74, 78-82.

[66] H. C. Hill, "The Development of Chicago as a Center of the Meat Packing Industry," *The Mississippi Valley Historical Review*, x (1923-1924), 254.

[67] Weld, "The Packing Industry," *loc.cit.*, 70.

[68] Hill, *loc.cit.*, 254; C. T. Leavitt, "Transportation and the Livestock Industry of the Middle West to 1860," *Agricultural History*, viii (1934), 23.

~~~~~~

FIGURE 1

Ice harvesting and storing in the eighties

The field has been prepared for the harvest by the scraper and the ice plane at the lower left. Strips of ice into which rectangular grooves have been cut are being poled through channels to the steam-powered elevators. Here they are being split into pieces of manageable size with iron bars. Note the long shafts and belts used to transmit the power.

From W. P. Blake, "Mining and Storing Ice," *Journal of the Franklin Institute*, LXXXVI 3rd s. (1883), 363.

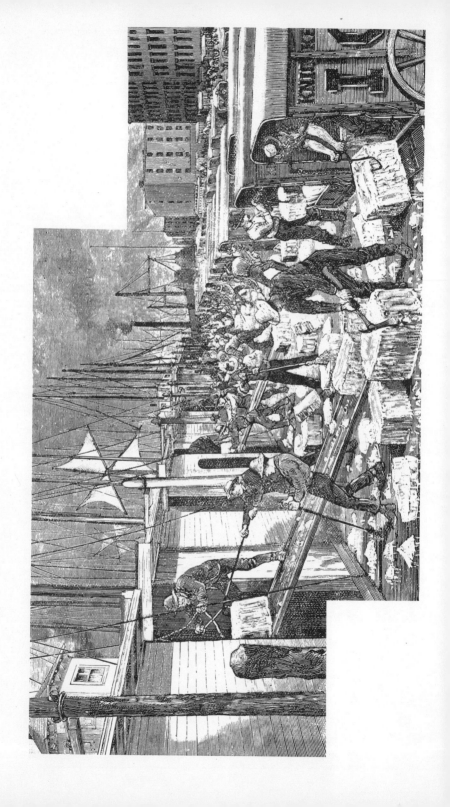

or mutton. Beef, however, was packed in increasing amounts west of Cincinnati.[69]

The industry was hampered by the absence of refrigeration. Curing everywhere was a winter operation, for only then were the temperatures low enough to enable the preservative qualities of salt or smoke to take effect before spoilage set in.[70] With the coming of the railroads, some fresh meat was shipped for considerable distances in the winter and much shorter distances during the cool of summer nights, but the risk of spoilage was a strict limitation and precluded western packers from sending fresh meat to eastern markets.[71] During the fifties attempts were made to remove these disadvantages by the application of refrigeration. Schooley was experimenting in Cincinnati in 1854 with a means of using ice to permit summer curing of pork, but he was not the first to do so.[72] Summer curing made possible by ice refrigeration seems to have begun in Chicago in 1858, but was not practiced on a large scale there until the early seventies.[73] Although there were experiments with refrigerator cars by 1857,[74] many years were to pass before the problem of shipping fresh meat for long distances was solved.

[69] Hill, *loc.cit.*, 254; Clemen, *op.cit.*, 79.
[70] Hill, *loc.cit.*, 271; Clemen, *op.cit.*, 110.
[71] Cummings, *The American Ice Industry*, 161-162; Weld, "The Packing Industry," *loc.cit.*, 70.
[72] Schooley, *op.cit.*, 1-5.   [73] Hill, *loc.cit.*, 271.
[74] "Transporting Produce," *loc.cit.*, 70.

Figure 2

Unloading Hudson River ice barges at New York, 1884

The ice is being removed from barges which were loaded at up-river points. The sailcloth windmills mounted on the deckhouses supplied power to pumps which removed meltwater from the holds. Economy of manpower apparently was no object, for there are four or five laborers visible for every wagon. At a modern ice plant essentially the same work is done by two men. One worker takes a three-hundred-pound block from storage and places it in a scoring machine, which delivers it down a chute to the loading dock, where the deliveryman, usually unassisted, stows it in his truck.

From *Harper's Weekly*, XXVIII (August 30, 1884), 562.

The modification of American diet contributed to the increased use of refrigeration between 1830 and 1860, while in turn refrigeration itself changed food habits among city-dwellers. In the home the refrigerator made possible economical eating of fresh foods, and by making them more palatable, stimulated their substitution for the traditional diet of salted, spiced, and smoked foods.[75] Beyond the home the first steps were taken in using refrigeration to close the gap between producer and consumer. The thirty years after 1860 were to see great progress in its application to this vital problem of food supply.

[75] For a complete discussion of the effect of the household refrigerator on diet see Cummings, *The American Ice Industry*, 140-145, 149, 151.

CHAPTER III

THE GROWING NEED FOR MECHANICAL
REFRIGERATION, 1860-1890

THE rapid expansion of American cities after the Civil War crowded into limited areas unprecedented multitudes that had to draw their fresh-food requirements from afar. The resources of the entire nation were needed to supply a great metropolis like New York. Between 1860 and 1890 refrigeration produced by natural ice was applied increasingly to the preservation of perishable foods.

The methods of harvesting and storing ice had reached a high state of perfection by 1860 because of the genius of Wyeth. No basic changes in technique were made during the next thirty years. Operations were so efficient that in 1883 the average cost of cutting the ice and packing it away was estimated to be only twenty-five cents a ton, while with an unlimited supply of good ice ten to twelve inches thick costs might run as low as twelve cents a ton.[1] During the seventies there were experiments with steam-powered circular saws, but these were too heavy and cumbersome to be successful.[2] Swing-guide-markers and ice-cutters with practical adjustable teeth were placed on the market in the eighties, but there were no radical changes in these basic tools.[3] There was, of course, a more general application of the power machinery that had been developed earlier. Icehouses continued to be improved slowly, and although they remained the same in principle, the greater demand for refrigeration resulted in the construction of huge storages that would hold sixty thou-

[1] W. P. Blake, "Mining and Storing Ice," *The Journal of the Franklin Institute*, LXXXVI 3rd s. (1883), 366-377. For descriptions of harvesting and storing operations in the eighties see *ibid.*, 355-356, and T. M. Prudden, "Our Ice-Supply and Its Dangers," *The Popular Science Monthly*, XXXII (1887), 671-676.

[2] H. Hall, "The Ice Industry of the United States," U. S. Department of the Interior, Census Office, *Tenth Census, 1880*, XXII (Washington, 1888), 17.

[3] W. E. Wood, "Twenty Five Years' Development in Harvesting and Housing Ice," *Ice and Refrigeration*, LI (1916), 190-191.

sand tons. The best houses were very efficient, for they would keep ice two or three years with no more than ten to twenty-five per cent wastage.[4] In spite of the efficient cutting and storing facilities, harvesting was still a ticklish task in areas where mild winters occurred. If the ice were only six inches thick, it was a great temptation to hold off in the hope that it would thicken by another two or three inches. But during the delay a freshet might spoil the crop. If all the ice were swept away, there still was a chance for a new formation, but if the dirt and debris carried down from upstream became mixed with the ice and the whole mass froze, the harvest probably was lost for the year.[5]

The export trade expanded briefly after 1860 and then dwindled away to practical extinction by the end of the century. In 1860 over 49,000 tons were sent out. By 1870 exports had grown to almost 66,000 tons, but by 1900 had dropped below 14,000.[6] In 1880 the company that continued Tudor's business after his death was forced to abandon its shipments to the East Indies by competition from manufactured ice in the ports to which ice had been sent from New England for almost half a century.[7] Other exporters, unable to compete with the machine, withdrew rapidly from their markets.[8]

After the Civil War Boston lost its pre-eminent position as a source of ice for the coastwise trade. In years of good winters during the early eighties only about 175,000 tons were shipped from Boston to both foreign and domestic ports, while the Maine harvest annually was over a million tons, only an insignificant part of which was consumed in the state itself. The great advantage of Maine was its dependable supply of good, thick ice. Harvesting there received an impetus during the Civil War, when James L. Cheeseman, who had an army contract to supply ice to Union troops in the southern states, began operations on the Kennebec River. Rapid expansion of

[4] Hall, *loc.cit.*, 9-10.   [5] Prudden, *loc.cit.*, 674.
[6] A. L. Hunt, "Manufactured Ice," U. S. Department of the Interior, Census Office, *Twelfth Census, 1900*, IX (Washington, 1902), 688.
[7] Hall, *loc.cit.*, 3.
[8] *Ice Trade Journal*, XI (May 1888), 2.

the Maine industry began in 1870, when there was an almost complete failure of the crop on the Hudson and the Schuylkill rivers. For several years Cheeseman, who got his start in the industry as a New York harvester and retailer, supplied ice for the export trade. The bulk of the cut in Maine, however, was sent by sea to New York, Philadelphia, Baltimore, Washington, and the southern ports. In the early eighties, sixty-five to seventy per cent was harvested on the Kennebec River, with the remainder cut on the Penobscot, the Cathance, and along the coast, principally at Booth Bay, Wiscasset, Damariscotta, and Portland. Especially large amounts were harvested during mild winters when the crop failed farther south. In the winter of 1879-1880, for example, about 1,300,-000 tons were gathered in Maine, while in the winter of 1889-1890 the harvest was over 3,000,000 tons.[9]

The coastal trade to the southern states suffered a decline in the decades after the Civil War because of the increasing competition encountered from manufactured ice. During the early years of the war shipments fell off, for the great ice-consuming ports of the South were blockaded.[10] The trade revived after the war, with shipments being made to Wilmington, Charleston, Savannah, Jacksonville, Appalachicola, Pensacola, Mobile, New Orleans, and Galveston. Savannah served as a rail gateway to Augusta, Macon, Atlanta, and other interior towns.[11] So rapid was the extension of ice-making machines, however, that by the early eighties the market for the natural product in the interior had been almost completely lost, and natural-ice firms were confined to the trade of the port cities.[12] Competition in the South from manufactured ice had become so great by 1888 that the *Ice Trade Journal*, the organ of the industry, predicted the coastwise traffic would become extinct in a few years unless its rapid decline could be checked.

[9] Hall, *loc.cit.*, 21-24; Hunt, *loc.cit.*, 688; *Ice Trade Journal*, XIII (Jan. 1890), 2.

[10] L. Weatherell, "The Ice Trade," U. S. Department of Agriculture, *Report of the Commissioner, 1863* (Washington, 1863), 449.

[11] Hall, *loc.cit.*, 35.

[12] *ibid.*, 5; Letter from Savannah, Ga., July 3, 1886, in *Ice Trade Journal*, x (Aug. 1886), 2.

The *Journal* suggested that ships hauling ice be equipped with refrigerating machinery in order to eliminate wastage from melting, for this would make possible lower prices and improve the competitive position of natural ice. The refrigerated ships, it was observed, could carry cargoes of perishable fruit on their return trip.[13] This suggestion was never tried, and ladings to the South continued to decline. By 1890 the natural-ice trade had almost disappeared from New Orleans, which for many years had been the leading market.[14]

The trade to the middle states, stimulated by the needs of New York, Philadelphia, Baltimore, and Washington, expanded after the Civil War. A growing use of ice in these cities caused greater dependence on other than local sources. Philadelphia relied regularly on Maine for about one-half of its requirements. The technique of the large-scale shipments to the middle states was carefully worked out and efficiently managed by big companies. The Knickerbocker Ice Company of Philadelphia, for example, devised an endless-chain conveyor for removing ice from the schooners and depositing it in wharfside warehouses.[15] This trade had a longer life than that to the South, for ice-manufacturing plants did not become important north of the Potomac as early as south. In 1889 there were in the southern states one hundred and sixty-five plants that made ice for sale in comparison with only fourteen in the middle states.[16] The trade to New York, Philadelphia, Baltimore, and Washington lasted well into the twentieth century, and as late as 1908 these cities received a large part of the 600,000 to 800,000 tons then harvested annually in Maine.[17]

[13] *Ice Trade Journal*, XII (Dec. 1888), 1.

[14] "Ice-Making in New Orleans," *Harper's Weekly*, XXXIV (Jan. 25, 1890), 67.

[15] Hall, *loc.cit.*, 19, 30.

[16] Hunt, *loc.cit.*, 677. Included as southern states were West Virginia, Virginia, North Carolina, South Carolina, Georgia, Florida, Kentucky, Tennessee, Alabama, Mississippi, Arkansas, Louisiana, Indian Territory, Oklahoma, and Texas. Included as middle states were New York, New Jersey, Pennsylvania, Delaware, Maryland, and the District of Columbia.

[17] S. S. Van der Vaart, "Growth and Present Status of the Refrig-

Towns of the interior South continued to receive ice from north of the Ohio River during the seventies and early eighties. Tennessee and Georgia obtained it by rail from the Ohio River region, while the lower Mississippi relied on the Missouri and upper Mississippi in a trade that centered at St. Louis. Much upper-river ice was cut on the Illinois. Some of it was sent as far as Texas.[18]

In the region north of the Potomac and the Ohio a steady growth marked the natural-ice industry between 1860 and 1890. The cities and towns there were able usually to exploit some local supply, and it became increasingly easy, where demand outran supply, to supplement local resources from distant points. The cities of the eastern seaboard could depend on obtaining Maine ice from coastal schooners. In the West railroads served commonly in ice transport. Rail shipments were unusual in the East, but in some cases roads acquired harvesting facilities to build up their freight traffic. In 1878 ice was railroaded successfully from Maine to Cincinnati.[19]

Boston continued to take its ice from the ponds that had served the city so many years. New York City cut some of its supply on lakes in the lower counties of the state, but its chief source was the Hudson River from Poughkeepsie to Troy, which could provide from 2,000,000 to 2,750,000 tons in a fairly good year. New York's needs usually were rounded out from the Lake Champlain region and from Maine. In the summer of 1880, when ice was scarce because of the failure of the winter crop, 257,000 tons were cut from the Kennebec River in addition to cargoes from other parts of Maine. Even this was not enough to make up the city's deficit, for 15,000 tons were imported from Canada and 18,000 from Norway. Ice was carried down the Hudson in barges built especially for the purpose. These scows, which were towed in fleets of

---

erating Industry in the United States," *Premier Congrès International du Froid*, iii (Paris [1908?]), 331.

18 Hall, *loc.cit.*, 34-35.

19 R. Maclay, "The Ice Industry," C. M. Depew, ed., *One Hundred Years of American Commerce*, ii (New York, 1895), 468.

six to twelve by steam tugs, could carry from four to eleven hundred tons stowed in holds and deckhouses. A small windmill installed on the top of the deckhouse drove a pump which removed meltwater from the hold. At some New York wharves steam power was available for unloading the blocks. On removal from the barge the ice was transferred to wagons for distribution.[20] After the Civil War Philadelphia expanded its local harvesting operations, using ponds in northeastern Pennsylvania and the Delaware and Lehigh rivers in addition to the Schuylkill. So great had the requirements of Philadelphia become that the city depended regularly on Maine for about one-half of its supply. Cincinnati drew mostly from natural or artificial ponds within a radius of one hundred miles from the city in Ohio and Indiana, and Cleveland depended on Lake Erie, the Cuyahoga and Rocky Rivers, and surrounding ponds. Chicago was able to obtain ice from nearby lakes in northern Indiana and Illinois and in Wisconsin, but the local harvest in the vicinity of St. Louis was so uncertain that it was necessary there to draw upon Illinois, Iowa, and Minnesota. On occasion St. Louis had to bring in ice from Toledo and even from Canada.[21]

San Francisco, the principal market for ice on the Pacific Coast, continued to receive supplies from Alaska after the Civil War, but with the construction of the Central Pacific Railroad, the resources of the Sierra Nevadas were opened. At first freight rates were high, permitting the Alaska product to compete, but eventually they were lowered, and about 1880 the Alaska enterprise collapsed. In 1890 the major part of San Francisco's requirements were met by natural ice from the mountains, with the balance supplied by manufactured ice, which had been in the picture since the seventies.[22]

It is difficult to generalize about prices, for they varied

[20] For an illustration of ice being removed from the barges and transferred to delivery wagons see Figure 2.

[21] Hall, *loc.cit.*, 17, 23-24, 26-27, 30-34; Maclay, *op.cit.*, 468.

[22] Hall, *loc.cit.*, 37; E. L. Keithahn, "Alaska Ice, Inc.," *Pacific Northwest Quarterly*, xxxvi (1945), 123-131; Letter from San Francisco, Calif., Aug. 5, 1890, in *Ice Trade Journal*, xiv (Sept. 1890).

from place to place, from year to year, and from month to month. When the harvest was bountiful, prices were low, but in years of scarcity they were almost prohibitive.[23] In southern ports, before manufactured ice became available, the regular retail price was twenty dollars a ton. The rate never went below ten dollars a ton and often rose to thirty. In years of yellow-fever epidemics, when demand was abnormal, prices often soared to sixty or seventy-five dollars a ton. The New York retail price was one dollar a hundred pounds during a short period in the summer of 1880, for the crop had failed the preceding winter. In the summer of 1881, however, the retail charge to small consumers ranged from four to eight dollars a ton. Responsible for the drop was the good harvest of 1880-1881.[24] Ice usually was sold to families not by the ton or hundredweight, but by flat seasonal or monthly rates for the delivery daily of a specified quantity, usually ten to thirty-six pounds.[25]

Large companies held leading positions in the business of distributing in cities, for considerable investments were required. About 1880, for example, the Knickerbocker Ice Company of Philadelphia had $300,000 invested in distribution facilities and employed eight hundred men. The big firms, however, faced competition from independent dealers and from peddlers, whose only capital might consist of a single wagon and team.[26] Notwithstanding the existence of large companies, the business was by no means stable. During seasons of shortage speculators appeared, tried to corner supplies, and made price fluctuations worse than they otherwise would have

[23] In 1890, a year in which there was a serious failure of the ice crop, a comedian in a Fay Templeton play at the 14th Street Theater, New York, won the applause of the house with a bit of byplay on the price of ice. After finding a drink of warm water unpalatable, he unlocked a safe, removed a two-inch piece of ice tied with a string, and worked it in the water for a few seconds. Then, without a word or smile, he locked the ice in the safe again. *Ice Trade Journal*, XIV (Oct. 1890), 8.

[24] Hall, *loc.cit.*, 27, 35.

[25] *Ice Trade Journal*, VII (May 1884), 2; XIII (June 1890), 5.

[26] Hall, *loc.cit.*, 27, 30, 33.

been,[27] while in years when ice was plentiful, business some-times was demoralized by price wars.[28]

Between 1860 and 1890 great progress was made in de-veloping equipment for using ice—refrigerators, cold-storage houses, and refrigerator cars. Many refrigerators continued to be mere ice chests, differing little from those of the eight-een-thirties. For example, the Davis refrigerator, patented by William Davis of Detroit in 1868, was a well-insulated, zinc-lined box, equipped with a galvanized-iron ice receptacle at one side.[29] Better models, however, with ice compartments at the top, were available.[30] The Cold-Blast refrigerator, patented in 1881 by Andrew J. Chase of Boston, was based on the correct principle of internal circulation. Chase realized that air, coming in contact with the ice, became drier and heavier and flowed to the bottom of the provision chamber, where it became warmer, absorbed moisture, and rose to pass again over the cold surfaces of the ice.[31] But the principle upon which refrigerators worked was still widely misunderstood. A Lowell, Massachusetts, newspaper, for example, advised its readers to keep ice consumption down by wrapping the block in cloths.[32]

A variety of systems was developed for using ice to cool storage rooms or buildings. During the sixties storage houses

---

[27] *Ice Trade Journal*, II (Oct. 1878), 2; XIV (Nov. 1890), 4.

[28] Such a year was 1883. The *Ice Trade Journal*, VIII (Aug. 1884), 2, printed a warning of the dangers of price wars:

> "Ah! long shall the ice men remember the day
> That led to this foolish and fatal affray;
> And look back with horror, the pale ghosts to see
> Of comrades who fell in the year '83.
> Let this be a warning to all who remain
> To guard in the future 'gainst the fate of the slain;
> Nor think that by quarrels or greed you'll grow rich,
> Such actions but lead, like the blind, to the ditch. . . ."

[29] *The Davis Refrigerator* (Boston [1870]), 3-4, 6-7. The Davis refrigerator was used on Pullman dining cars.

[30] E. M. Bacon, *The Growth of Household Conveniences in the United States from 1865 to 1900* (Unpublished doctoral dissertation, Radcliffe College), 94-95.

[31] [A. J. Chase], *The Cold-Blast Refrigerator* (Boston [1881?]), 6.

[32] *Ice Trade Journal*, XXI (July 1898), 4.

of the type designed by Benjamin Nyce were built at cities as widely separated as Indianapolis, Cleveland, and New York. Although some houses constructed on the Nyce principle were in service as late as 1907, the Nyce system was neglected after a few years, when improved designs for ice-cooled storages were introduced.[33]

A common improved type, of which the Jackson system was an example, featured an overhead ice bunker so arranged that the air of the provision chamber was cooled and partially dried by circulation over the ice surface. Temperatures as low as thirty-five to thirty-eight degrees Fahrenheit were possible with such storages, which combined the benefits of air circulation with the advantages of cheapness and simplicity. Overhead-bunker systems were adapted to buildings of several stories by massing the ice at the top and using shafts to carry the cooled air to lower floors. Systems were tried in which the bunker was placed at the side of the room to be cooled, but these encountered the problem of inadequate air circulation.[34] In some large cold-storage houses the air was forced over the ice and through the provision chambers by fans.[35]

Since the lowest temperature obtainable from ice alone was about thirty-five degrees Fahrenheit, there were early attempts to gain lower levels by mixing salt with the ice. A certain amount of heat, called the latent heat of solution, was required for the ice and salt to mix to form a liquid; since this heat was taken from the mixture itself, the temperature of the mixture fell below the freezing point of water to a degree that depended chiefly on its relative proportions.[36] This principle was

[33] T. L. Rankin, "Prof. Benj. M. Nyce," *Ice and Refrigeration*, VI (1894), 405; W. A. Taylor, "The Influence of Refrigeration on the Fruit Industry," U. S. Department of Agriculture, *Yearbook, 1900* (Washington, 1901), 565; M. Cooper, *Practical Cold Storage . . .*, 2nd edn. (Chicago, 1914), 344; M. Cooper, "Cold Storage By Means of Ice," *Transactions of the American Society of Refrigerating Engineers*, III (1907), 45.

[34] Cooper, "Cold Storage By Means of Ice," *loc.cit.*, 46-47; M. B. C., "Cold Storage," *Garden and Forest*, VII (1894), 352.

[35] *Ice Trade Journal*, XII (Feb. 1889), 8; F. A. Horne, "Development of the Cold Storage Industry," *Ice and Refrigeration*, LI (1916), 170.

[36] J. T. Bowen, *The Application of Refrigeration to the Handling*

employed about 1865 by A. and M. Robbins, dealers at Fulton Market in New York. They suspended V-shaped galvanized-iron tanks from the ceiling of a storage room in such a way that the tanks could be filled with ice and salt from the room above. With this arrangement it was possible to attain temperatures sufficiently low to freeze large quantities of poultry and game.[37] Ice-and-salt mixtures, necessary where freezing work was to be done, were used as adjuncts to other systems as well as independently. In some cases the freezing mixtures were placed directly in the provision rooms; in others the tanks were located in the upper part of the storage house so that the brine could flow by gravity through pipes in the rooms to be cooled.[38]

The improvement of methods for cooling storage space with ice was accompanied by the growth of the cold-storage industry. Some of the earliest firms operated storages constructed on the Nyce plan, and the houses Nyce built in the sixties at Covington, Kentucky and at Indianapolis may have been the first profitable commercial refrigerated warehouses.[39] His houses were built in other cities. They were introduced in Boston by the Massachusetts Fruit Preserving Company, which was incorporated in 1866 for "the purpose of carrying on the business of preserving fruits, foreign and domestic, in the city of Boston. . . ."[40] Cold-storage warehousing is said to have been placed on a commercial basis in New York in 1875,[41] starting as a branch of the general-warehousing industry. At first old buildings and lofts were used, but gradually separate structures were erected, several stories in height.

---

of Milk (U. S. Department of Agriculture, *Bulletin No. 98*) (Washington, 1914), 14.

[37] Taylor, *loc.cit.*, 565.

[38] Cooper, "Cold Storage By Means of Ice," *loc.cit.*, 47; M. E. Pennington, "Fifty Years of Refrigeration In the Egg and Poultry Industry," *Ice and Refrigeration*, CI (1941), 45; M. B. C., *loc.cit.*, 352.

[39] Rankin, *loc.cit.*, 405; Taylor, *loc.cit.*, 565.

[40] *Massachusetts Fruit Preserving Company* (Boston, 1866), 3, 7-8. George B. Loring, Commissioner of Agriculture in the Garfield and Arthur administrations, was president of the company.

[41] M. B. C., *loc.cit.*, 352.

Sawdust and shavings were the principal insulating materials.[42] One cold-storage house in the Middle West, the largest in the world refrigerated by ice and salt, had a total capacity of 1,800,000 cubic feet.[43]

Practical refrigerator cars—the solution to the problem of overland transportation of perishables—were developed in the quarter-century that followed the Civil War. The idea went back at least to 1842, when the Boston and Albany Railroad proposed to run refrigerator cars so that Michigan game might be carried East in return for westward shipments of fresh cod. These plans did not mature,[44] but about 1848 unsuccessful experiments were made in forwarding dressed veal and lamb in primitive refrigerator cars on the Concord and Nashua Railroad, while in 1851 butter was moved successfully in refrigerator cars from northern New York to Boston.[45] In the summer of 1857 fresh meats and poultry were being brought to New York City from the "Western States" in "a new description of rail car, fitted up on the principle of a refrigerator."[46] Experiments in transporting fresh meat from Chicago to the East were made in the early sixties by the Michigan Central Railroad. Box cars were equipped with ice bins built about three feet above the floor at each end. Trial shipments showed that these cars lacked efficiency and con-

[42] Horne, *loc.cit.*, 170; *Ice Trade Journal*, XII (Feb. 1889), 8. Hay, cottonseed hulls, fibrous mineral materials, cinders, charcoal, air spaces between double walls, and pitch and asphalt poured between masonry walls also served as insulation. "What the Insulants Have Contributed," *Refrigerating Engineering*, XXVIII (1934), 313.

[43] I. C. Franklin, "The Service of Cold Storage in the Conservation of Foodstuffs," U. S. Department of Agriculture, *Yearbook . . . , 1917* (Washington, 1918), 364.

[44] A serious obstacle to this project was the existence of many short rail lines with tracks of varying gauges. These made long-distance through-shipments impossible. Effective steps to overcome this obstacle were taken in the late fifties and early sixties. R. O. Cummings, *The American and His Food* (Chicago, 1940), 60-61.

[45] Letter from J. Wilder, Bristol, Tenn., to the Editor of *The Railroad Gazette*, May 1, 1886, in *The Railroad Gazette*, XVIII (May 14, 1886), 325.

[46] "Transporting Produce," *The Scientific American*, XIII (Nov. 7, 1857), 70.

venience.[47] About the same time as the Michigan Central attempts the Pennsylvania Railroad allocated thirty box cars to experimental work directed by W. W. Chandler, head of the Star Union Transportation Company, a fast-freight line. These were insulated with sawdust, but the only provision for refrigeration was a box of ice placed on the floor in the middle of the car. Later this method was modified by suspending a container for ice from the ceiling at each end. The principal defects of this, as of the Michigan Central car, were that no arrangements were made for air circulation or for icing from the outside.[48]

Many patents for refrigerator cars were issued in the late sixties and early seventies, but a large number of the designs were too complicated to be practical.[49] In November 1867 J. B. Sutherland of Detroit was issued the first United States patent. Sutherland's car, although it does not seem to have been used to any great extent, was much like modern refrigerator equipment, for it had ice bunkers at each end and provision for air circulation.[50] The car used most extensively in the early seventies was that patented in 1868 by another resident of Detroit, William Davis, who was interested primarily in the transport of fish. An early version of the Davis car may have been cooled by vertical cylinders, one at each corner, filled with an ice-and-salt mixture, but by 1870 wedge-shaped metal tanks were substituted. These were fitted along the sides of the car from top to bottom and so designed that they could be filled through hatches in the roof. They were arranged with the intent of promoting air circulation, but the

[47] Taylor, *loc.cit.*, 574; Federal Trade Commission, *Food Investigation. Report on . . . Private Car Lines, June 27, 1919* (Washington, 1920), 26.

[48] Federal Trade Commission, *op.cit.*, 26; Cummings, *The American and His Food*, 61; L. D. H. Weld, "Private Freight Cars and American Railways" (Faculty of Political Science of Columbia University, eds., *Studies in History, Economics and Public Law*, xxxi, Number 1) (New York, 1908), 12, states that these experiments occurred about 1857.

[49] *Ice and Refrigeration*, i (1891), 133-134; xxi (1901), 207-208.

[50] Federal Trade Commission, *op.cit.*, 26; U. S. Department of the Interior, Patent Office, *Annual Report of the Commissioner of Patents for the Year 1867* (Washington, 1868), ii, 1386.

numerous attempts that were made to improve the car indicated that the circulation actually achieved left much to be desired.[51] Another of the more important early plans was that patented by Joel Tiffany in 1877. It featured an overhead galvanized-iron ice tank that formed the roof of the provision compartment and a novel arrangement for introducing outside air that had been cooled by passing it through tubes at the bottom of the tank.[52] The car patented by James H. Wickes in 1877 was unusual among early models in that it had an ice compartment at one end and relied for air circulation on a fan which took its power from the axle.[53]

Experimentation on a commercial basis was to a large extent the work of meat-packers of the Middle West. They had every incentive to undertake this work, for if equipment could be developed that would preserve fresh meat during the long rail haul to the Atlantic seaboard, they would be able to slaughter for the fresh as well as the preserved-meat market of eastern cities. G. H. Hammond, a Detroit packer, was probably the first to ship meat commercially in refrigerator cars. Hammond sent beef to Boston in 1869 in a Davis car and continued to operate on a small scale.[54] In the winters of

[51] Federal Trade Commission, *op.cit.*, 26; D. W. Davis, quoted by J. M. Culp, "Report No. 1, Question XVI, Perishable Goods," International Railway Congress Association, *Proceedings, Eighth Session, 1910*, III (Brussels, 1912), XVI/11; Letter, Parker Earle, quoted by F. S. Earle, "Development of the Trucking Interests," U. S. Department of Agriculture, *Yearbook, 1900*, 444; "Transportation of Fresh Meats and Fruits, Etc., Through Long Distances—The Davis Refrigerator Car," *Scientific American*, XXIII n. s. (Nov. 12, 1870), 312. For an exterior view of the Davis car see *The Davis Refrigerator*, back cover. A description of the car by the inventor's son is D. W. Davis, "Davis Refrigerator Car," *Ice and Refrigeration*, VII (1894), 166-167.

[52] "The Tiffany Refrigerator Car," *The Railroad Gazette*, IX (July 13, 1877), 311; *Ice and Refrigeration*, XXI (1901), 208.

[53] *Ice and Refrigeration*, XXI (1901), 208.

[54] Newspaper extracts in *The Davis Refrigerator*, 10-11; Davis, *loc.cit.*, 166; L. F. Swift, in collaboration with A. Van Vlissingen, Jr., *The Yankee of the Yards; The Biography of Gustavus Franklin Swift* (Chicago, 1927), 186. A shipment of fresh beef was sent under refrigeration from Denison, Texas to New York in 1873. W. R. Woolrich, "Mechanical Refrigeration—Its American Birthright," *Refrigerating Engineering*, LIII (1947), 307.

1874 and 1875 Nelson E. Morris sent frozen beef to the East in ordinary box cars, but this was not satisfactory, for it could be done in the winter months only, and even then there was danger of loss from sudden temperature changes.[55]

Gustavus F. Swift, who moved his New England packing business to Chicago in the middle seventies, was most influential. Swift, keenly aware of the potentialities of the fresh-meat trade to the East, tried in vain to induce the railroads running between Chicago and the Atlantic coast to furnish refrigerator cars for his use. The railroads were unwilling to provide highly specialized and expensive equipment of this type, especially when it was still in the experimental stage; furthermore, they had no desire to encourage anything that might interfere with their profitable traffic in live animals. Despairing of assistance from the livestock-transporting lines, Swift turned to the Grand Trunk, which refused to build any refrigerator cars, but agreed to haul such as he furnished himself. Swift experimented with several existing cars and established meat shipments to the East with one that had overhead ice tanks. This was not wholly satisfactory, since moisture condensing on the tanks dripped on the meat hung below. His efforts to find a better car were successful when about 1881 he obtained one designed by Andrew J. Chase, the Boston inventor of the Cold-Blast refrigerator. In this car the necessary air circulation was attained by placing the ice-containers, which could be filled through roof hatches, at each end. With

[55] Federal Trade Commission, *op.cit.*, 28.

FIGURE 3

Delivering ice in New York, 1884

At the left is illustrated delivery to domestic consumers. In this neighborhood, apparently, icemen did not enter the kitchen, but sold their wares to the housewife or her maid at the curb. At the right is delivery to some commercial establishment, possibly a saloon. Though ice is available, the butcher a few numbers down the street still allows his meat to hang unrefrigerated in the open air.

From *Harper's Weekly*, xxviii (August 30, 1884), 562.

DELIVERING UP TOWN.

FILLING A CELLAR.

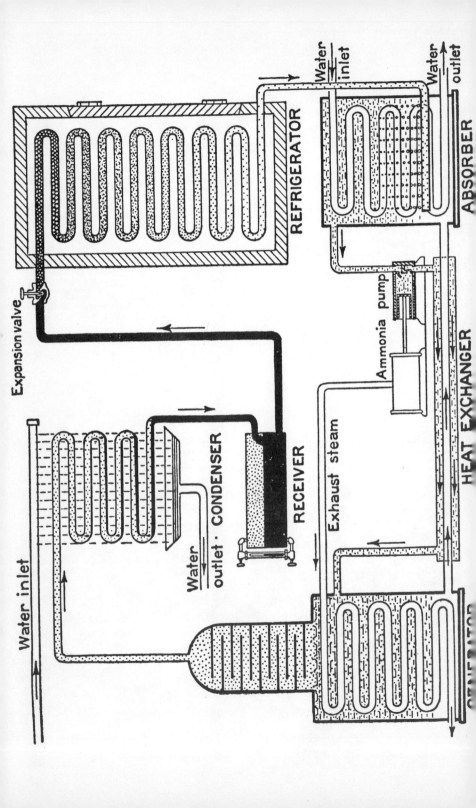

such location of the ice the danger of moisture dripping on the meat was removed. As Swift demonstrated the practicability of refrigerator cars, other packers followed his example and built their own. Although the railroads did not build cars themselves, they soon were competing for a traffic they once had spurned.[56] By about 1881 the business of shipping dressed beef from Chicago to the East had become firmly established on a remunerative basis.[57] Equipment continued to be improved, and by the end of the decade Philip D. Armour was using the Hamilton car, which was chilled by four galvanized-iron cans at each end. These cans were loaded from the roof with a mixture of crushed ice and salt. A natural circulation

[56] *ibid.*, 28-30; Swift, *op.cit.*, 187-190; [Chase], *op.cit.*, 2-3, 9-10.

[57] Philip D. Armour, testimony in Select Committee on Transportation and Sale of Meat Products, "Report," *Senate Report*, 51 Cong., 1 sess., no. 829, p. 432.

FIGURE 4

An ammonia-absorption refrigerating machine

"In the evaporator, the liquid ammonia vaporizes, the heat required for this process being absorbed from the substance surrounding the evaporator coils. The necessary low pressure in the evaporator results from the rapid absorption of the anhydrous ammonia gas by the weak aqua ammonia in the absorber. This absorption of the gas by the weak solution generates heat, which is carried off by the cooling water circulating through coils in the absorber. The ammonia solution, made stronger by the refrigerant absorbed, is pumped into the generator. There it is heated, usually by low-pressure steam. Under the influence of the higher temperature and the consequent high pressure the ammonia gas is driven out. The weakened solution is then returned to the absorber. In the heat exchanger, the weak solution coming from the generator through the inner space gives up part of its heat to the stronger solution flowing to the generator through the outer space, and heat is conserved. The ammonia gas released in the generator then passes into the condenser, where its heat is taken out by the cooling water and it returns to liquid form. It flows next into the receiver, and thence through an expansion valve into the evaporator to absorb more heat from the refrigerator."

From J. T. Bowen, *Refrigeration in the Handling, Processing, and Storing of Milk and Milk Products* (U. S. Department of Agriculture, Miscellaneous Publication No. 138) (Washington, 1932), 8-9.

was induced as the warm air from the top of the car came in contact with the surface of the cans, was cooled, became heavier, and flowed out along the floor.[58]

The work of meat-packers was paralleled by the efforts of certain fruit-growers and dealers to find a car suitable for fruit. As early as 1868 peaches were carried from the Middle West to the East in Davis cars,[59] and in October 1870 one loaded with grapes, peaches, and pears from California arrived in New York with its contents in good condition.[60] One of the most active experimenters was Parker Earle, a strawberry-grower of Union County, Illinois. Earle, who had experimented about 1866 with forwarding strawberries by express in refrigerator chests, made rather discouraging trials in 1869 with a Michigan Central car that had been built to carry dairy products. About 1872, according to his own account, he used a Tiffany car successfully, although its ice capacity was so small that he had to cool the fruit before loading it. Not satisfied with this, Earle continued his efforts to find suitable equipment.[61] Others experimented with shipping fruit, and in the eighties satisfactory cars were developed.[62] Air circulation achieved by placing ice bunkers at each end became the rule. It was with one of these improved models, invented by Carlton B. Hutchins of Detroit, that regular refrigerated hauls of fruit were begun from California late in the eighties.[63]

[58] *Ice and Refrigeration*, I (1891), 134; Culp, *loc.cit.*, xvi/12.
[59] Newspaper extracts printed in *The Davis Refrigerator*, 10-11.
[60] "Transportation of Fresh Meats and Fruits, Etc., Through Long Distances—The Davis Refrigerator Car," *loc.cit.*, 312.
[61] Letter, P. Earle, quoted by Earle, *loc.cit.*, 444-445.
[62] Federal Trade Commission, *op.cit.*, 27; E. F. McPike, "The Refrigerator Car—Retrospective and Prospective," *Transactions of the American Society of Refrigerating Engineers*, ix (1913), 127-128; "The Tiffany Refrigerator Car," *loc.cit.*, 311; *Ice Trade Journal*, iii (Oct. 1879), 4; Earle, *loc.cit.*, 446.
[63] C. E. Russell, *The Greatest Trust in the World* (New York, 1905), 34-37. Although many accounts of the evolution of the refrigerator car have been written, none has been found that is adequate, particularly for the important decade between 1880 and 1890. Before the complete story of the refrigerator car can be told there must be close examination of the patents.

The amount of ice used in the United States increased rapidly after the Civil War. In 1856 New York needed 100,000 tons for all purposes; in 1860 Boston required 85,000 tons and New Orleans 24,000.[64] In the year from October 1, 1879 to September 30, 1880, a year in which the supply was somewhat below normal because of a partial failure of the crop, the New York consumption was 956,559; that of Boston was 381,588. In the case of New Orleans the increase was less spectacular; only 31,500 tons were sold there. In Chicago in 1879-1880, 576,687 tons were consumed; in Brooklyn and Philadelphia over 300,000 tons; in Cincinnati and St. Louis over 200,000. The total in the United States for this year was estimated at from 5,000,000 to 5,250,000 tons.[65]

A growing use of refrigeration in the home underlay the increased ice consumption. Forty-five per cent of the ice sold in New York in 1879-1880 went to private families. In Boston and Chicago they bought thirty-three per cent. About half that sold in Baltimore and Philadelphia was purchased for the home, and in New Orleans, where ice consumption was low, sixty-nine per cent went to domestic buyers.[66] Ice continued to be employed for many medical purposes,[67] but the volume utilized in this way was insignificant. In the home it was used to cool beverages, and the increasing presence of refrigerators indicates that more food was being preserved by ice

[64] R. O. Cummings, *The American Ice Industry* (Unpublished doctoral dissertation, Harvard University), 188.

[65] Hall, *loc.cit.*, 5-6, 38-41.

[66] *ibid.*, 38-41.

[67] The medical uses of ice are summarized in the *Ice Trade Journal*, VII (April 1884), 2, as follows: "Swallowed by the patient it is deemed a specific in gastritis, gastric fevers, and in all inflamatory affections of the larynx and bronchia. Externally, it is applied to the head with highly favorable results in acute mania and brain fever, to the temporal and carotid arteries in diphtheria and scarlet fever; to the spine in case of cholera, yellow and typhoid fevers; and over the intestines in cholera and other diseases of these organs." Large quantities of ice were ordered by the federal government for use in army hospitals during the Civil War. For information on alleged profiteering and other irregularities in connection with this see "Report of the Joint Committee on the Conduct of the War," Section on "Ice Contracts," *Senate Report*, 38 Cong., 2 sess., no. 142.

than before.[68] It is difficult to estimate how many people had refrigerators. One writer asserted in 1884 that they were as common as stoves or sewing machines in all but the poorest tenements, but there is no doubt that boxes or tubs often served as substitutes, and that poorer families used ice only in small quantities in the hottest weather if at all.[69] In rural areas wells and cellars continued to be the principal means of preserving fresh foods for the family table.

The needs of a growing number of dealers in perishable foods, particularly those who handled meat, fish, poultry, and butter, were served by ice refrigeration. Public eating and drinking places—hotels and restaurants, saloons and soda fountains—used more ice than ever before.[70]

Brewers of lager beer found improved methods of meeting their refrigeration requirements after 1860. The earliest way of using ice to cool the hot wort was to permit it to flow from preliminary cooling vats at the top of the brewery through copper coils submerged in cold water to which ice was added in the summer. The process was improved in 1864 by the introduction of the countercurrent cooler, which consisted of two pipes, one inside the other, through which the hot wort and ice water flowed in opposite directions. To remove the heat generated during fermentation, ice originally was piled around the fermenting tubs, but a more effective means of heat transfer was needed. This was attained by employing "swimmers," cone-shaped vessels of sheet tin or copper which were filled with ice and allowed to float on the surface of the fermenting wort. A later method was to pump water which had been ice-cooled through coils immersed in the tub. A cool environment for the fermentation and storage phases of lager-beer production had first been achieved by locating these operations in great underground cellars and by applying ice directly to the beer-containers. This was unsatisfactory, not only because of the inadequate control of temperature afforded,

[68] G. M. Towle, *American Society* (London, 1870), I, 268; Bacon, *op.cit.*, 95.

[69] J. Ralph, "The Ice Industry of New York," *Harper's Weekly*, XXVIII (Aug. 30, 1884), 565; Bacon, *op.cit.*, 94.

[70] Ralph, *loc.cit.*, 565; Hall, *loc.cit.*, 38-41.

but also because of the limits imposed by nature on the size and location of cellars. To make available more suitable space, great structures were built above ground with overhead ice bunkers which made it possible to refrigerate large rooms with a natural circulation of cool air. Many brewers used ice-and-salt mixtures to produce temperatures lower than could be obtained with ice alone.[71] So important was ice to the brewing industry that a million dollars was spent for it annually by 1864.[72] By 1877 brewers were the largest single class of consumers in the country, several of them using twenty to thirty thousand tons each year.[73]

Ice refrigeration was nowhere more important between 1860 and 1890 than in the meat-packing industry. It was applied to preservation of fresh meats during rail transportation and also to their preservation in the packing plant itself. It was used, beginning in the late fifties, for summer curing of meats; by 1870 it was being employed in the larger plants for preserving fresh meats as well. Space for meat chilling and storing usually was cooled by ice placed in overhead lofts, which were connected with the space to be refrigerated by flues which made possible a natural circulation of cold air. To attain temperatures below freezing, some packers installed systems in which mixtures of crushed ice and salt could be used. Construction details of packing plants varied because methods were regarded as trade secrets, and each firm worked

[71] G. J. Patitz, "Progress in Brewery Refrigeration," *Ice and Refrigeration*, LI, (1916), 167; A. J. Dixon, *Manual of Ice-Making and Refrigerating Machines: A Treatise on the Theory and Practice of Cold-Production by Mechanical Means* (St. Louis, 1894), 74; J. C. Goosman, "History of Refrigeration," *Ice and Refrigeration*, LXIX (1925), 149; J. E. Siebel, "Refrigeration and the Fermenting (Brewing) Industry in the United States," *Premier Congrès International du Froid*, III (Paris [1908?]), 72-73. J. P. Arnold and F. Penman, *History of the Brewing Industry and Brewing Science in America* (Chicago, 1933), 92.

[72] Editorial, *New-York Daily Tribune*, Nov. 24, 1864.

[73] *Ice Trade Journal*, I (Oct. 1877), 2; Hall, *loc.cit.*, 5. A factor contributing to the rise of Milwaukee as a brewing center was its bountiful and cheap supply of natural ice, which gave its brewers an important advantage over their competitors in Cincinnati and St. Louis. T. C. Cochran, *The Pabst Brewing Company; The History of an American Business* (New York, 1948), 77-78.

out its own problems. So great were the amounts of ice required that storages for it often covered one-half the area of the plant. Some refrigerating machines were installed in the eighties, but natural-ice refrigeration was the only method used generally until about 1890.[74]

The development of practical methods of using ice in packing plants worked revolutionary changes. Fresh meat no longer had to be distributed immediately after slaughter, but could be cooled promptly and kept in storage for short periods pending shipment. Ice refrigeration was particularly important in the production of cured-meat products, for it was a means of preventing spoiling until the preservatives could penetrate to all parts of the flesh. Until late in the sixties curing was a seasonal operation, carried on only in months when temperatures could be depended on that were low enough to check decay. Refrigeration not only made curing a year-around industry, but it also improved the quality of the meat. With operations spread over the entire year, the haste and carelessness that had accompanied crowding all curing into the few winter months were eliminated. Largely because of the stimulus given by processing throughout the year, the annual hog pack increased eighty-six per cent between 1872-1873 and 1879-1880.

Refrigeration in the packing plant had an important effect on stock raising, and particularly on hog raising, for pork was the principal product preserved by curing. As soon as packing plants could function the entire year, farmers no longer had to market their hogs during the winter, but could sell them whenever they reached their best marketing weight. No longer was it necessary to hold back and feed already mature animals for an entire summer waiting for the opening of the curing season. The ability to market at any time of the year also benefited the farmer by giving him some control, at least, over prices. If hogs were low and corn were cheap, he

[74] A. Cushman, "The Packing Plant and Its Equipment," *The Packing Industry* (Chicago, 1924), 101, 107-108; R. A. Clemen, *The American Livestock and Meat Industry* (New York, 1923), 215-216; Swift, *op.cit.*, 196.

could delay marketing for a time in the hope that a more favorable price situation would develop.[75]

The refrigerator car was the essential factor in the startling changes that took place in American meat packing during the eighteen-seventies and eighties. Shipments of dressed beef from Chicago to the East, a stable feature of the industry after 1880, grew steadily in volume. There were great advantages in forwarding fresh meat rather than livestock. In dressed-meat shipments no freight was paid on the thirty-five per cent of the animal that was waste. Rail transportation resulted in the death of some livestock and serious loss of weight and deterioration in quality of meat among those that survived. Not only was it less expensive to care for dressed meat than live animals, but dressed meat contributed to further economies by permitting the transportation of non-perishable animal by-products in ordinary rolling stock at lower rates than were possible when the animals, by-products and all, were forwarded on the hoof in stock cars.[76]

In spite of these economic advantages, the dressed-meat trade encountered serious opposition from the railroads, the public, and the local butchers and slaughterers. Rail lines with a vested interest in livestock traffic resisted it, for it not only gave them less tonnage to haul, but also endangered investments in loading docks and feeding stations. Until a compromise was reached, they charged rates on dressed meat high enough to yield them the same return as on livestock. Since the roads continued reluctant to supply refrigerated rolling stock, packers built their own and formed freight organizations to manage it. Railroads hauled the packer cars, paying a rental charge based on mileage. Public prejudice was a serious obstacle for dressed meat brought in from afar in re-

[75] Cushman, *loc.cit.*, 102-103; "The Hog Crop and Its Product," *The Merchants' Magazine*, LXI (1869), 447-448; H. C. McCarty, "Slaughtering and Meat Packing," *Twelfth Census, 1900*, IX, 388.

[76] Federal Trade Commission, *op.cit.*, 36; Clemen, *op.cit.*, 238. A contemporary discussion of the waste involved in shipping live animals can be found in D. E. Somes, *Mr. D. E. Somes' Plan for Ventilating, Cooling and Heating the Capitol, and for Purifying and Moistening the Air* ([Washington?] 1867), 8-9.

frigerator cars. Local slaughterers did all they could to encourage this prejudice, for the new competition threatened their livelihood. But they also fought directly. Usually distributors as well as killers, they often refused to handle Chicago meat. By exerting pressure through their associations, they obtained the passage of state and municipal restrictive measures. The dressed-meat interests counterattacked by taking their antagonists into partnership or by establishing branch houses to market their products. The opposition to the dressed-meat trade continued vigorously throughout the eighties, but collapsed early in the next decade, when local interests found it impossible to compete in price with meat butchered at central points and brought to market under refrigeration.[77]

The refrigerator car enabled western packers to supply fresh meat to the growing urban population of the East. By making possible this solution to meat-supply problems, it affected fundamentally the location and organization of the packing industry. It permitted a westward movement which brought the packing plants nearer the supply of livestock. The number of cattle handled at eastern cities declined relatively, while the business of western slaughterers boomed. The census statistics between 1880 and 1900 told the story of the shift to the West. In 1880 the value of meat products from slaughtering and packing at Boston was $7,096,777, at New York $29,297,527, at Chicago $85,324,371, and at Kansas City $965,000. By 1890 the value of operations at Boston had fallen to $2,782,823. In New York the value had increased to $50,251,504, but this increase was outstripped by that of the western centers, for Chicago products were worth $203,606,402 and those of Kansas City $39,927,192. The trend was even more evident by the end of the century. In 1900 Boston's business had declined to $1,392,010 and New York's to $38,752,586. Meanwhile the value of Chicago products rose to $256,527,949 and those of Kansas City to $73,787,771.[78]

[77] Clemen, *op.cit.*, 237-251; M. Keir, *Manufacturing* (New York, 1928), 269-270.
[78] Weld, *loc.cit.*, 44-45; McCarty, *loc.cit.*, 391-392.

One of the striking effects of the development of a practical refrigerator-car service was an increasing trend toward centralization of the meat-packing industry. The growth of large meat-curing centers, first at Cincinnati and then at Chicago, had been the result largely of transportation factors, but slaughter of livestock for sale as fresh meat had remained essentially a local industry until a practical refrigerator car was invented.[79] The statistics recording the rising volume at western packing centers between 1880 and 1900 reflected a decline in the number of animals killed locally and marked the centralization of slaughter for the fresh-meat trade.[80] There were other important factors—the development of a new source of livestock supply in the trans-Mississippi West, the extension of railroads, and the appearance of leaders of great organizational genius—but without the refrigerator car centralization would have been impossible.[81]

An important element in the success of the large centralized packers in the competition with local slaughterers was the effective distribution system which the packers devised. They established refrigerated branch houses through which they could supply fresh meats to retailers as easily as could local abattoirs. To meet the needs of outlying districts where the volume of trade did not justify branch houses, "peddler car" routes were established. Over these routes refrigerator cars were dispatched from central packing plants or branch houses to deliver small quantities of meat products at the various stations along the way.[82] These basic changes in the organization of distribution would have been impossible without the refrigerator car.

In its indirect effects the car also was important in transforming the meat-packing industry. By centralizing slaughter

[79] H. C. Hill, "The Development of Chicago as a Center of the Meat Packing Industry," *The Mississippi Valley Historical Review*, x (1923-1924), 259-263.

[80] Federal Trade Commission, *op.cit.*, 34; McCarty, *loc.cit.*, 387-388. The tendency toward centralization was particularly strong between 1890 and 1900.

[81] Clemen, *op.cit.*, 6-7, 222, 224.

[82] *ibid.*, 389-390, 400-403; Federal Trade Commission, *op.cit.*, 37-38.

for the fresh-meat trade, it made possible the utilization of by-products, with benefits that were passed on to the consumer in the form of lower meat prices.[83] The refrigerator car was partly responsible for the increasing concentration of control in the industry, for large business organizations with great resources of capital were necessary to distribute on a national scale.[84] Since the cars were not furnished by the railroads, but were supplied by the packers themselves, it became increasingly difficult for packers who did not have the capital necessary to acquire their own to participate in the large-scale distribution of fresh meat.[85]

The effect of the development of a satisfactory car reached out to the livestock-producer on the grasslands west of the Missouri. By rendering possible an efficient and cost-saving means of selling fresh meat nationally, it provided an almost limitless market for the stockmen of the western ranges.[86]

If the refrigerator car, its bunkers filled with natural ice, enabled American packers to exploit the national market for meat, ships refrigerated with natural ice permitted them to pioneer in distributing on an international scale. Consignments of livestock to Great Britain, which began in the seventies, were limited by the inherent difficulty and inefficiency involved in the transport of live animals. Some attempts were made to send dressed meat to Britain without refrigeration, but these were unsuccessful. Refrigerated shipments in quantity began in 1875 with the experiments of Timothy C. Eastman of New York. Eastman employed an arrangement invented by a fellow New Yorker, John J. Bate. The carcasses were stowed in a special compartment between decks and were chilled by air which was circulated through a thirty-ton ice bunker and the meat compartment by a steam-powered fan.

[83] J. O. Armour, *The Packers, the Private Car Lines, and the People* (Philadelphia, 1906), 24; H. Sloan, "Modern Developments in Packing House Refrigeration," *Ice and Refrigeration*, LXXVI (1929), 105.

[84] L. R. Edminster, Section on "History and American Developments" under "Meat Packing and Slaughtering," *Encyclopaedia of the Social Sciences*, x, 246.

[85] Federal Trade Commission, *op.cit.*, 18-19.

[86] Weld, *loc.cit.*, 45; M. J. Adams, "The Refrigerator Car," *Ice and Refrigeration*, LVIII (1920), 241.

Many followed Eastman into the export trade, and before long all lines from New York and Philadelphia to England and Scotland carried dressed meat under refrigeration. Various systems were employed. One invented by a Dr. Craven involved pumping cold brine through pipes that ran between the rows of meat; others used various methods of forced-air circulation.[87] About 1881 twenty-five steamships were equipped with systems designed by Andrew J. Chase.[88] The export of meat under refrigeration from the United States to the British Isles, which increased from 109,500 pounds of beef in 1874 to 72,427,200 in 1880, marked the real beginning of the tremendous overseas meat trade to Great Britain. Up to the end of 1880 only four hundred carcasses of mutton had been shipped from Australia to England, but 120,000 tons of fresh beef had been sent there from the United States.[89] But natural-ice refrigeration was not used for long. In 1879 an Anchor Line vessel was fitted with a refrigerating machine, and at least as early as 1891 mechanical refrigeration was used exclusively for transatlantic shipments.[90]

In the catching and distributing of fish, ice became more and more important in the three decades after 1860. On fishing vessels its use was continued. Methods of freezing with mixtures of ice and salt were introduced. But the most important contribution of ice was in distribution. The bulk of the catch of American fishermen came to be marketed fresh, a development furthered by the appearance of speedier transportation facilities both at sea and on shore.[91] Railroading improved so much that in 1884 a carload of fresh salmon was shipped from

[87] Clemen, *op.cit.*, 271-273, 275-278.

[88] [Chase], *op.cit.*, 3. For a contemporary description of the export meat trade see J. MacDonald, *Food from the Far West; or, American Agriculture, with Special Reference to the Beef Production and Importation of Dead Meat from America to Great Britain* (New York [1878?]), 5-7.

[89] J. T. Critchell and J. Raymond, *A History of the Frozen Meat Trade*, 2nd edn. (London, 1912), 13, 423.

[90] *ibid.*, 336-337; "Coal Ammonia for Refrigeration," *Scientific American*, LXIV (April 18, 1891), 241.

[91] C. H. Stevenson, "The Preservation of Fishery Products for Food," U. S. Fish Commission, *Bulletin*, XVIII (1898), 358.

The Dalles, Oregon, to New York City. High rates kept the experiment from being repeated at that time, but in a few years regular movements from the west coast began.[92] Procedures for packing in ice varied. In some cases fish were dressed first; in others they were packed round, which generally was the best method. Shipping containers were boxes and barrels of varying sizes in which the fish were placed with alternating layers of crushed ice.[93]

Packing fish in ice retarded, but did not prevent decomposition. The fish acquired a musty taste and lost flavor. Under the best circumstances they could be kept in good condition only for two weeks or so. If they were frozen, however, they could be kept much longer. This made possible holding fish over from periods of glut to periods of scarcity.[94] In March 1861 Enoch Piper of Camden, Maine, was granted a patent on a method of freezing and storing fish. Piper froze them by setting pans filled with ice and salt on fish which were placed on racks in an insulated compartment. When frozen, the fish were covered with a material like gutta percha or were dipped several times in cold water to cover them with a thin coating of ice. Then they were stored in insulated rooms refrigerated by vertical metal tubes filled with brine. Piper established a plant near Batħurst, New Brunswick, from which he sent salmon to New York, where they were held for market in a storage room of his design. After three or four years he lost exclusive rights to his method, which his competitors then applied to fish other than salmon. Piper's method was improved by William Davis, who patented in 1869 a system for freezing fish by placing them in metal boxes which then were

[92] *Ice and Refrigeration*, IX (1895), 159.

[93] Stevenson, *loc.cit.*, 359-362.

[94] *ibid.*, 362; C. H. Stevenson, "Freezing and Storing Fish," *Ice and Refrigeration*, XVIII (1900), 99. Fish should be frozen as soon as possible for good results. Unfortunately, some dealers would make every effort to sell fish unfrozen, and if unable to do so, they would freeze them just in time to prevent spoilage and total loss. This resulted in an inferior frozen product and happened often enough to lead many consumers to believe that freezing itself was the cause of the poor quality. H. F. Taylor, "Science in the Distribution of Fish," *Scientific American*, CXL (1929), 251.

buried in an ice-and-salt mixture. To assure efficient heat transfer from the fish to the refrigerant, the boxes were constructed so that their tops and bottoms were always in contact with the fish. The practice of preserving by freezing spread from the seacoast to the Great Lakes. By 1892 ice-and-salt freezers were to be found in almost all of the fishing ports of the Great Lakes as well as in New York and along the New England shore. The freezing method invented by Davis and the storage system devised by Piper were used without basic modification until 1892, when mechanical refrigeration was first employed in a plant designed exclusively for freezing fish.[95]

In the processing and distributing of dairy products ice refrigeration came to play an important part. Spring water and deep cellars were the age-old method of providing low temperatures, but gradually a more extensive use of ice was made where it was available. It commonly was added to the water that cooled the cans of milk or cream,[96] and in the eighties well-equipped dairies in the northern United States had cold-storage rooms refrigerated by ice placed in overhead bunkers so arranged that air circulated over it and through the chamber.[97] Although the seventies and eighties saw an expanding use of refrigeration to keep milk en route to market, a more striking development of those years was the transportation of butter and cheese for great distances in refrigerator cars. Before 1875 butter and cheese were being shipped in them to eastern cities from the Middle West, enabling dairymen there, who had the advantage of relatively cheap lands, to

[95] Stevenson, "The Preservation of Fishery Products for Food," *loc.cit.*, 371-373. Some frozen fish were stored in mechanically refrigerated warehouses before 1892.

[96] E. Wiest, "The Butter Industry in the United States: An Economic Study of Butter and Oleomargarine" (Faculty of Political Science of Columbia University, eds., *Studies in History, Economics and Public Law*, LXIX, Number 2) (New York, 1916), 14; H. E. Alvord, "Dairy Development in the United States," U. S. Department of Agriculture, *Yearbook*, *1899* (Washington, 1900), 387-388, 390.

[97] H. Myrick, "Associated Dairying in New England," U. S. Department of Agriculture, *Fourth and Fifth Annual Reports of the Bureau of Animal Industry for the Years 1887 and 1888* (Washington, 1889), 378-380.

compete with the East.[98] In the eighties there were large and regular movements of butter, and areas were being turned to dairying that could not have been so used without dependable means of refrigerated transport.[99] Butter was one of the products stored in the ice-cooled warehouses that appeared in the seaboard cities.[100]

The preservation of eggs before the day of cold storage was accomplished by a variety of chemicals, including water glass, vaseline, and limewater. In the eighties storing in ice-cooled rooms displaced other methods. Unfortunately, eggs were stored too frequently only as a last resort. There was no selection of especially sound eggs for refrigerated preservation, and inferior grades were kept which gave the cold-storage product a bad reputation.[101] Statistics are not available, but the quantities stored were not large enough to prevent great scarcities during the fall and winter which made prices prohibitively high.[102]

Poultry also was kept by refrigeration. In 1865, A. and M. Robbins, dealers at Fulton Market, New York, preserved poultry by freezing it in an underground room refrigerated by metal tanks that contained a mixture of ice and salt. In 1870 six carloads of chickens were shipped to New York from Wisconsin, where they had been frozen by air blown through pipes immersed in a tank filled with brine.[103] Other experiments were made from time to time, but freezing was not important commercially as long as it was necessary to depend

[98] *Ice Trade Journal*, III (Oct. 1879), 4; III (Nov. 1879), 6; F. Merk, *Economic History of Wisconsin During the Civil War Decade* (State Historical Society of Wisconsin, *Studies*, I) (Madison, 1916), 29-30.

[99] H. H. Wing, "The Dairy Industry of the United States," U. S. Department of Agriculture, *Fourth and Fifth Annual Reports of the Bureau of Animal Industry for the Years 1887 and 1888*, 400-401.

[100] Horne, *loc.cit.*, 170.

[101] *Twelfth Census, 1900*, v, ccxxvi; M. A. Jull et al., "The Poultry Industry," U. S. Department of Agriculture, *Agriculture Yearbook, 1924* (Washington, 1925), 386.

[102] M. E. Pennington, "Fifty Years of Refrigeration In the Egg and Poultry Industry," *Ice and Refrigeration*, CI (1941), 47.

[103] *ibid.*, 45; D. K. Tressler and C. P. Evers, *The Freezing Preservation of Foods*, 2nd edn. (New York, 1947), 2.

upon ice and salt.[104] Refrigerated transportation made feasible the establishment of killing plants in the Middle West, near the areas where poultry raising was concentrating. In 1890 two and one-half million pounds of dressed fowl were shipped from Kansas City alone.[105]

The distribution of fruits and vegetables to the expanding urban population received little assistance from ice refrigeration before 1890. Some refrigerated storage of fruit was done on a commercial basis even before 1870,[106] but the volume held in this way was not great. As late as 1890 comparatively few apples, the fruit later to be preserved in greatest volume, were held under refrigeration.[107] Since little advantage was taken of the preservative effects of cold, fruits had to be sold shortly after they were harvested. The efforts of farmers to sell during a brief period resulted in glutted markets and ruinously low prices. In a few weeks, when the season had passed, scarcity and high prices prevailed.[108]

After the Civil War the dwellers in American cities wanted increasing amounts of fruits and vegetables. Their demands could not be met by the areas adjacent to the cities, which could provide fresh foods in sufficient volume only during the brief period they were locally in season. Besides, near urban centers increased land values raised the costs of agricultural production.[109] As a result it was necessary to bring fruits and

[104] *Ice and Refrigeration*, LXXVIII (1930), 545.
[105] *Twelfth Census, 1900*, v, ccxxvi.   [106] Taylor, *loc.cit.*, 565.
[107] Cooper, *Practical Cold Storage*, 341.
[108] G. H. Powell, "Relation of Cold Storage to Commercial Orcharding," *Ice and Refrigeration*, XXV (1903), 240; E. L. Overholser, "History of Fruit Storage and Refrigeration in the United States," *Better Fruit*, XXIX (Aug. 1934), 8. Some apples—varieties that ripened slowly and that had thick protective skins—were kept in pits, cellars, or insulated and ventilated buildings. This made possible some control over the period in which they could be marketed.
[109] E. G. Ward, Jr., and E. S. Holmes, Jr., *Rates of Charge for Transporting Garden Truck, with Notes on the Growth of the Industry* (U. S. Department of Agriculture, Division of Statistics, *Miscellaneous Series—Bulletin No. 21*) (Washington, 1901), 9; J. W. Lloyd, "Truck-Growing," L. H. Bailey, ed., *Cyclopedia of American Agriculture; A Popular Survey of Agricultural Conditions, Practices and Ideals in the United States and Canada*, II (New York, 1907), 653.

vegetables from producing areas that were often hundreds of miles away.

Persistent efforts were made during the quarter-century after the Civil War to apply refrigeration to the preservation of these products while they were in transit. Successful and regular shipments in refrigerator cars began in the eighties, but it was not until early in the next decade that a volume trade on a nationwide scale began.[110]

Until refrigerator cars were used generally for bringing fruits and vegetables to cities from distant areas, high-speed transportation was relied upon to reach the market before decay set in. Prior to 1885 all produce grown in the southeastern states for the large eastern cities was transported in fast coastwise vessels. The first all-rail movement of garden truck from Norfolk to New York took place in 1885, from eastern North Carolina in 1887, and from Charleston in 1888.[111] Farther north, in the vicinity of New York, shipments by railroad began in the eighteen-forties.[112] Rail hauling began earlier from southern states west of the Appalachians than in the Southeast, since water transportation was not available. Mississippi peaches were sent north in 1866, and in the seventies tomatoes and strawberries were forwarded.[113] Fruits and vegetables at first went in small lots by express, but as volume increased, carload shipments became necessary. Initially box or cattle cars were employed, but these were unsatisfactory. The one excluded the air too completely, while the other offered insufficient protection from the weather. Neither had the springs necessary to prevent injury to the load. Eventually special equipment was developed with springs, air brakes, and numerous wire-covered openings for ventilation. It was desirable to have the cars hauled to market as rapidly as possible. In some instances they were attached to passenger trains, but

[110] Federal Trade Commission, *op.cit.*, 27; M. C. B., "The California Fruit-supply in New York," *Garden and Forest*, VI (1893), 432.

[111] A. Oemler, "Truck Farming," U. S. Department of Agriculture, *Report of the Commissioner . . . , 1885* (Washington, 1885), 604; *Twelfth Census, 1900*, VI, 304.

[112] Cummings, *The American and His Food*, 57-58.

[113] *Twelfth Census, 1900*, VI, 304.

as the traffic developed, railroads that moved large quantities of produce provided special trains that ran on fast schedules.[114] Specialized areas of production appeared under the stimulus of these methods of transportation. In the East the Chesapeake peninsula and the hinterlands of Norfolk, Charleston, and Savannah were important. In the West the eastern shore of Lake Michigan, southern Illinois, parts of Tennessee and Mississippi, and the areas around Mobile and New Orleans were among the specialized centers.[115] But fast transportation alone was inadequate. Only the best-carrying varieties could be shipped any distance, and frequently even they arrived in bad condition. Choice Florida tomatoes sometimes decayed on the wharves at Savannah while waiting to be loaded on a New York-bound steamer.[116] It was a risky business to forward even the best California fruit in ventilated cars on express schedules. When such fruit arrived in Chicago, it had to be disposed of quickly before it became unsalable, and very little of it could be sent on to New York.[117] Not only was reliance on speedy transportation hazardous; it was also expensive because it was costly to move solid trains of produce almost as fast as passenger trains.[118] By the eighties there was a serious need for extensive use of refrigerator cars throughout the United States.

In the years 1860 to 1890 natural-ice refrigeration in its varied applications was a basis for progress in the solution of the food-supply problems that faced the growing urban economy of the United States. But the potentialities of natural ice as a refrigerant were limited, and its limitations became more and more apparent as its use increased. In the southern states the shortcomings were evident early, since there the uncertainty and cost of supply were severe handicaps. The ice consumed in the South had to be brought in. Freight rates on it were high, for it was a bulky and highly perishable com-

---

[114] Earle, *loc.cit.*, 441-442; A. W. McKay et al., "Marketing Fruits and Vegetables," U. S. Department of Agriculture, *Agriculture Yearbook, 1925* (Washington, 1926), 684.
[115] Earle, *loc.cit.*, 439-440.   [116] Oemler, *loc.cit.*, 586.
[117] *Twelfth Census, 1900*, VI, 305.
[118] McKay et al., *loc.cit.*, 684.

modity. So great were transportation costs that the regular retail price in southern ports was twenty dollars a ton, while in towns away from the coast prices were even higher. The result was that ice consumption was small. For example, Columbus, Georgia, which with its adjacent towns and villages had a population of twenty thousand, about 1880 used annually only fifteen hundred tons, and this included ice wasted during distribution.[119] When warm winters in the North resulted in small harvests, the cost in the South was especially high and shipments uncertain.[120] In the northern states the supply usually was abundant, but even there shortages and high prices followed warm winters such as that of 1879-1880. A particularly serious situation developed after the seasons of 1888-1889 and 1889-1890, when there were two successive partial failures of the ice crop.[121]

The lack of a cheap and dependable means of refrigeration was a handicap to the South not only in the home but also in industry. It was a factor in the backwardness of the section in such pursuits as meat packing, dairying, and brewing. More serious still, it interfered with the diversification of agriculture by restricting the use that could be made of refrigerator cars for marketing fruits and vegetables. A few cars were tried in the South in the late seventies,[122] but it was not until about 1890, when ice-making plants were widely distributed throughout the section, that the refrigerator car could become an important adjunct to agriculture there.[123]

Ice was not a satisfactory refrigerant even in the North where the supply usually was cheap and dependable. Its limitations became especially apparent at an early stage in the brewing industry. Since large quantities of it were required, its

[119] Hall, *loc.cit.*, 35; *Ice Trade Journal*, III (May 1880), 4.

[120] W. J. Rushton, "Early Days of the Manufacture of Ice," *Ice and Refrigeration*, LI (1916), 151.

[121] *Ice Trade Journal*, III (April 1880), 8; XII (May 1889), 5; XIII (March 1890), 4.

[122] McPike, *loc.cit.*, 127-128; Weld, *loc.cit.*, 52-53.

[123] Weld, *loc.cit.*, 27; W. A. Sherman, *Merchandising Fruits and Vegetables; A New Billion Dollar Industry* (Chicago, 1928), 34; Letter, L. M. Rhodes, Commissioner of the Florida State Marketing Bureau, quoted by Sherman, 33-34.

bulk and inconvenience were disadvantages. Beer was stored on the ground floor of structures built with a high upper story in which the ice was placed. Since extremely heavy construction was necessary, these buildings were expensive. Storages also were needed to house the ice required for cooling the wort. Added to these costs and that of the ice itself was the cost of labor employed to handle it. Moreover, ice was not a satisfactory means of refrigerating fermentation and storage rooms because the air that circulated over it and through the space below was moist enough to favor the development of fungi which might infect the beer.[124]

In the meat-packing industry as well, ice was found to be an unsatisfactory refrigerant. Packing plants had to make large investments to assure an adequate ice supply each year.[125] So great were the amounts called for that in many cases houses for storing it covered over one-half of the plant area. The air in coolers refrigerated with ice was too damp, and the range of temperatures that could be produced was too limited for packing-plant specifications.[126]

In the dairy industry, where cleanliness was at a premium, ice was found unsatisfactory for many purposes. Natural ice frequently contained dirt and vegetation that made it unsanitary. Even when it was pure, the moisture involved in its use produced an unwholesome environment in refrigerated compartments. Furthermore, the unevenness of the temperatures that could be obtained with it was an obstacle to the production of high-quality butter and cheese.[127]

As a refrigerant for cold-storage warehouses ice was inadequate. Its bulk and inconvenience, as in other industries, were objectionable. Since it was only slightly colder than the

[124] Patitz, *loc.cit.*, 167; F. E. Matthews, *Elementary Mechanical Refrigeration; A Simple and Non-Technical Treatise* (New York, 1912), 14-15.

[125] Swift, *op.cit.*, 191.

[126] Cushman, *loc.cit.*, 107, 114-115. For G. F. Swift's difficulties with ice refrigeration see Swift, *op.cit.*, 196-199.

[127] Charles E. Hart, paper presented at convention of Wisconsin Buttermakers' Association at La Crosse, Wis., Feb. 2, 1911, quoted by *Ice and Refrigeration*, XL (1911), 135; H. M. Brandt, "Dairy Refrigeration," *Ice and Refrigeration*, XII (1897), 442.

air it cooled, only a small amount of moisture was condensed on its surfaces, and the air in the refrigerated chambers remained too damp for proper preservation of many foods. Circulation often was poor, for it varied in direct proportion to the difference in temperature between the ice and the air. As the differential became smaller, circulation was less vigorous. Furthermore, it was difficult to achieve the precise temperature control necessary if foods were to be stored under optimum conditions. Although temperatures lower than the freezing point of water could be obtained by adding salt,[128] even these were not low enough for some commodities. The long-term storage of butter, for example, required five to eight degrees below zero Fahrenheit, a level difficult to obtain with ice and salt.[129]

By 1890 it thus was clear that though refrigeration produced by natural ice had done much to supply great city populations with fresh foods,[130] if the vast agricultural and industrial resources of the United States were to be exploited fully, and if perishable foods were to be used with a minimum of waste, a means of refrigeration more dependable and satisfactory than natural ice was needed. Increasingly between 1860 and 1890 men turned to machines that were being developed to produce refrigeration.

[128] E. T. Skinkle, "A Modern Small Cold Storage Warehouse," *Ice and Refrigeration*, XIX (1900), 156; J. A. Moyer and R. U. Fittz, *Refrigeration, Including Household Automatic Refrigerating Machines* (New York, 1928), 348; Matthews, *op.cit.*, 15.

[129] O. Erf, "Refrigeration of Dairy Products," L. H. Bailey, ed., *Cyclopedia of American Agriculture*, 2nd edn., III (New York, 1910), 246.

[130] After 1860 there was a tremendous expansion in the amount of food preserved by canning. Cummings, *The American and His Food*, 69. Canned food was a factor of the first importance in providing for the requirements of city populations and in preserving food generally, but in spite of its virtues, canned food was not a substitute for fresh.

CHAPTER IV

THE INVENTION OF MECHANICAL
REFRIGERATION, 1755-1880

MAN had known for centuries that the evaporation of water produced cooling effects. At first he did not recognize or understand the phenomenon, but he knew that any portion of his body that became wet felt cold as it dried in the air. At least as early as the second century evaporation was employed in Egypt to chill jars of water, and it was used in ancient India to make ice.[1] It was not surprising that the first attempts to produce refrigeration mechanically depended on the cooling effects of the evaporation of water. In 1755 William Cullen, a Scottish physician and teacher of medicine, obtained temperatures sufficiently low to make ice. He accomplished this by reducing the pressure on water in a closed container with an air pump, for at a very low pressure the liquid evaporated violently or boiled at a low temperature. The heat required for a portion of the water to change from liquid to vapor was taken from the water itself. So much heat was absorbed that at least part of the water remaining turned to ice. Partly because Cullen found it difficult to maintain a pressure low enough to create temperatures that would make ice,[2] his work, a laboratory experiment, found no practical application. About 1810 another Scot, Sir John Leslie, went a step beyond. He placed two saucers, one containing water and the other sulphuric acid, in a vessel from which air had been evacuated. As the water evaporated, the sulphuric acid, which had an affinity for it, maintained the low pressure by absorbing the water vapor. This increased the speed of freezing.[3] An improved apparatus was devised in 1824 by Vallance,

---

[1] A. Neuberger, *The Technical Arts and Sciences of the Ancients*, H. L. Brose, tr. (New York, 1930), 123.

[2] J. C. Goosman, "History of Refrigeration," *Ice and Refrigeration*, LXVII (1924), 329; J. F. Nickerson, "The Development of Refrigeration in the United States," *Ice and Refrigeration*, XLIX (1915), 170.

[3] J. A. Ewing, *The Mechanical Production of Cold*, 2nd edn. (Cambridge, 1921), 47-48; Goosman, *loc.cit.*, LXVII (1924), 329.

who employed a more efficient method of utilizing the absorbent qualities of the acid.[4]

A major obstacle to the production of refrigeration by evaporating water was the difficulty of producing and working with extremely low pressures.[5] It was not long before men tried more volatile liquids, which could be evaporated under practical pressures. In 1805 Oliver Evans, a talented American engineer, suggested that ether be used to make ice. According to his plan, ether was to be placed in a vessel submerged in the water to be frozen. As vaporization took place vigorously in a vacuum produced by a steam-powered pump, the ether would absorb heat from the water, which would freeze. The ether vapor was then to be compressed into another vessel submerged in water. There, subjected to pressure and the cooling influence of the water, it would condense, giving off as it assumed liquid form the heat it had absorbed earlier. The liquid ether then was ready to be used once more.[6] Although Evans apparently did not realize that a vacuum was unnecessary to obtain satisfactory results with ether, his plan showed a clear understanding of the principles involved in the vapor-compression system of mechanical refrigeration. Evans did not carry this device beyond the planning stage, but in 1834 an American residing in England, Jacob Perkins,[7] constructed and patented a vapor-compression machine. Perkins made ice by evaporating under reduced pressure a volatile fluid obtained by the destructive distillation of indiarubber. Vaporization was effected inside a hollow-walled basin which contained the water to be frozen. The vapor was constantly condensed so that the fluid could be

[4] Goosman, *loc.cit.*, LXVII (1924), 329; A. J. Dixon, *Manual of Ice-Making and Refrigerating Machines* (St. Louis, 1894), 10.

[5] Goosman, *loc.cit.*, LXVII (1924), 329.

[6] O. Evans, *The Young Steam Engineer's Guide* (Philadelphia, 1805), 137-139.

[7] Perkins was born in Newburyport, Massachusetts, in 1766. The inventor of a machine to cut and head nails, he went to England in 1818, where he achieved business success as a plate-maker and banknote-printer and continued to exercise his talents as an inventor. For a biography see G. and D. Bathe, *Jacob Perkins, His Inventions, His Times, & His Contemporaries* (Philadelphia, 1943).

used repeatedly. This machine had a compressor, a condenser, an evaporator, and a cock between the condenser and the evaporator—all the basic parts of a modern system. It could produce a small quantity of ice, but it was not used commercially.[8]

Great inventive accomplishment marked the thirty years after 1850. Stimulated by growing need and aided by progress in science and technology, inventors devoted their attention to the problem of devising practical mechanical refrigeration. The need for it was felt in several countries. In Great Britain, where the seriousness of the problem of supplying food to the industrial population became acute about 1860,[9] inventors turned to mechanically produced low temperatures to preserve meat during ocean transportation. In Australia the desirability of finding a means of exporting meat was a factor. On the Continent the principal impetus came from the brewing industry. In the southern part of the United States mechanical refrigeration was necessary if there were to be a dependable and inexpensive supply of ice.

Fortunately, at the time when the need became imperative, science and technology had developed sufficiently to make possible its satisfaction. Materials more suitable than water and ether were made available by Faraday, Thilorier, and others, who had demonstrated that certain substances formerly thought to exist only in gaseous form—ammonia and carbon dioxide, for example—could be liquefied. The theoretical background necessary for mechanical refrigeration was provided by Rumford and Davy, who had explained the nature of heat, and by Kelvin, Joule, and Rankine, who were continuing the work begun by Sadi Carnot in formulating the science of thermodynamics. A suitable source of power, the steam engine, had long been ready, and by the eighteen-seventies the

---

[8] F. Bramwell, "Ice-Making," *Journal of the Society of Arts*, xxxi (Dec. 8, 1882), 76-77; Goosman, *loc.cit.*, lxvii (1924), 429. Ewing, *op.cit.*, 82-83, has a discussion of the Perkins machine based on Bramwell's account. Drawings of the machine can be found in Bramwell, 77, Ewing, 82, and Bathe, *op.cit.*, Plate xxxvi.

[9] J. T. Critchell and J. Raymond, *A History of the Frozen Meat Trade*, 2nd edn. (London, 1912), 4-5.

design and construction of machinery had so improved that equipment could be built capable of working with the pressures involved in the vapor-compression process.[10]

The refrigerating machines which appeared between 1850 and 1880 could be classified according to the substance, called a refrigerant, they used to produce low temperatures. They fell into two broad classes—those that used air and those that used a material which could be vaporized and liquefied alternately.[11]

Machines employing air as a refrigerant, called compressed-air or cold-air machines, played a significant role in the history of refrigeration. They were based on the reverse of the phenomenon of heating that occurred when air was compressed —the fact that air became cool as it expanded against resistance. This behavior had long been recognized. It was observed as early as the middle of the eighteenth century in connection with the operation of an air pump. The Cornish engineer and inventor, Richard Trevithick, who lived until 1833, constructed engines in which expanding air was used to convert water to ice,[12] and in 1846 an American, John Dutton, obtained a patent for making ice by the expansion of air.[13] The real development of the cold-air machine, however, began with the one which Dr. John Gorrie of Appalachicola, Florida, patented in England in August 1850 and in the United States in May 1851.[14]

[10] M. W. Travers, "Liquefaction of Gases," *Encyclopedia Britannica* (Chicago, 1946), XIV, 172-173; A. R. Leask, *Refrigerating Machinery. Its Principles and Management* (London, 1895), 12-14, 17; "Steps in the History of Refrigerating Systems," *Refrigerating Engineering*, XXI (1931), 346; Goosman, *loc.cit.*, LXIX (1925), 100.

[11] Ewing, *op.cit.*, 30-31, discusses the classification of refrigerating machines.

[12] "Scientific Gossip," *The Athenaeum*, Dec. 15, 1849, 1277-1278; Goosman, *loc.cit.*, LXVII (1924), 330.

[13] U. S. Department of State, Patent Office, *Report of the Commissioner of Patents for the Year 1846* [Washington, 1847], 148, 295.

[14] U. S. Department of the Interior, Patent Office, *Report of the Commissioner of Patents for the Year 1851* (Washington, 1852), I, 76.

Gorrie, a public-spirited physician,[15] was motivated by his desire to prevent the ravages of malaria. He had studied the disease closely—as he had every opportunity. to do in Appalachicola—and concluded that it was caused by a volatile oil which emanated from the surface of the earth when organic decomposition, moisture, and heat were present. Gorrie believed controlling the generation of this oil impracticable, but he thought it could be removed from the air supplied to a room in a dwelling or hospital. He had noticed that malaria was not contracted during the day, and he reasoned it could be prevented by remaining at night in a room in which the atmosphere had been purified. Since it was a well-known fact that malaria was checked by the advent of cold weather, particularly of frost, he decided to try to purify the atmosphere of a limited space by "the refrigerative, condensing and disinfecting agency of ICE." He proposed to do this by suspending close to the ceiling of a room an ornamental basin filled with ice. Over this container he made an opening in the ceiling from which a pipe was to run between the ceiling and the floor above to the chimney. Another pipe, which could be opened and closed by a valve, was to be placed at the level of the floor to connect the room with the outside. The room, sealed except for the pipes, was to be ventilated by air drawn down the chimney and expelled through the outlet. Gorrie believed that when the air flowed down the chimney, which was lined with carbon in the form of soot, the malaria in it would tend to decompose because of the affinity of carbon for "vapors and organic oils." As the air passed over the ice, its vapor would be condensed and "other volatile matters and extraneous gases" absorbed.[16] By 1844 Gorrie had succeeded

[15] R. E. Mier, *John Gorrie: Florida Medical Pioneer and Harbinger of Air Conditioning* (Unpublished master's dissertation, John B. Stetson University) is the only reliable account of the facts of Gorrie's life. A brief summary is R. E. Mier, "More About Dr. John Gorrie and Refrigeration," *The Florida Historical Quarterly*, XXVI (1947-1948), 167-173.

[16] J. Gorrie, "On the Nature of Malaria, and Prevention of its Morbid Agency," *The New Orleans Medical and Surgical Journal*, XI (1854-1855), 619-620, 753-760, 767.

in lowering the temperature of a hospital room,[17] presumably with such an arrangement.

To obtain the cheap, dependable supply of ice necessary if this contrivance were to be used generally in Appalachicola or any other southern town, Gorrie tried to find a practical method of manufacturing ice. By 1845 he had constructed a cold-air machine. His invention was described in the *Scientific American* in 1849, and he seems to have produced ice with a model at a public demonstration in 1850. A unit of his design was working in England in 1854.[18] In Gorrie's invention air at atmospheric pressure was compressed, an operation accompanied by heating. To remove this heat, water was injected into the compressing cylinder by a small pump. The air then was allowed to pass into a receiver surrounded by cold water, where it was cooled further and separated from the water that had been introduced. When the air had cooled sufficiently, it was allowed to expand against a piston in a cylinder surrounded by salt-water brine. As it did this, it became very cold. Brine was injected into the cylinder, cooled by the air, and then drawn off and circulated around containers of fresh water to be frozen. Part of the power expended in compressing was recovered by using the force exerted by the expanding air to aid in compressing more air.[19] Gorrie's work was not carried beyond the experimental stage, for he failed to get financial backing in the United States,[20] and a machine built in England performed unsatisfactorily.[21] Never-

[17] Mier, *op.cit.*, 51.

[18] *ibid.*, 53, 57-58, 64; J. C. C., "Ice Made by Mechanical Power," *Scientific American*, v (Sept. 22, 1849), 3; J. J. Coleman, "Air Refrigerating Machinery and its Applications," *Minutes of Proceedings of the Institution of Civil Engineers*, LXVIII (Feb. 14, 1882), 152; "Ice-Making Machine," *Journal of the Society of Arts*, II (Feb. 24, 1854), 250.

[19] U. S. Department of the Interior, Patent Office, *op.cit.*, I, 163-164; Gorrie, *loc.cit.*, 754n; J. C. C., *loc.cit.*, 3; Coleman, *loc.cit.*, 152-153; "Ice-Making Machine," *loc.cit.*, 250. A model of Gorrie's machine is on exhibit at the National Museum, Washington, D.C. A diagram is in *Ice and Refrigeration*, XXI (1901), 47-48.

[20] Mier, *op.cit.*, 68-69, 72, 75. For Gorrie's own account of why his invention was not well received see Gorrie, *loc.cit.*, 754n.

[21] A. C. Kirk, "On the Mechanical Production of Cold," *Minutes of*

theless, his accomplishments encouraged other inventors.[22] Refrigerating machines in which the refrigerant was cold air could be divided into two types—closed cycle and open cycle. In the closed cycle, air confined to the machine at a pressure higher than that of the atmosphere was used repeatedly in a sequence of operations. In the open cycle, air was drawn into the machine at atmospheric pressure and when cooled was discharged directly into the space to be refrigerated.[23]

Of the closed-cycle machines the earliest to achieve commercial success were the inventions of Europeans. In 1862 a Scot, Dr. Alexander C. Kirk, led the way. Kirk was the manager of a paraffin-oil works at Bathgate. The owners asked him to find a means of producing refrigeration for use in making paraffin. They had tried a vapor-compression system which had proved a fire hazard. Kirk was able to develop on the basis of Gorrie's work a practical machine which met their requirements. He also designed several similar units for ice making.[24] Another European, Franz Windhausen of Brunswick, Germany, invented a closed-cycle machine for

---

Proceedings of the Institution of Civil Engineers, xxxvii (Jan. 20, 1874), 246. Sir William Siemens, who was asked to examine this machine, criticized, it in 1857 among other reasons because the expanded air was allowed to escape to the atmosphere at low temperatures instead of being used to cool the fresh air being drawn in for compression. C. W. Siemens, "Report on the performance of Newton's Patent Refrigerating Apparatus, with suggestions for improving the same," *Minutes of Proceedings of the Institution of Civil Engineers*, lxviii (1881-1882), 179-186. A summary of the Siemens report may be found in Ewing, *op.cit.*, 32-33.

[22] Dixon, *op.cit.*, 16-17; Kirk, *loc.cit.*, 246. Gorrie finally received a rather surprising amount of recognition. In 1914 Florida placed his statue in Statuary Hall in the Capitol at Washington, and in 1935 the Dr. John Gorrie Memorial Fund, endorsed by the governor and legislature of Florida and the Florida Medical Association, was launched to fight cancer. *Ice and Refrigeration*, xlvi (1914), 311; lxxxix (1935), 84. Gorrie, a man of considerable vision, was aware of the benefits that would come from the successful application of refrigeration to what is now called air conditioning and to food preservation. Gorrie, *loc.cit.*, 768-769.

[23] Ewing, *op.cit.*, 31, 39.

[24] Kirk, *loc.cit.*, 245-246; Ewing, *op.cit.*, 34-36.

which he received in 1870 a United States patent and which was used in German and American breweries.[25] An American, Leicester Allen, the editor of the *American Artisan*, entered the field in 1879 with an invention which underwent several modifications before it proved satisfactory.[26]

Open-cycle machines, which were to be used more widely than the closed, were given theoretical outline early in the fifties by both Kelvin and Rankine, but it was not until 1873 that a Frenchman, Paul Giffard, invented one. A mechanism similar to Giffard's was patented in Great Britain in 1877 by Joseph J. Coleman and Henry and James Bell. The Bell-Coleman machine was designed to provide a means better than natural ice for refrigerating meat in shipment to Great Britain. It was used to preserve meat imports from the United States in 1879, from Australia in 1880, and from New Zealand in 1882. About the same time T. B. Lightfoot also contributed to the development of open-cycle equipment by improving Giffard's invention.[27]

Destined to become vastly more important than cold-air machines were those which used as a refrigerant liquids capable of being alternately vaporized and liquefied. Around refrigerants of this type two kinds of systems developed, vapor absorption and vapor compression. Both had as their basis the same natural phenomena: first, when liquids vaporized they absorbed heat from their surroundings, leaving them refrigerated; second, the liquids could be made to evaporate violently or boil and subsequently to condense or liquefy again at temperatures that could be raised or lowered by increasing or decreasing the pressure.

Vapor-absorption machines were of two varieties: ammonia absorption and water-vapor absorption. In the ammonia type the pressure that regulated the vaporization and liquefaction was produced by the interaction of ammonia and water, which

[25] Nickerson, *loc.cit.*, 171.

[26] H. B. Roelker, "The Allen Dense Air Refrigerating Machine," *Transactions of the American Society of Refrigerating Engineers*, II (1906), 52, 54; Goosman, *loc.cit.*, LXVII (1924), 429.

[27] Ewing, *op.cit.*, 34, 38-39; E. C. Smith, "Pioneers of Refrigeration," *Nature*, CLI (April 10, 1943), 413.

tended to combine when both were cold but to separate when heated. No pressure pumps were necessary. When a mixture of ammonia and water was heated in a vessel called a generator, the ammonia was driven off as vapor at high pressure. This vapor passed to a condenser, where it was liquefied and the heat removed by circulating water or air. Then it was admitted to an evaporator, where it drew heat from the surroundings as it vaporized rapidly. The evaporator was connected with a reservoir of water known as an absorber. Since both ammonia and water were now cold, they tended to unite in the absorber. So rapid was the absorption of the ammonia vapor forming in the evaporator by the water in the absorber that the low pressure necessary for evaporation was maintained.[28]

The water-vapor-absorption machine employed the interaction of water and sulphuric acid. It was based on the same principle as the ammonia absorption, in which the ammonia evaporated, took heat from its surroundings, and then was condensed and re-evaporated. In the water-vapor-absorption variety, however, the evaporation from a quantity of water took heat from and froze the water itself. When one quantity was frozen and the resulting ice removed, more water was admitted and the process repeated indefinitely. No condensing operation was needed. A pump created the initial low pressure required and assisted the sulphuric acid in maintaining it. Since the water had to be evaporated at an extremely low pressure to attain the freezing temperature, water-vapor-absorption machines were also known as vacuum machines.[29]

In the evolution of the ammonia-absorption system the key figure was a French engineer, Ferdinand P. E. Carré. About 1860 he invented a crude apparatus based on the affinity of ammonia for water. In spite of its limitations, it was an im-

---

[28] Ewing, *op.cit.*, 46-47. In practice the operation of ammonia-absorption machines was considerably more complex than this description indicates. See Figure 4. The high and low pressures described in connection with both vapor-compression and ammonia-absorption machines were relative, for with most refrigerants the "low" pressure that had to be maintained in the evaporator was higher than that of the atmosphere.

[29] *ibid.*, 47-48.

portant step. Its chief defect was that it had only two basic parts, each of which had to perform two functions; one was both generator and absorber, while the other was both condenser and evaporator. Since these functions could only be performed alternately, the refrigerating action was intermittent. By providing a separate unit to handle each of the four basic operations, Carré succeeded in 1862 in making a model that refrigerated continuously. Inventors in France, Germany, England, Australia, and the United States soon made further improvements and refinements.[30]

In the development of the water-vapor-absorption machine the pioneer work was done by Leslie and Vallance. In 1850 Edmond Carré made a practical water-vapor-absorption apparatus for producing ice on a limited scale. His device, in which a hand pump was used, could freeze a small amount of water in about five minutes. The Carré machine was used widely in Paris for a while, but it suffered from a serious disadvantage in that the sulphuric acid quickly became diluted with water and lost its absorbing power.[31] The real inventor of the small, hand-operated machine was H. A. Fleuss, who devised a particularly effective pump.[32] A system capable of making ice on a comparatively large scale was constructed in 1878 by Franz Windhausen, who did not confine his activities to any single type of apparatus. In order to make its operation continuous, he drove the water vapor from the sulphuric acid by applying heat, an operation which restored the absorbing power of the acid. This machine, which was installed in an English dairy, could make about twelve tons of ice in twenty-four hours. Further attempts at improvement were made, but this type never became important commercially, for its efficiency was low and the ice produced poor in appearance and quality.[33]

[30] Goosman, *loc.cit.*, LXVIII (1925), 71; LXIX (1925), 99-100; Ewing, *op.cit.*, 52-53; *Ice and Refrigeration*, XXI (1901), 49; Nickerson, *loc.cit.*, 172. A contemporary description of a Carré machine can be found in "Manufacture of Ice," *American Artisan*, X n. s. (Feb. 9, 1870), 89-90. A diagram is in *Ice and Refrigeration*, XXI (1901), 49.

[31] Goosman, *loc.cit.*, LXVII (1924), 329.

[32] *ibid.*, LXVII (1924), 329; Ewing, *op.cit.*, 50.

[33] Ewing, *op.cit.*, 48, 51; Goosman, *loc.cit.*, LXVII (1924), 330.

The vapor-compression system, which was to become by far the more important, differed from the vapor-absorption in that the low pressure needed for violent evaporation was produced by mechanical means. It was created in a vessel called an evaporator by a pump called a compressor. The compressor had a dual function. In addition to furnishing low pressure to the evaporator, it provided the high pressure required for liquefaction of the vapor in a condenser. As vaporizing and condensing occurred, the heat absorbed in the process of evaporation was removed in the condenser.[34]

One of the earliest of the compression machines was that patented in 1853 by Alexander C. Twining, an American inventor, engineer, astronomer, and college professor.[35] Twining, who believed manufactured ice would prove more convenient and economical than natural ice, particularly in the South, began experimenting in 1848. By 1850 he had succeeded in making small quantities, and by 1855 he had a machine erected in Cleveland, Ohio, which could produce close to a ton a day. Twining's invention had the familiar elements of the vapor-compression system—compressor, evaporator, and condenser—but it featured an unusual method of applying the refrigeration to the process of ice making. Instead of using the evaporation of a small amount of ether to cool brine that circulated around containers holding the water to be frozen, Twining employed a large amount of ether which was cooled by the evaporation of a part of it. At a low temperature the refrigerant itself was circulated around cast-iron vessels filled with water.[36] Twining was approached by New

[34] J. T. Bowen, *Refrigeration in the Handling, Processing, and Storing of Milk and Milk Products* (U. S. Department of Agriculture, *Miscellaneous Publication No. 138*) (Washington, 1932), 2-3. An excellent description of the vapor-compression process is included in E. L. Carpenter and M. Tucker, *Farm and Community Refrigeration* (University of Tennessee Engineering Experiment Station, *Bulletin No. 12*) (Knoxville, 1936), 15. For a diagram of a vapor-compression machine see Figure 5.

[35] C. W. Mitman, "Alexander Catlin Twining," *Dictionary of American Biography*, XIX, 83-84. Mitman is in error in classifying Twining's invention as an absorption process.

[36] A. C. Twining, *The Manufacture of Ice on a Commercial Scale*

Orleans promoters interested in ice manufacturing, and he supplied them with a detailed estimate of initial and operating costs, but his machine was never developed commercially. He attributed this to accidents, defects in construction, inability to obtain financial backing, and the Civil War, which prevented its introduction to the South. In an 1870 plea for extension of his patent he insisted that his was the basic invention and that the compression and absorption machines that had appeared in Europe were only the natural consequences of his work.[37]

In the period of Twining's experiments James Harrison, a Scot, was at work in Australia on machinery for ice production. In 1855 he obtained an Australian patent and in 1856

---

and with Commercial Economy, by Steam or Water Power . . . (New Haven, 1857), 4-6, 8-9; Goosman, *loc.cit.*, LXVII (1924), 430; Eli W. Blake, Esq., and Prof. J. D. Dana, quoted by Twining, *op.cit.*, 12. A diagram is in *Ice and Refrigeration*, XXI (1901), 46.

[37] Goosman, *loc.cit.*, LXVII (1924), 430; Twining, *op.cit.*, 8; A. C. Twining, *The Fundamental Ice-Making Invention; Patented November 8, 1853; In a Plea for Extension, Before the Committee on Patents in the House of Representatives of the United States* . . . [Washington, 1870], 2-3, 14-21.

FIGURE 5

A vapor-compression refrigerating machine

"In the evaporator (the coils in the refrigerator) the liquid boils and in the process absorbs heat from the surrounding medium. The compressor is a specially designed pump that takes the gas from the evaporator coils and compresses it into the condenser coils, reducing its volume and increasing its temperature. The condenser consists of coils of pipe over or through which water or air flows to absorb the heat from the gas, which is thereby liquefied. In some systems the cooling water passes through an inner tube, and the gas from the compressor through the annular space between the inner and the outer pipes. From the condenser the refrigerant passes first to a liquid receiver, and then through a throttling or expansion valve into the evaporator coils, to repeat the process of transferring heat from the refrigerator to the water flowing through the condenser. . . ."

From J. T. Bowen, *Refrigeration in the Handling, Processing, and Storing of Milk and Milk Products* (U. S. Department of Agriculture, Miscellaneous Publication No. 138) (Washington, 1932), 3.

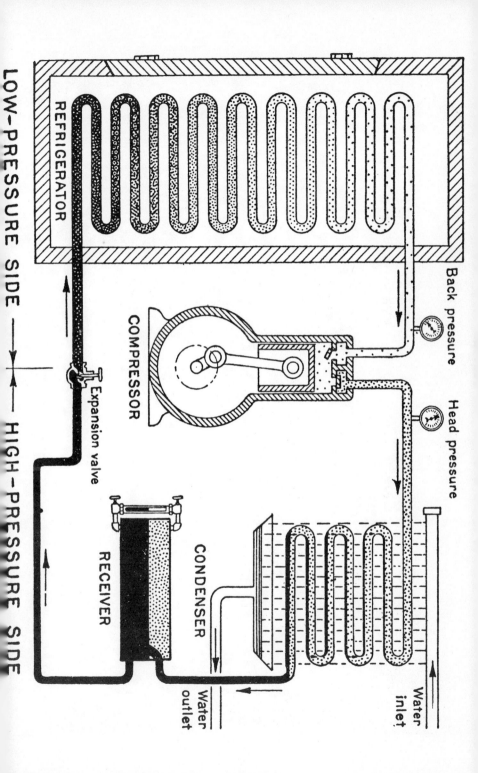

LOW-PRESSURE SIDE ← → HIGH-PRESSURE SIDE

REFRIGERATOR

Back pressure

COMPRESSOR

Head pressure

Expansion valve

RECEIVER

CONDENSER

Water outlet

Water inlet

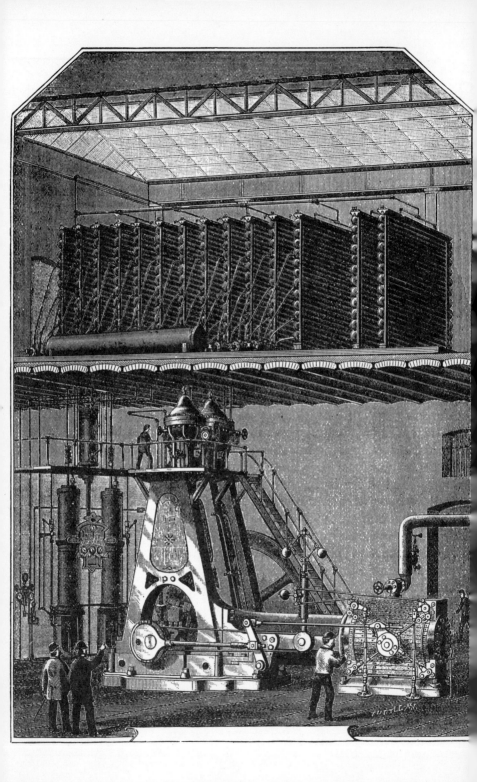

and 1857 English patents. A few of his machines, which used sulphuric ether, were erected in Australia in the early fifties, but apparently were too small in scale to be successful. In 1857 Harrison visited England and was able to have several large models constructed by expert builders. One of these was installed in a paraffin-oil works at Bathgate, Scotland, another was erected in an English brewery, and a third was shipped back to Australia. Soon a Sydney firm undertook manufacture of the Harrison machine, and before long several were in operation in Australia.[38]

A number of other inventors experimented with compression machines that used ether or its compounds. A resident of Maryland was reported to have built such an apparatus about 1859, and a certain Andrew Muhl developed one in San Antonio, Texas, about 1867. The Maryland device failed when tried on a large scale, while only a few of the Muhl units were ever constructed.[39] In France F. P. E. Carré built an ether-compression machine, and Charles Tellier, a versatile pioneer of mechanical refrigeration and its application, constructed one whose refrigerant was methyl ether.[40] In Ger-

[38] Critchell and Raymond, *op.cit.*, 22-24; Smith, *loc.cit.*, 412-413; Goosman, *loc.cit.*, LXVII (1924), 430; "Ice-Making Machine," *The Illustrated London News*, XXXII (May 29, 1858), 546.

[39] "Artificial Ice," *The Builder*, XX (April 19, 1862), 272; W. R. Woolrich, "Mechanical Refrigeration—Its American Birthright," *Refrigerating Engineering*, LIII (1947), 250.

[40] Goosman, *loc.cit.*, LXVII (1924), 430; C. Tellier, *Histoire d'une Invention Moderne, le Frigorifique* (Paris [1910]) is the inventor's own account of his work.

FIGURE 6

An ammonia-compression machine installed in an ice plant, 1891

The two compression cylinders are at the top of the A-shaped frame. The water-cooled condenser is in the upper room. The ammonia liquefied in the condenser collected in the horizontal cylinder, passed down into one of the vertical cylinders to the left of the compressor, whence it entered the evaporating coils in the freezing tank. This equipment, manufactured by De La Vergne, could make 130 tons of ice a day.

From F. Fernald, "Ice-Making and Machine Refrigeration," *The Popular Science Monthly*, XXXIX (1891), 24.

many Carl Linde, financed by brewers, finished a methyl-ether unit in 1874. Just a few years earlier—in 1870 and 1871—Linde had paved the way for great improvements in refrigerating machinery by demonstrating how its thermo-dynamic efficiency could be calculated and increased.[41]

Inventors of compression machines also experimented with ammonia, the refrigerant that was to be used most widely for a great many years. In the sixties Tellier developed an ammonia machine.[42] Of the American workers in this field the most important was David Boyle, a Scot who arrived in Mobile, Alabama, shortly before the Civil War. Boyle patented his in 1872 and by October, 1873 had one at work making ice in Jefferson, Texas. The contribution of Boyle, who made very satisfactory equipment, has been underestimated,[43] but never-theless, the most important figure in the development of am-monia-compression machines probably was Linde, who in 1876 obtained a patent for one which was installed in a Trieste brewery the following year. Linde's model, which became very popular, was particularly excellent in its mechanical de-tails.[44] The use of ammonia in compression machines was an important forward step, for in addition to a thermodynamic advantage which it possessed over other practical refrigerants, the pressures it required were easy to produce, and machines which used it could be small in size.[45]

Late in the sixties Professor P. H. Van der Weyde of Phila-delphia patented a compression machine which featured a refrigerant composed of petroleum products. Machines of this type did not become important, though several, built by Daniel L. Holden, a consulting engineer who purchased the Van der Weyde patent rights, were erected in southern cities.[46] In

[41] Goosman, *loc.cit.*, LXVIII (1925), 137, 336; LXIX (1925), 99.

[42] Tellier, *op.cit.*, 51-55, 257; Critchell and Raymond, *op.cit.*, 27.

[43] Goosman, *loc.cit.*, LXIX (1925), 100-101; Woolrich, *loc.cit.*, 305-306.

[44] Goosman, *loc.cit.*, LXVIII (1925), 413; J. H. Awberry, "Carl von Linde: A Pioneer of 'Deep' Refrigeration," *Nature*, CXLIX (June 6, 1942), 630.

[45] Ewing, *op.cit.*, 86.

[46] Nickerson, *loc.cit.*, 171; Goosman, *loc.cit.*, LXIX (1925), 99; Woolrich, *loc.cit.*, 246.

1875 Raoul P. Pictet, a professor of physics at the University of Geneva, introduced compression equipment which employed sulphurous acid. This became well known in France and Switzerland, but was not used much in other countries.[47] Carbon-dioxide machines, which because of the harmlessness of the gas became important in installations where safety was the primary concern, were not used extensively until the eighties and nineties. Although an American, Thadeus S. C. Lowe, devised a mechanism which produced ice in 1866, the development of equipment employing carbon dioxide was more closely associated with the later work of Windhausen and the English firm that took up his patents.[48]

Europeans provided most of the theoretical background for the development of mechanical refrigeration, but Americans participated vigorously in the widespread inventive activity of the three decades between 1850 and 1880.[49] That their role was not larger was due perhaps to the fact that in the northern states the bountiful natural-ice resources were not yet clearly inadequate.

[47] Nickerson, *loc.cit.*, 171; Ewing, *op.cit.*, 84; "Raoul Pictet's Sulphurous Acid Ice-Machine," *Nature*, xiii (March 30, 1876), 433-434.

[48] Ewing, *op.cit.*, 94; Woolrich, *loc.cit.*, 246, 248.

[49] W. R. Woolrich, a prominent American refrigerating engineer and Dean of the College of Engineering at the University of Texas, believes firmly that American contributions have not been sufficiently emphasized. One reason for this, he thinks, is that inventors in the United States were not as active in writing about their work as were their European contemporaries. Woolrich, *loc.cit.*, 196.

CHAPTER V

THE EXTENSION OF MECHANICAL
REFRIGERATION, 1860-1890

IN the United States the first refrigerating machines to be operated on a commercial basis were located in the South, where a cheap, dependable ice supply was needed. A small Carré ammonia-absorption machine is said to have been run through the blockade to Augusta, Georgia, in 1863, but not to have been a success. By 1865 three others had been brought to New Orleans, but not until one of these had been shipped to San Antonio, Texas, where it was modified, was successful operation achieved.[1] Carré-designed equipment built at Gretna, Louisiana, was installed in 1868 by the Louisiana Ice Manufacturing Company of New Orleans.[2] By 1869 four plants in the United States were manufacturing ice for sale—two in Louisiana, one in Tennessee, and one in Texas.[3]

During the seventies ice-making equipment was still in a period of trial and experiment. Early in the decade plants were set up in some of the larger cities, usually where there was little competition from natural ice. From these centers they spread gradually to neighboring communities and by 1880 were beginning to threaten the New England ice trade to southern ports. By 1879 the number of establishments in the southern states that manufactured ice for sale had increased to twenty-nine. Texas had twelve, Georgia eight, Louisiana four, Alabama three, and Kentucky and Arkansas each had one.[4] The size of these concerns varied greatly. One

[1] Daniel L. Holden, quoted in *Ice and Refrigeration*, xxi (1901), 50.
[2] J. F. Nickerson, "The Development of Refrigeration in the United States," *Ice and Refrigeration*, xlix (1915), 171.
[3] U. S. Department of the Interior, Census Office, *Ninth Census* (Washington, 1872), iii, 523, 571, 573. The twelve-month period on which census statistics on manufactures were based varied somewhat. In the interests of uniformity and clarity all statistics are assumed to be for the calendar year preceding the census year.
[4] A. L. Hunt, "Manufactured Ice," U. S. Department of the Interior, Census Office, *Twelfth Census, 1900*, ix (Washington, 1902), 676-677; Editorial, *Ice Trade Journal*, iii (Sept. 1879), 4.

plant in Austin, Texas, could make only two and one-half tons a day, while that of the Louisiana Ice Company in New Orleans, the largest in the United States at the time, could turn out one hundred and eighteen tons every twenty-four hours.[5]

Between 1880 and 1890 ice manufacturing was extended throughout the South. In the early eighties the natural product remained dominant in seaports, for it still could compete in price, and it enabled dealers to avoid the expenditures involved in manufacturing experiments.[6] But this situation did not last long, and before 1890 competition had forced the natural-ice business into a real decline.[7] Plant construction in the South began to increase steadily in 1883 and in 1888 to accelerate rapidly.[8] By 1889 one hundred and sixty-five plants were in operation. Fifty-three were located in Texas, eighteen in Alabama, sixteen in Georgia, thirteen in Tennessee, and twelve in Kentucky. No southern state was without several.[9] The ammonia-absorption system predominated during the seventies, but between 1880 and 1890 ammonia-compression installations became more common.

The quality of manufactured ice improved greatly between 1870 and 1890. In the first plants the water to be frozen was placed in metal cans suspended in tanks containing chilled brine. The ice obtained was opaque, for particles of air present in the water were trapped in the ice, breaking up its usual crystalline structure.[10] A transparent product was demanded

[5] *Ice Trade Journal*, II (July 1879), 2; IV (Jan. 1881), 5.

[6] H. Hall, "The Ice Industry of the United States," U. S. Department of the Interior, Census Office, *Tenth Census, 1880*, XXII (Washington, 1888), 36.

[7] *Ice Trade Journal*, XII (Dec. 1888), 1.

[8] W. J. Rushton, "Early Days of the Manufacture of Ice," *Ice and Refrigeration*, LI (1916), 152; *Ice and Refrigeration*, II (1892), 118. In 1888 there was a marked increase in the number of references in the *Ice Trade Journal* to the construction of ice plants.

[9] Hunt, *loc.cit.*, 677.

[10] Nickerson, *loc.cit.*, 171; Van R. H. Greene, "Modern Systems for Manufacture of Raw Water Can Ice," *Ice and Refrigeration*, LXII (1922), 287. Not all early ice plants used the can system. In a few cases ice was made by spraying water over refrigerated pipes that were arranged vertically. When enough ice had accumulated to fill the space

by the public, however, and before 1870 the manufacture of it became possible as a result of the discovery that distillation of the water before freezing drove off the air.[11] Soon many refinements were effected in this method, known as the distilled-water can system. Distilled water was obtained by condensing steam, part of which, before condensation, was used in compression systems for power. When cooled and filtered, the water was introduced into cans in a brine tank to be frozen. The brine, kept circulating by pumps, was chilled by evaporator coils placed between the rows of cans. When the water was frozen, the cans were lifted from the tank by a crane, dipped in a bath of tepid water, and then tipped over so that the ice slid out easily. The cake was clear except for a cloudy core, which was caused by the air the water absorbed while freezing.[12]

In 1890 the distilled-water can system was used most widely,[13] but another method—the plate system—was employed in some important establishments. This was an attempt to obtain transparent ice by imitating nature's method. Water, constantly circulated, was frozen from one side only, not from four sides at once as in the can arrangement. As freezing progressed, air and other undesirable elements were expelled into the unfrozen water, which was removed when sufficient ice had formed. With the plate method distilled water was not needed.[14] Plate-system plants, it was said, could produce more

---

between the pipes, it was removed. J. C. Goosman, "History of Refrigeration," *Ice and Refrigeration*, LXX (1926), 314; Nickerson, *loc.cit.*, 171.

[11] Daniel L. Holden, quoted in *Ice and Refrigeration*, XXI (1901), 50; F. V. De Coppet, "Pioneer Ice Making," *Ice and Refrigeration*, II (1892), 37; A. J. Dixon, *Manual of Ice-Making and Refrigerating Machines* (St. Louis, 1894), 66.

[12] Dixon, *op.cit.*, 62, 68-69; F. Fernald, "Ice-Making and Machine Refrigeration," *The Popular Science Monthly*, XXXIX (1891), 23; "Ice-Making in New Orleans," *Harper's Weekly*, XXXIV (Jan. 25, 1890), 67.

[13] Dixon, *op.cit.*, 62; J. E. Starr, "Refrigeration Twenty-Five Years Ago," *Ice and Refrigeration*, LI (1916), 143.

[14] Dixon, *op.cit.*, 63-64; E. Penney, "Plate versus Can Systems," *Ice and Refrigeration*, XXIX (1905), 259; G. Braungart, Jr., "Crystal Ice from Raw Water," *Ice and Refrigeration*, XXXVIII (1910), 142; Goosman, *loc.cit.*, LXX (1926), 314.

ice per ton of coal than could those using the can system, but their initial cost was much higher, and they made ice in large, unwieldy plates which had to be sawed into blocks of manageable size.[15]

Although most of the refrigerating machinery in the South was installed in ice-manufacturing plants, some was used in other activities. Early in the seventies mechanical refrigeration was employed in Texas meat-packing plants.[16] In 1869 one of Charles Tellier's methyl-ether-compression machines was erected in the brewery of George Merz in New Orleans. David Boyle, noted for his work with ammonia-compression equipment, obtained a contract about 1878 to refrigerate a Louisville, Kentucky, brewery,[17] and Thomas L. Rankin, an early experimenter, built absorption machines for several breweries after 1879.[18] In the eighties a considerable number of breweries in the South were refrigerated mechanically.[19]

On the Pacific coast ice manufacturing did not develop as rapidly. A plant was erected at Los Angeles in 1869 and one at San Francisco in 1870. The following year manufactured ice was introduced in Portland, Oregon,[20] but according to the Tenth Census there were only five firms on the Pacific coast, all in California, that made it for sale in 1879. A decade later the number had increased to thirteen, seven of which were in

[15] L. K. Doelling, "Twenty Five Years Evolution of Refrigeration," *Ice and Refrigeration*, LI (1916), 160; Braungart, *loc.cit.*, 142; H. D. Pownall, "The Development of Raw Water Ice Making," *Ice and Refrigeration*, XLIII (1912), 229.

[16] Goosman, *loc.cit.*, LXIX (1925), 99; R. A. Clemen, *The American Livestock and Meat Industry* (New York, 1923), 216.

[17] E. T. Skinkle, *Practical Ice Making and Refrigerating: A Plain, Common Sense Series of Papers on the Construction and Operation of Ice Making and Refrigerating Plants and Machinery* (Chicago, 1897), 8.

[18] J. P. Arnold and F. Penman, *History of the Brewing Industry and Brewing Science in America* (Chicago, 1933), 93-94. For accounts of Rankin's career see Goosman, *loc.cit.*, LXIX (1925), 100, and W. R. Woolrich, "Mechanical Refrigeration—Its American Birthright," *Refrigerating Engineering*, LIII (1947), 306-307.

[19] Advertisement, Blymer Ice Machine Company, *Ice Trade Journal*, XIV (Dec. 1890), 2; *Ice Trade Journal*, III (May 1880), 3; Arnold and Penman, *op.cit.*, 95.

[20] Nickerson, *loc.cit.*, 171.

California, four in Oregon, and two in Washington.[21] As late as 1890 San Francisco, where consumption was sixty thousand tons a year, depended for the major part of her needs on ice from the Sierra Nevadas.[22]

In the northern United States, where ice was in cheap and abundant natural supply, very little machinery was erected for its manufacture. In 1879 there was only one plant in Missouri making it for sale and none in the states north of the Potomac and the Ohio. A decade later ice factories were still completely absent from New England, though fourteen were in operation in the middle states and twenty-three in the central states, chiefly in Ohio, Illinois, Pennsylvania, and Maryland.[23] The most important installations of refrigerating machines in the North before 1890 were not in ice plants, but in breweries and packing houses.

Brewing was the first activity in the northern states to use mechanical refrigeration extensively. A Carré absorption machine was installed by S. Liebmann's Sons Brewing Company in Brooklyn in 1870. Later in the seventies machines were adopted by others in the industry throughout the North— in New York, Philadelphia, Cincinnati, Chicago, Milwaukee, St. Louis, and elsewhere.[24] Increasing numbers were used in the early eighties,[25] and by the end of the decade their advantages and practicability had been demonstrated thoroughly.[26] By 1891 nearly every brewery was equipped with refrigerating machines. It was claimed that more were employed by brewers at that time than by all other users.[27] Both com-

[21] Hunt, *loc.cit.*, 677.

[22] Letter from San Francisco, Calif., Aug. 5, 1890, in *Ice Trade Journal*, xiv (Sept. 1890).

[23] Hunt, *loc.cit.*, 677.

[24] Arnold and Penman, *op.cit.*, 93-94; Skinkle, *op.cit.*, 7-10.

[25] *Ice Trade Journal*, v (March 1882), 4.

[26] G. J. Patitz, "Progress in Brewery Refrigeration," *Ice and Refrigeration*, li (1916), 168. The Best (later Pabst) brewery in Milwaukee was slow to adopt mechanical refrigeration; not until 1879 did Captain Frederick Pabst install the first unit, a 150-ton Boyle machine. In the eighties additional equipment was purchased at a rate that made natural ice unnecessary after 1888. T. C. Cochran, *The Pabst Brewing Company* (New York, 1948), 107-109.

[27] Fernald, *loc.cit.*, 28.

pression and ammonia-absorption systems were installed, but compression equipment was in greater favor.

In breweries mechanical refrigeration first was used to chill the water which cooled the hot wort and removed the heat generated during fermentation. In the early days brewers hesitated to make themselves dependent on machinery to cool their cellars, for they feared the interruption of refrigeration that would result from a breakdown and that would mean the loss of all beer stored there. They preferred to rely on ice which they manufactured. But before long experiments were made in cooling fermenting rooms by pumping chilled brine through pipes. When these experiments proved successful, the system was extended to storage cellars.[28]

Mechanical refrigeration was earliest adopted on a large scale by the brewing industry because of its inherent advantages. It freed brewers from dependence on natural ice with its uncertainties of supply and price. It reduced labor costs and permitted more advantageous use of the space that had been required for ice storage. It enabled brewers to make a better product, since temperatures could be controlled more exactly. It maintained a more intense cold that shortened the time necessary for beer to ripen and helped reduce the amount of foaming waste involved in filling tanks, kegs, and bottles. In fermentation and storage rooms, which no longer had to be located in deep cellars, it provided a dry environment which contributed to cleanliness, checked the growth of molds, and made possible healthier conditions for workers.[29]

In a few meat-packing plants refrigerating machinery was employed in the early seventies, but not until 1880 was it introduced by a Chicago packer.[30] The eighties were years of experimental installations. By 1890 mechanical refrigeration had proved to be both practical and economical. But in general meat-packers were less quick to accept it than brewers.

[28] Patitz, *loc.cit.*, 168; Goosman, *loc.cit.*, LXIX (1925), 149.

[29] Arnold and Penman, *op.cit.*, 92-93; Patitz, *loc.cit.*, 168; J. E. Siebel, "Refrigeration and the Fermenting (Brewing) Industry in the United States," *Premier Congrès International du Froid*, III (Paris [1908?]), 74-75; Cochran, *op.cit.*, 97-98, 107, 109-110.

[30] Clemen, *op.cit.*, 216.

So important a processor as Swift did not install it until 1887, and it was not used generally until about 1890.[31]

The assignment given refrigerating machines was to chill salt brine which was circulated through the space to be cooled. Sometimes the brine was pumped through open troughs, but more frequently it was forced through pipes banked in the overhead bunkers formerly used for ice or suspended from the ceiling or walls of the storage rooms. In some cases brine was not circulated; instead, the refrigerant was evaporated in the pipes rather than in a separate evaporator. An advantage of this method was its lower initial cost.[32]

Mechanical refrigeration proved more satisfactory than ice for producing the low temperatures required. It made huge ice-storage houses unnecessary and made it possible to eliminate or reduce in size the bunkers over refrigerated space. It provided a dry environment, and it produced temperatures sufficiently low to reduce materially the time needed to cool carcasses. When ice was used to chill hogs, forty-eight to seventy-two hours were necessary. After the introduction of the refrigerating machine the time was lowered to from ten to sixteen hours.[33]

The extension of mechanical refrigeration to cold-storage warehouses, where it offered advantages comparable to those in breweries and in packing plants, occurred in northern cities in the eighties. A pioneer in this development was the Mechanical Refrigerating Company of Boston, which opened a warehouse about 1881. In 1882 an East St. Louis, Illinois, canning concern cooled its storage space mechanically. Machinery was used to refrigerate a Baltimore warehouse in 1886 and one in Chicago in 1889,[34] but not until the early

[31] A. Cushman, "The Packing Plant and Its Equipment," *The Packing Industry* (Chicago, 1924), 101, 114; L. F. Swift, with A. Van Vlissingen, Jr., *The Yankee of the Yards* (Chicago, 1927), 200-201.

[32] Cushman, *loc.cit.*, 115-116.

[33] F. H. Boyer, "Abattoir Refrigeration," *Ice and Refrigeration*, xx (1901), 195.

[34] Taylor, *loc.cit.*, 568-569; Goosman, *loc.cit.*, lxix (1925), 205; lxx (1926), 314.

nineties was it used generally for cold storage in the cities of the North.[35]

The manufacture of refrigerating machines in the United States before 1880 was mainly the work of a number of able engineers, some of whom entered the field as a result of their efforts to improve the early equipment imported from Europe. One of the first of these pioneers was Daniel L. Holden, who had served as chief of the Magnetic Signal Corps in the Confederate Army. While attempting to make ice in San Antonio with a Carré absorption machine, he discovered that its operation could be accelerated if a steam coil were used to heat the solution in the generator, and that transparent ice could be made from distilled water. In addition to his work with absorption machinery, Holden built and installed a number of compression units.[36] Even more important among these early inventors and builders was David Boyle, who supplied ice plants and breweries. In 1875, unable to keep pace with orders, he interested the Crane Brothers Manufacturing Company of Chicago in constructing his ammonia-compression machine, and in 1878 he organized the Boyle Ice Machine Company.[37]

Difficult problems faced the first manufacturers. Since it was not easy to construct pumps and joints capable of working with gases at high pressures, leakage of the refrigerant was a common occurrence. The danger of explosion always existed, especially when ether was used. Frequently operating costs of an installation proved too high, and sometimes a machine that worked satisfactorily in the North would fail in the South.[38] The trials of the manufacturers, serious enough

[35] F. G. Urner, quoted by G. K. Holmes, *Cold Storage and Prices* (U. S. Department of Agriculture, Bureau of Statistics, *Bulletin 101*) (Washington, 1913), 8; F. A. Horne, "Development of the Cold Storage Industry," *Ice and Refrigeration*, LI (1916), 170.

[36] Goosman, *loc.cit.*, LXIX (1925), 99; *1925 Ice and Refrigeration Blue Book and Buyer's Guide: A Directory of the Ice Making, Cold Storage, Refrigerating and Auxiliary Trades* . . . (Chicago, 1926), 14; Woolrich, *loc.cit.*, 199, 246.

[37] Goosman, *loc.cit.*, LXIX (1925), 100-101; *1925 Ice and Refrigeration Blue Book and Buyer's Guide*, 15; Skinkle, *op.cit.*, 7-10; *Ice Trade Journal*, XV (Aug. 1891), 4; Woolrich, *loc.cit.*, 305-306.

[38] Hall, *loc.cit.*, 20.

when of a technical nature alone, often were complicated by patent troubles.[39]

About 1880 leadership passed from individual mechanics to companies which could command greater financial and technical resources. Some of the pioneers—David Boyle, for example—were able to make the transition to large-scale operation, but most of the firms of the eighties were comparative latecomers. Although some organizations were ill-prepared and soon failed, several of the concerns still held in 1950 positions of prominence.[40]

Ammonia-compression machines occupied the attention of some of the most progressive manufacturers of the eighties. The De La Vergne and Mixer Refrigerating Company, a New York City enterprise organized in 1881, had its origin in the interest of John C. De La Vergne, a brewer, in more practical equipment for his plant. This company was noted especially for the system of oil sealing with which its compressors were furnished to prevent ammonia leakage.[41] The Frick Company of Waynesboro, Pennsylvania, a builder of steam engines since the early fifties, became interested in refrigerating machinery in 1882, when it was asked to convert an old steam engine into an ammonia-compressor. By 1886 the firm was well established in its new activity.[42] The Vilter Manufacturing Company of Milwaukee, which originated in a small machine shop, began in 1882 to make compression equipment which soon was characterized by improved valves

[39] *Ice Trade Journal*, I (April 1878), 1. Little evidence, however, has been found to suggest that patents had a marked effect either in stimulating or impeding progress in the art of refrigeration. For a full account of the most spectacular patent litigation in the history of the industry see Special Supplement to *Electric Refrigeration News*, reprinted from the issues of March 27 and April 10, 1929.

[40] Goosman, *loc.cit.*, LXIX (1925), 148, 203.

[41] *ibid.*, LXIX (1925), 203-205, 267; Doelling, *loc.cit.*, 159; J. A. Ewing, *The Mechanical Production of Cold*, 2nd edn. (Cambridge, 1921), 90-94. See Figure 6 for an illustration of a De La Vergne machine.

[42] Goosman, *loc.cit.*, LXX (1926), 124, 198, 373; T. Mitchell, "Fifty Years Development in Refrigerating and Ice-Making Machinery," *Ice and Refrigeration*, LXXXII (1932), 247. See Mitchell, 247, for illustrations of Frick machines built in the eighties.

and condensers and by excellent gas-tight fittings. Among other prominent manufacturers was the Fred W. Wolf Company of Chicago. Wolf, a designer of breweries and sugar factories and a builder of brewing equipment, obtained in 1881 the American and Canadian rights to the patents of Carl Linde.[43]

Ammonia-absorption machinery was made by several American companies in the eighties. One of the most successful types was that designed by H. D. Stratton, a resident of Columbus, Georgia. Built first by the Columbus Iron Works Company, it was used widely in ice plants throughout the South.[44] The Blymer Ice Machine Company of Cincinnati[45] and the Henry Vogt Machine Company of Louisville were among the firms that specialized in the manufacture of absorption systems.[46] At least one British absorption machine, the Pontifex, was made in the United States.[47]

The companies which assumed leadership were individualistic in their methods. Since there was no formal training and since there were no journals devoted to refrigerating engineering, technical knowledge frequently was inadequate. The only test of a machine was whether or not it worked. Manufacturers guarded jealously whatever they considered the secret of their success and did all they could to prevent their equipment from being imitated or, so it seemed, even from being repaired by others.[48]

By 1890 dependable refrigerating machinery—both absorption and compression—had been developed. Durable con-

---

[43] Goosman, *loc.cit.*, LXVIII (1925), 414; LXIX (1925), 268-269, 372-373.

[44] *Ice Trade Journal*, IX (May 1886), 1; C. T. Baker, "Twenty-Five Years of Ice Manufacture in the South," *Southern Power Journal*, XLVII (June 1929), 131; Woolrich, *loc.cit.*, 307-308. See Figure 7 for an illustration of this machine.

[45] A list of the installations made by this company elsewhere than in ice plants is included in its advertisement in the *Ice Trade Journal*, XIV (Dec. 1890), 2.

[46] "What the Refrigerating Machine Companies Have Contributed," *Refrigerating Engineering*, XXVIII (1934), 300.

[47] *ibid.*, 299; *Ice Trade Journal*, VIII (Sept. 1884), 5.

[48] Goosman, *loc.cit.*, LXIX (1925), 203.

struction was the rule, but the problem of low efficiency remained to be solved.[49]

By 1888 some of the leading manufacturers feared that the greater part of their market had been supplied.[50] These fears were short-lived, however, for the natural-ice crop was perhaps twenty per cent below normal in the winters of 1888-1889 and 1889-1890. Failures had been experienced in single years, but never before had they occurred in two successive winters.[51] The serious shortages in the following summers, a striking demonstration of the inadequacy of natural ice, gave rise to an unprecedented demand for refrigerating machinery, not only for ice making, but for all industries which required low temperatures.[52] This led to a new era in mechanical refrigeration.

[49] Doelling, *loc.cit.*, 159, 161. Ammonia-compression and absorption machines of 1891 are described by Fernald, *loc.cit.*, 22-23, 25, 29. Long working lives were characteristic of these early machines. In 1942 a compressor built by the Arctic Ice Machine Company in 1883 was still used at the meat-packing plant of John Morrell and Co., Ottumwa, Iowa. *Ice and Refrigeration*, CII (1942), 153.

[50] L. Block, "Recollections of a Quarter Century Experience," *Ice and Refrigeration*, LI (1916), 146; "What the Refrigerating Machine Companies Have Contributed," *loc.cit.*, 298; Nickerson, *loc.cit.*, 172.

[51] *Ice Trade Journal*, XII (May 1889), 5; XIII (March 1890), 4.

[52] Editorial, *Ice Trade Journal*, XIV (June 1891), 4. So great was the demand that some manufacturers failed as a result of their inability to make deliveries they had guaranteed. *Ice Trade Journal*, XV (Nov. 1891), 4; Nickerson, *loc.cit.*, 172.

CHAPTER VI

TECHNICAL PROGRESS
1890-1917

STEADY technical progress in the field of mechanical refrigeration marked the years after 1890. Revolutionary changes were not the rule, but many improvements, the work of inventors in several countries, were made in the design of machinery.

In compression machines the principal refrigerants were ammonia and carbon dioxide or, as it was called frequently, carbonic-acid gas. Ammonia was by far the more important in spite of its toxicity and its corrosive effect on all non-ferrous metals except aluminum, for the pressures it required were easy to provide and its thermal cycle was efficient.[1] The use of carbon dioxide was limited to ships, restaurants, hotels, and other places where safety was particularly important.[2]

Improvements in ammonia-compression systems were made in the three basic parts—the compressor, the condenser, and the evaporator.[3] Compressor operating speeds were increased, in some cases from forty to two hundred and twenty revolutions a minute.[4] The multiple-effect compressor, designed by Gardner T. Voorhees, a Boston refrigerating engineer, made possible greater efficiency during low-temperature work by enabling the compressor to handle a greater weight of vapor at each stroke.[5] Electricity rather than steam was used more

[1] W. H. Carrier, "How Air Conditioning Has Advanced Refrigeration," *Mechanical Engineering*, LXV (1943), 332-333.

[2] J. F. Nickerson, "The Development of Refrigeration in the United States," *Ice and Refrigeration*, XLIX (1915), 173; F. Wittenmeier, "Development of Carbon Dioxide Refrigerating Machines," *Ice and Refrigeration*, LI (1916), 165.

[3] Improvements were not limited to ammonia-compression systems; the efficiency of carbon-dioxide equipment was increased significantly.

[4] L. K. Doelling, "Twenty Five Years Evolution of Refrigeration," *Ice and Refrigeration*, LI (1916), 161.

[5] Voorhees' invention also made it possible for a compressor to maintain evaporators at different pressures. G. T. Voorhees, "Multiple Effect Compressors," *Transactions of the American Society of Refrigerating Engineers*, II (1906), 97-111; "Recent Improvements in

frequently as a source of power.[6] Condenser efficiency was enhanced by the adoption of the double-pipe design, which featured one pipe inside the other. The refrigerant flowed through one, while cooling water passed through the other in the opposite direction.[7] In condensers perhaps the most important development was the "flooded" system, which took advantage of the fact that heat transfer from the refrigerant to the cooling water was more rapid when the refrigerant was in the form of a liquid than when it was a gas.[8] Evaporator performance was improved by the gravity-feed or "flooded" method, in which the coils were supplied automatically with the proper quantity of refrigerant. The evaporator was kept practically filled with liquid refrigerant, a condition which speeded the flow of heat through its walls.[9]

the Refrigerating Industry," *Scientific American*, CIX (Oct. 4, 1913), 261; Nickerson, *loc.cit.*, 173. Another method of increasing efficiency when it was necessary to provide very low temperatures—the two-stage compressor—was not used extensively before 1917. H. Sloan, "Low Temperatures by Means of Multi-Stage and Other Compression Systems," *Refrigerating Engineering*, XLV (1943), 419-420.

[6] When only induction electric motors were available, power had to be transmitted from the motor to the compressor by belt, for induction motors could not be operated slowly enough to permit them to be connected directly. Belt drive was expensive, for it required too much floor space, and there was excessive loss of power in transmission. With the appearance of the synchronous motor, which could operate at slow speeds, direct connection with all its economies became practical. C. N. Drake, "Direct Connected Compressors," *Ice and Refrigeration*, LVIII (1920), 267-268.

[7] Nickerson, *loc.cit.*, 173.

[8] *ibid.*, 173; "Recent Improvements in the Refrigerating Industry," *loc.cit.*, 261, 272.

[9] "Recent Improvements in the Refrigerating Industry," *loc.cit.*,

FIGURE 7

The Stratton Absorption Ice Machine

Absorption machinery manufactured by H. D. Stratton and Company, Columbus, Georgia, was adopted by many of the ice plants built in the South during the eighteen-eighties. In the installation pictured the brine tank in which the cans of distilled water were frozen is at the left.

From Advertisement, H. D. Stratton and Company, Columbus, Georgia, *Ice Trade Journal*, XI (January 1888), 8.

Freezing Bath.   Gas Exchanger. Distilled Water Tank.   Ammonia Pump   Gas Collector   Absorber   Condenser   Air Compressor

Ammonia-absorption machines, which were used more frequently than compression equipment to provide very low temperatures because of their greater efficiency in such work,[10] were improved between 1890 and 1917. Early in the century appeared the exhaust-steam type, which, since it needed only low-pressure steam in its generator coil, could be operated with exhaust steam that otherwise would go to waste.[11]

Progress was made in developing small machines for other than domestic purposes. The equipment designed for ice-making establishments, breweries, and packing plants was too large for the needs of a variety of commercial activities. Early in the nineties it was clear that machinery of small capacity which did not have to be operated by a skilled mechanic was needed for markets, dairies, and similar small users. The first attempts to manufacture such equipment failed, but before the end of the century small machines were available. By 1917 units with capacities ranging from one-fourth to three or four tons were used widely.[12] They did not require highly

---

272; Nickerson, *loc.cit.*, 173-174; T. O. Vilter, "The Gravity Feed System," *Ice and Refrigeration*, XLII (1912), 21-22.

[10] Sloan, *loc.cit.*, 419.

[11] "What the Refrigerating Machine Companies Have Contributed," *Refrigerating Engineering*, XXVIII (1934), 299; Nickerson, *loc.cit.*, 173.

[12] Doelling, *loc.cit.*, 160; "Medium and Small Independent Refrigerating Plants," *Scientific American*, LXXVIII (Feb. 12, 1898), 104-105. A ton of refrigeration is the removal of heat at the rate of 288,000 B.t.u. every twenty-four hours.

FIGURE 8

An ice plant of the eighteen-nineties

This is a diagram of a model ice plant designed by the Hercules Iron Works of Aurora, Illinois. Refrigeration was produced by a steam-powered ammonia-compression machine, and the distilled water frozen was obtained by condensing steam generated in the boiler at the left. This illustration explains why the ice factory captured the imagination of all who saw it operate. Coal was brought in at one end of the plant; blocks of crystal-clear ice were removed at the other.

From Advertisement, The Hercules Iron Works, *Ice and Refrigeration*, II (1892), 411.

skilled attention, but they were not automatic.[13] Their development was limited by a number of factors. The supervision and care required for those with capacities of less than five tons cost practically as much as for machines of from five to ten tons. Power costs were heavy. If steam had to be generated especially for a small plant, the expense was almost prohibitive, and rates for electricity were high. Various devices were used to lower these costs. Brine tanks were employed which reduced the hours of operation by storing up refrigeration. Internal-combustion engines, which provided relatively inexpensive power, were used frequently where a cheap supply of steam was not available and where power needs were small.[14]

Another method of filling small commercial requirements was pipe-line refrigeration. As the name implied, chilled brine or liquid ammonia was pumped from a central plant to a large number of small users dispersed usually throughout a metropolitan market district. In 1890 a successful pipe line was laid in St. Louis, and shortly afterward one was built by the Quincy Market in Boston. Before many years had passed similar service was provided in New York, Brooklyn, Baltimore, Philadelphia, Kansas City, Los Angeles, and several other cities. Pipe-line installations, however, did not expand as rapidly as other applications of mechanical refrigeration, for there were relatively few localities in which a large number of consumers could be found concentrated within a limited area around a central plant.[15]

Household refrigerating machines appeared long before 1917. They were proposed at least as early as 1887,[16] and several were on the market by 1895. In 1916 over two dozen

[13] Editorial, *Ice and Refrigeration*, L (1916), 327; J. E. Starr, "Refrigeration Twenty-Five Years Ago," *Ice and Refrigeration*, LI (1916), 144.

[14] F. E. Matthews, *Elementary Mechanical Refrigeration* (New York, 1912), 15-16.

[15] R. H. Tait, "Development of Pipe Line Refrigeration," *Ice and Refrigeration*, LI (1916), 174-176; Nickerson, *loc.cit.*, 174. For the varied uses of the refrigeration provided by the pipe line in St. Louis see *Ice and Refrigeration*, II (1892), 21-22.

[16] *Ice Trade Journal*, X (Feb. 1887), 4.

different makes, mostly compression models, were available.[17] Perhaps the most significant of these was the "Audiffren," a sulphur-dioxide-compression appliance which embodied a basic principle of later, more practical machines. Developed in France, it was manufactured in the United States after 1911 by the General Electric Company. Except for the motor, all parts, including the compressor itself, were contained in an hermetically sealed unit. The enclosure of the compressor eliminated the necessity for a tight stuffing box, a device which was used in conventional models to prevent leakage along the piston rod, but which caused so much friction that it decreased materially the efficiency of very small machines.[18] None of the household equipment developed before 1917 was satisfactory. No machine was fully automatic, and the initial cost of all was high.[19]

The appearance of means for the exchange of technical information was a significant factor in the progress made in mechanical refrigeration. The printed word played an important part. *Ice and Refrigeration*, the first periodical in the field, began publication in Chicago in 1891, to be followed after some years by other journals, both in the United States and abroad. Books appeared which made technical data available to engineers, cold-storage men, brewers, packers, and others.[20] Not only the first periodicals, but the first books as well were of American origin.[21] Trade and technical associations were established which served as clearing houses for information. The most important of these was the American Society of Refrigerating Engineers, founded in 1904. These

[17] *Ice and Refrigeration*, IX (1895), 325; LI (1916), 203.

[18] "The Audiffren Refrigerating Machine," *Scientific American Supplement*, LXVI (Aug. 22, 1908), 119; "What the Refrigerating Machine Companies Have Contributed," *loc.cit.*, 304.

[19] P. Neff, "The Domestic Refrigerating Machine," *Ice and Refrigeration*, XLIX (1915), 144; G. T. Voorhees, "Observations of a Refrigerating Engineer," *Ice and Refrigeration*, LI (1916), 150.

[20] J. E. Siebel's *Compend of Mechanical Refrigeration . . .* , published in Chicago, was in its second edition by 1896.

[21] See A. J. Wallis-Tayler, *Refrigeration, Cold Storage and Ice-Making* (London, 1902), preface, for British recognition of the preeminent position of American publications on refrigeration.

groups tended to break down the secretive methods that had characterized the manufacture of machinery and to make possible steps toward greater standardization.[22] The benefits of association were extended beyond national borders by the international congresses of refrigeration which met at Paris in 1908, Vienna in 1910, and Chicago in 1913, and by the formation of the International Association of Refrigeration.

[22] Nickerson, *loc.cit.*, 172-173; H. Sloan, "Refrigeration Then and Now," *Refrigerating Engineering*, xxxix (1940), 10.

CHAPTER VII

THE ICE INDUSTRY AND DOMESTIC
REFRIGERATION, 1890-1917

THE methods of manufacturing ice underwent important changes between 1890 and 1917. All processes benefited by the development of improved refrigerating machinery. The distilled-water can system, by which most ice was made, was increased in efficiency. For chilling the brine in which the cans of distilled water were frozen, special coolers were developed that proved more effective than the old method of placing the evaporator coils in the brine tanks themselves. Better insulation around the tanks reduced the waste of refrigeration. The removal of ice from the tanks was speeded by the introduction of power hoists which could handle several blocks at a time.[1]

This system, in spite of its improvement, had certain disadvantages. One of the drawbacks was the expense of distillation. Steam had to be generated especially to make distilled water, for adequate quantities could not be obtained by condensing that used for power. This was expensive enough, but if other power than steam were used, distillation was even more costly.[2] Great economies would be possible if a practical way were discovered to make clear ice from water that had not been distilled.

The plate system was an attempt to accomplish this, but it never was adopted extensively, for its disadvantages could not be overcome.[3]

[1] J. F. Nickerson, "The Development of Refrigeration in the United States," *Ice and Refrigeration*, XLIX (1915), 174; L. K. Doelling, "Twenty Five Years Evolution of Refrigeration," *Ice and Refrigeration*, LI (1916), 160; "Recent Improvements in the Refrigerating Industry," *Scientific American*, CIX (Oct. 4, 1913), 272.

[2] "Recent Improvements in the Refrigerating Industry," *loc.cit.*, 273; J. C. Goosman, "History of Refrigeration," *Ice and Refrigeration*, LXX (1926), 314.

[3] In 1909 eight per cent of the ice manufactured for sale was made by the plate system. By 1919 the per cent so made had declined to five. U. S. Department of Commerce, Bureau of the Census, *Thirteenth Census of the United States Taken in the Year 1910* (Washington, 1912-1914), x, 443; U. S. Department of Commerce, Bureau of the

The raw-water can system, which became important in the second decade of the twentieth century, was another method of making clear ice by freezing undistilled water in cans. By its use the advantages of convenient shape could be achieved and the costs of distillation avoided. It was recognized long before 1890 that the air which made ice opaque could be removed not only by distilling the water, but also by agitating it during freezing so that the air particles adhering to the surface of the forming ice were continuously washed away. At an early date several means of accomplishing this were devised. In Europe a system was developed in which cans containing undistilled or raw water were shaken during the freezing process. Since the apparatus required was bulky and expensive, this method was not suitable for the large plants of the United States. Another way was developed in Europe in which agitation was accomplished by rods suspended in the water. Satisfactory when small cans were used, it was ill-suited to large ones.[4] Agitation obtained by blowing air through the water was the basis of a process patented in the United States in 1877 by T. B. E. Turrettini of Geneva, Switzerland. This proved to be the key to a practical plan for making clear ice with undistilled water.[5]

The Turrettini idea developed slowly for many years,[6] but by 1917 several variations of it were available. In all of them the water was agitated by air introduced into the cans through pipes. While the water froze, particles of air were washed from the surface of the ice, which as it formed rejected substances in solution or suspension. Both air and other matter were concentrated in the unfrozen water. When all but a

---

Census, *Fourteenth Census of the United States Taken in the Year 1920* (Washington, 1921-1923), x, 964. For disadvantages of the system see *supra*, 89.

[4] G. Braungart, Jr., "Crystal Ice from Raw Water," *Ice and Refrigeration*, xxxviii (1910), 143.

[5] V. H. Becker, "The Ice Factory of the Future," *Ice and Refrigeration*, xxxviii (1910), 56.

[6] Becker, *ibid.*, 56, believes the Turrettini patent checked its development until the distilled water system became so well established that it could not be dislodged.

slender center section of the block was frozen, agitation was stopped. In some cases the remaining water was allowed to freeze, but since this resulted in a white, opaque core, it was more common for the residual water to be sucked out and fresh, either distilled or not, substituted for it.[7] Transparent ice with only a slight core could be made by this method at low cost, for distillation and the expensive equipment it required were eliminated. Steam engines could be replaced with electric motors and internal-combustion engines, which not only had lower initial costs, but also required less space, less labor, and were more convenient.[8] The raw-water can system was not used to any great extent before about 1914, but in the next few years it became important.[9] In 1919 thirty-five per cent of the ice manufactured for sale by the can system was made from undistilled water.[10]

Steam continued to be the principal type of power employed in the manufacture of ice, but its use diminished rapidly after 1914. In 1899 ninety-six per cent of the power in the industry was generated by it. In 1914 ninety per cent was so obtained, but by 1919 only seventy-one per cent.[11] This sudden drop was due to the substitution of electric power by raw-water plants.

Ammonia-compression equipment came more and more to dominate the field. In 1899 it had been adopted by seventy-four per cent of the establishments manufacturing ice for sale. Ten years later it was used by about eighty-six per cent.[12]

Ice-making plants multiplied during the eighteen-nineties. In the southern states so many had been constructed by 1890

---

[7] Braungart, *loc.cit.*, 143; *Ice and Refrigeration*, XLIII (1912), 115, 118; XLVI (1914), 307. A system had been devised in 1916 in which undesirable elements concentrating in the unfrozen water were removed automatically. This reduced costs and increased capacity by speeding operations. *ibid.*, L (1916), 93.

[8] Doelling, *loc.cit.*, 161.

[9] J. E. Starr, "Refrigeration Twenty-Five Years Ago," *Ice and Refrigeration*, LI (1916), 143.

[10] *Fourteenth Census, 1920*, x, 964.

[11] *Thirteenth Census, 1910*, x, 441; *Fourteenth Census, 1920*, x, 960.

[12] *Thirteenth Census, 1910*, x, 445; *Fourteenth Census, 1920*, x, 963.

that natural ice was no longer a significant item of trade.[13] Before long a problem of overcapacity arose. In 1897 the productive capacity of ice factories in southern cities was said to be two to four times the amount needed.[14] By the end of the decade 386 concerns, an increase of 134 per cent in ten years, were manufacturing ice for sale.[15]

In the northern states the construction of plants boomed after the natural-ice shortages of 1889 and 1890. Some of the new establishments failed in 1891, when there was a good harvest in most places,[16] but the machine with its dependable output soon demonstrated that it was in the North to stay. Shortly, plants erected in cities were expanded to take care of enlarged demand. In 1893, despite a bountiful harvest and low prices, the manufactured product was able to compete successfully.[17] By the turn of the century there were 169 concerns that manufactured ice for sale in the middle states, 152 in the central states, and seven in New England. None were in operation in Maine, New Hampshire, Vermont, Massachusetts, Michigan, Wisconsin, or Minnesota, for in these states nature favored the distribution of natural ice and discouraged the machine-made product. In 1889 only seventeen per cent of the plants in the country were located in the middle, central, and New England states, but ten years later forty-four per cent were operating there. So great was the expansion of the industry in the North that in 1899 the five states that led in the production of manufactured ice were Pennsylvania, New York, Missouri, Illinois, and Ohio. Two

[13] Editorial, *Ice Trade Journal*, xv (Oct. 1891), 4.

[14] Special contributor in *Ice and Refrigeration*, xii (1897), 16-17.

[15] A. L. Hunt, "Manufactured Ice," U. S. Department of the Interior, Census Office, *Twelfth Census, 1900*, ix (Washington, 1902), 677. The statistics of the Twelfth Census were criticized by the ice industry. The census included only establishments that manufactured ice for sale; it did not include the ice plants of breweries, packing houses, cold-storage warehouses, and similar concerns. Pointing out that many of these activities that made ice primarily for their own use entered the open market at certain seasons, *Ice and Refrigeration* claimed that the Census did not give a true picture of the size of the industry. Editorial, xxiii (1902), 17.

[16] Editorial, *Ice Trade Journal*, xv (Oct. 1891), 4.

[17] *Ice and Refrigeration*, vi (1894), 11, 91; ix (1895), 304-305.

northern states, Indiana and New Jersey, held seventh and ninth places.[18]

In the western states forty plants were producing by 1899, almost half of them in Kansas, but none in the Dakotas, Montana, Idaho, Wyoming, and Nevada. Thirty-three were working in the states bordering the Pacific, chiefly in California.[19]

In the twentieth century plants spread to all parts of the country.[20] There were 775 in 1899. By 1909 their number had increased to 2,004 and by 1919 to 2,867.[21]

More efficient methods of harvesting and housing natural ice were introduced after 1890. Certain improvements would have been made in the ordinary course of events, but competition from manufactured ice undoubtedly sped their evolution. The outstanding development in harvesting was the perfection of field and basin saws, powered by gasoline engines or electric motors. The field saw with its circular blade was designed to do the same job as the ice-cutter, but more quickly and economically. The basin saw, which had several circular blades mounted on a shaft, was located near the foot of the elevator which raised the ice to storage. By its use large floats of ice could be separated into blocks more efficiently than had been possible before. In housing, probably the most significant development was an improved elevator which was faster and cost less than the older inclined-plane type.[22]

In the first decade of the twentieth century the natural-ice industry was still a big business. In New England, New York, Pennsylvania, and in the states bordering the Great Lakes, harvesting on a large scale continued. In 1907 it was estimated that between fourteen and fifteen million tons were consumed each year in the United States. This was almost three times the estimated consumption of 1880. But the industry

[18] Hunt, *loc.cit.*, 677, 684.

[19] *ibid.*, 677. See Figure 9 for a map showing the distribution of ice-making plants.

[20] In some of the more northern states plants made slow progress. Maine, for example, had only one in 1942. *Ice and Refrigeration*, CIII (1942), 95.

[21] *Fourteenth Census, 1920*, x, 956.

[22] W. E. Wood, "Twenty Five Years' Development in Harvesting and Housing Ice," *Ice and Refrigeration*, LI (1916), 191-193.

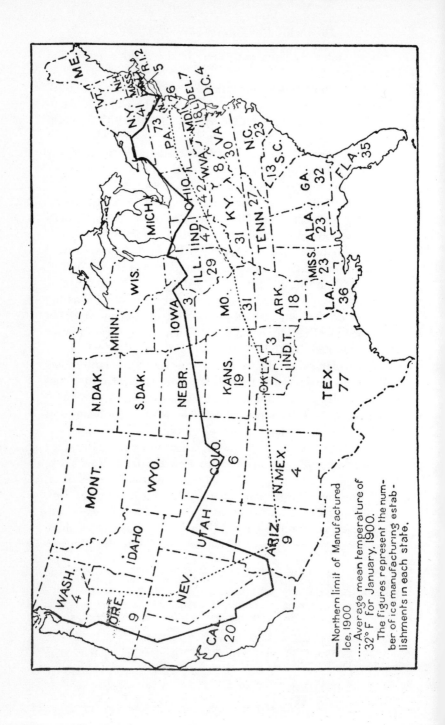

ME.

VT. N.H. MASS. 2
CONN. R.I. 2
5

N.Y. 41
PA. 73
N.J. 26
DEL. 7
MD. 18
D.C. 4
W.VA. 8
VA. 30
N.C. 23
S.C. 13
OHIO 42
IND. 47
KY. 31
TENN. 27
GA. 32
FLA. 35
ALA. 23
MISS. 23
LA. 36
ILL. 29
IOWA 3
MO. 31
ARK. 18
OKLA. 7
IND.T. 3
TEX. 77
MICH.

WIS.

MINN.
N.DAK.
S.DAK.
NEBR.
KANS. 19
COLO. 6
N.MEX. 4
MONT.
IDAHO
WYO.
UTAH
NEV.
ARIZ. 9
WASH. 4
ORE. 9
CAL. 20

—— Northern limit of Manufactured
Ice, 1900.

.... Average mean temperature of
32° F for January, 1900.

The figures represent the number of ice manufacturing establishments in each state.

was not healthy. The harvest decreased steadily.[23] In New York City in 1901 the per capita consumption of natural ice had been about 1,120 pounds; in 1909 it was only 426 pounds.[24]

After 1910 the industry lost ground rapidly. In 1914 shipments from Maine practically had been abandoned, and the capacity of Hudson River icehouses was twenty-five per cent less than it had been ten years earlier.[25] The quantity manufactured by the American Ice Company of New York surpassed the amount of natural ice it harvested in 1914 and three years later accounted for more than three-fourths of the sales of the company.[26] Nevertheless, in the North natural ice still was an important factor. Although in 1914 it constituted only six per cent of all consumed in St. Louis and nine per cent in Cincinnati, in Chicago it accounted for sixty-seven per cent, in Detroit for seventy-five, and in Boston for eighty-five. Even in the middle states it remained important, making up thirty-one per cent of that sold in Baltimore, forty per cent in Philadelphia, and forty-five in New York.[27]

In the nineteen-twenties the natural-ice industry ceased to be a big business. In the winter of 1917-1918 over 2,500,000 tons had been harvested on the Hudson River, but in 1923-

[23] S. S. Van der Vaart, "Growth and Present Status of the Refrigerating Industry in the United States," *Premier Congrès International du Froid*, III (Paris [1908?]), 331-332.

[24] H. W. Bahrenburg, "President's Address," *Proceedings of the First Regular Meeting of the Natural Ice Association of America . . .* (New York [1910?]), 7.

[25] Dwelley-Nichols Engineering Co., "Resume of Report on the Ice Situation," *Ice and Refrigeration*, XLVI (1914), 270.

[26] *Ice and Refrigeration*, LII (1917), 322.

[27] *1915 Ice and Refrigeration Blue Book . . .* (Chicago, 1915), 13.

FIGURE 9

The distribution of ice-manufacturing plants, 1900

This map shows clearly the close relation that existed between the availability of natural ice and the introduction of ice-manufacturing establishments.

From A. L. Hunt, "Manufactured Ice," U. S. Department of the Interior, Census Office, *Twelfth Census of the United States, Taken in the Year 1900*, IX (Washington, 1902), 686.

1924 less than 250,000 tons were cut there. By 1925 New York City was no longer dependent on natural ice, even during prolonged hot spells, and the storage houses on the Hudson were being demolished or sold to mushroom-growers. Harvesting continued in the more northern states and in some mountain areas—northeastern Pennsylvania, for example—but on a small scale. In 1949 the crop was estimated at only 317,500 tons.[28]

Producers of natural ice did not give up their old supremacy without a struggle, especially in areas where natural and manufactured-ice interests both could compete. The issue was decided on grounds of relative price and dependability, but claims and counterclaims as to the quality of the rival products marked the fray.[29]

An issue in this rivalry was the purity of natural ice. Its wholesomeness began to be questioned sharply in the eighties, when numerous pamphlets and newspaper and magazine articles appeared connecting its use in beverages with the prevalence of certain diseases, particularly typhoid fever.[30] Residents of New York City were warned of the dangers of ice cut from the Hudson just below Albany, inasmuch as sewage from Albany and Troy, where typhoid was frequent when ice was forming, was emptied directly into the river.[31] By the nineties these warnings had resulted in widespread municipal supervision. St. Joseph, Missouri, for example, licensed dealers, requiring them to make a statement as to the source of their product and the purposes for which it was sold. Harvesting was prohibited in certain condemned areas,

[28] *Ice and Refrigeration*, LXVI (1924), 375; LXVIII (1925), 465; XCII (1937), 225; CXIX (Aug. 1950), 27.

[29] In 1897 natural-ice interests claimed that the high rates charged by the railroads for hauling ice were largely responsible for the inroads of the machine. Editorial, *Ice Trade Journal*, XX (May 1897), 4. Losing sight of the fact that ice produced refrigeration by melting, both sides on occasion claimed their product would melt more slowly than the other. *ibid.*, XIV (March 1891), 1.

[30] *Ice Trade Journal*, X (July 1887), 2.

[31] T. M. Prudden, "Our Ice Supply," *The Popular Science Monthly*, XXXII (1887), 679-681; Prudden, "Ice and Ice-Making," *Harper's New Monthly Magazine*, LXXXV (1892), 383.

and provision was made for frequent chemical analysis of samples.[32]

The attention given to purity was part of the growing concern, stimulated by knowledge of the germ theory of disease and by European example, which American cities began to feel for the safety of their water supplies in the late eighties.[33] Manufactured-ice interests did all they could to convince the public that the natural product was a danger to health and that security lay only in ice made from potable water. Before the days of the raw-water plant, ice-manufacturers stressed particularly the purification assured by distillation.[34] Natural-ice men attributed to their competitors the public interest in the regulation of their product. In the convention of the Empire State Ice Association in 1907 they charged that their competitors, not the residents of New York City, were responsible for requests for legislation to regulate harvesting on the Hudson.[35]

[32] *Ice and Refrigeration*, IV (1893), 388. The *Ice Trade Journal*, xx (April 1897), 4, objected strenuously to the prosecution of ice men for cutting in places prohibited by local boards of health. In an 1897 editorial it declared: "This is a usurpation of power which does not belong to them, and which cannot be delegated to them by any city ordinance. They have no legislative functions whatever, their duties being exclusively executive. The Board may legally suppress a public nuisance but they cannot take upon themselves to decide whether the act of cutting ice on one's own property is a nuisance or not. It is a judicial act, and it is not competent even for a legislative body to make a law binding on an individual, declaring him a criminal, without evidence of any kind, and without the authority of a State Law. These arbitrary methods of Health Boards are entirely irregular and need repressing just as much, or more, than the ice trade does." Health authorities were concerned principally with natural ice, but the purity of manufactured ice as well came to be supervised. For the regulations of a board of health of Louisiana, for example, see W. D. Bigelow, with N. A. Parkinson, *Food Legislation During the Year Ended June 30, 1908* (U. S. Department of Agriculture, Bureau of Chemistry, *Bulletin No. 121*) (Washington, 1909), 28-29.

[33] A. M. Schlesinger, *The Rise of the City, 1878-1898* (A. M. Schlesinger and D. R. Fox, eds., *A History of American Life*, x) (New York, 1933), 104.

[34] *Ice Trade Journal*, xvii (Jan. 1893), 4; Editorial, *Ice and Refrigeration*, xvii (1899), 99. Natural-ice interests reciprocated by attacking the purity of manufactured ice. See W. J. Snyder, "The Ice Business or Ice from a Scientific Standpoint," *Ice and Refrigeration*, xxxviii (1910), 255.

[35] N. B. Van Derzee, "Address to the Fifth Annual Convention of

The danger of contracting disease from polluted ice probably was slight. As ice crystals formed on the surface of a lake or river, they expelled most of the foreign matter in suspension or solution, including bacteria, into the water still unfrozen. Such self-purification, of course, was impossible when the ice was flooded in order to increase its thickness, and when filth was deposited upon it by surface drainage. Even if the ice were contaminated, storage tended to purify it, for continued low temperatures killed almost all of the typhoid bacilli present and weakened the few that survived. In general the scientists who investigated the matter felt that there was little danger from natural ice, but advised against taking it from polluted waters and against consumption of any that might have been formed by flooding. It was significant that although New York City and Lawrence and Lowell, Massachusetts, used ice from sewage-laden streams, they had low death rates from typhoid fever, and no typhoid epidemics were traced conclusively to ice.[36]

---

the Empire State Ice Association," *Ice and Refrigeration*, XXXII (1907), 298.

[36] W. T. Sedgwick and C.-E. A. Winslow, "Experiments on the Effect of Freezing and Other Low Temperatures upon the Viability of the Bacillus of Typhoid Fever, with Considerations Regarding Ice as a Vehicle of Infectious Disease," *Memoirs of the American Academy of Arts and Sciences*, XII.—No. v (Aug. 1902), 519-520; M. J. Rosenau, L. L. Lumsden, and J. H. Kastle, *Report on the Origin and Prevalence of Typhoid Fever in the District of Columbia* (U. S. Treasury Department, Public Health and Marine-Hospital Service, Hygienic Laboratory, *Bulletin No. 35*) (Washington, 1907), 22-23; "Report of Investigations of the State Board of Health upon the Pollution of Ice Supplies," *Twenty-first Annual Report of the State Board of Health of Massachusetts* (Boston, 1890), 145, 152-153; G. C. Whipple, "Stream Pollution and Sanitation [o]f Natural Ice," *Proceedings of the Fourth Annual Convention of the Natural Ice Association of America* . . . (New York [1912?]), 174-175, 177, 179; M. J. Rosenau, *Conservation of Our Natural Ice Business* (New York, n.d.), 3; E. H. Porter, "Report on Contamination and Self-Purification of Ice," *Ice and Refrigeration*, XXXII (1907), 299-300. Icemen were sometimes blamed for epidemics for which they certainly were not responsible. In 1893 A. J. Connors, an Evanston, Illinois, dealer, was arrested and charged with responsibility for an epidemic of malarial complaints! *Ice Trade Journal*, XVII (Aug. 1893), 4. Rosenau, Lums-

Regardless of the reality of the infection danger, the pollution of water from which ice was cut was a factor contributing to the extension of the manufactured commodity.[37] Some uncertainty was inevitable when ice from rivers and lakes was used, but unless gross carelessness were tolerated, no doubts could exist as to the purity of ice from the refrigerating machine. The public, more conscious of germ dangers than ever before, preferred the machine-made product.

In the design of ice refrigerators for home use, improvements came slowly. Refrigerators based on the principle of air circulation were common, but the quality of most was very poor. They were not designed by engineers; their size and shape were determined largely by the efforts of manufacturers to use economically the commercial sizes of lumber and sheet metal. The temperatures obtained, which fluctuated with the size of the ice charge, were insufficiently low for proper food preservation.[38] In many of the cheaper types the insulation was hardly worthy of the name, consisting only of one or two sheets of paper. Poor insulation allowed moisture to condense within the wall, corroding the metal lining and furnishing a favorable medium for the growth of bacteria. One investigator estimated that inefficient insulation in low-priced units resulted in the waste of at least eighty per cent of the ice used.[39]

---

den, and Kastle, *op.cit.*, 22, found it was possible that typhoid infection might occasionally be present in manufactured ice sold in Washington, D.C. They discovered that urine and excrement were carried sometimes on the shoes of workmen to the top of the tanks and dropped into the freezing water. Sedgwick and Winslow, *loc.cit.*, 520, pointed out that ice manufactured from impure water might be dangerous, for the manufactured product usually was consumed quickly, and there was little time for purification in storage.

[37] F. E. Matthews, *Elementary Mechanical Refrigeration* (New York, 1912), 14.

[38] M. E. Pennington, "The Construction of Household Refrigerators," *Ice and Refrigeration*, LXXIV (1928), 521; C. F. Holske, "Domestic and Commercial Ice Refrigerators," *Refrigerating Engineering*, XXXII (1936), 150-151.

[39] J. R. Williams, "A Study of Refrigeration in the Home, and the Efficiency of Household Refrigerators," *Proceedings Third International Congress of Refrigeration*, III [Chicago, 1914?], 9-20; J. R.

Ignorance on the part of the public concerning the essentials of a good refrigerator was the basic reason for the low level of quality that prevailed. Manufacturers, who soon discovered that consumers did not appreciate the necessity for good construction and insulation and that cheap units sold well, had no incentive for making better equipment.[40] The ice industry, which might have been expected to educate the public as to the best methods of using its product, took little interest. Shortsightedly, it concerned itself only with selling ice. One member of the Natural Ice Association of America probably expressed the attitude of many of his fellows when he declared: "The only refrigerator that I ever recommended to a customer was any old kind of refrigerator that was always kept full of ice; any refrigerator that is kept well filled is a good refrigerator."[41]

The quantity of ice used in the United States increased rapidly after 1890. In 1914 over 21,000,000 tons of manufactured ice alone were consumed, four times the estimated consumption of natural ice in 1880.[42] More and more was required by the nation's rapidly growing cities. Philadelphia, Baltimore, and Chicago used over five times as much in 1914 as in 1880, and in the same period the amount consumed in New Orleans increased almost thirteenfold.[43]

The growing consumption reflected an expansion in domestic refrigeration. Its extent is difficult to gauge, but some indication is afforded by the growth of the refrigerator-manufacturing industry. The value of its products in 1889 was $4,513,616, an increase of almost $2,775,000 over 1879. By 1899 this had grown to $5,317,886, by 1909 to $10,-

Williams, "Use of Ice and Other Means of Preserving Food in Homes," *Scientific American Supplement*, LXXVII (Jan. 3, 1914), 6-7.

[40] Holske, *loc.cit.*, 150.

[41] *Proceedings of the Eighth Annual Meeting of the Natural Ice Association of America* . . . (New York [1916?]), 135.

[42] *Fourteenth Census, 1920*, x, 963; H. Hall, "The Ice Industry of the United States," U. S. Department of the Interior, Census Office, *Tenth Census, 1880*, XXII (Washington, 1888), 5. No figures exist for the amount manufactured in 1880, but it had not yet reached large proportions.

[43] *1915 Ice and Refrigeration Blue Book*, 13; Hall, *loc.cit.*, 38-41.

689,275, and by 1919 to $26,048,808.[44] Nevertheless, a large number of urban homes—probably more than half—bought no ice.[45] Few farm families used it even when it could be had for the harvesting.[46] Aware of the advantages of refrigeration, both federal and state departments of agriculture tried to awaken farmers to the benefits they could obtain from ice.[47]

In commercial food-handling establishments ice usage became more general than it had been before 1890, but many retailers did not maintain the low temperatures for perishables that were essential if the food were to reach the consumer in good condition. Food-dealers no longer relied exclusively on ice refrigerators, for the small machine was beginning to offer in this field successful competition.[48]

The ice industry was troubled by a demoralizing competitive situation. The principal abuse was price cutting, which appeared in an aggravated form in the nineties. Its basic cause was the entry of more operators into the industry than the business warranted.[49] Conditions seem to have been particularly bad in the South, where shortly after 1890 excess ice-making capacity existed in many localities.[50] Salesmen of refrigerating machinery, with extravagant claims as to the low costs possible with their equipment, frequently made bad

[44] *Thirteenth Census, 1910*, VIII, 426; *Fourteenth Census, 1920*, VIII, 167.

[45] J. R. Williams, a doctor, made a study of 519 homes in Rochester, New York, in 1912 and found that sixty-nine per cent of them used ice. Very likely this figure was too high for the city as a whole, for the homes he studied do not seem to have constituted a proper sample. Williams, "Use of Ice and Other Means of Preserving Food in Homes," *loc.cit.*, 6. Estimates by members of the ice industry in the middle twenties indicated that at least half of the urban homes in the United States did not take ice, even in summer. L. C. Smith, "Because It Is Right," *Ice and Refrigeration*, LXX (1926), 268; Advertisement of Rochester Ice and Cold Storage Utilities, Inc. reproduced in *Ice and Refrigeration*, LXXI (1926), 302.

[46] L. C. Corbett, *Ice Houses* (U. S. Department of Agriculture, *Farmers' Bulletin 475*) (Washington, 1911), 20.

[47] See *ibid.* and W. L. Nelson, "Ice on the Farm," Missouri State Board of Agriculture, *Monthly Bulletin*, XIII (Sept. 1915), 3-19.

[48] Editorial, *Ice and Refrigeration*, L (1916), 327.

[49] H. D. Norvell, "Controlling the Ice Industry in Any Locality Legitimately and Lawfully," *Ice and Refrigeration*, LXIII (1922), 369.

[50] Special contributor in *Ice and Refrigeration*, XII (1897), 16-17.

situations worse by inducing installations in areas where already there were too many.[51] Brewers, ice-cream manufacturers, and others who used refrigeration in connection with their main line sometimes entered the ice business, selling at prices regular manufacturers could not meet profitably. The confusion was intensified by the free-lance deliverymen, or peddlers, who operated in most localities. Often notoriously dishonest, these worthies made price cutting their specialty.[52]

One of the first attempts to improve conditions was the establishment of trade associations, which brought members of the industry together for the discussion of common problems. The Southern Ice Exchange, formed in 1889, was followed by a large number of other groups, all on a geographic basis. Organization on a nation-wide scale was achieved during the first World War with the appearance of the National Association of Ice Industries.[53] These associations were attacked in the press as trusts and as attempts to raise prices,[54] but such charges were in the main unjustified. True, these groups were interested in reducing the evils of price cutting, but their methods apparently were confined to exhortation.

Different from the trade associations were the ice exchanges formed in some cities—New York, Buffalo, and Chicago, for example—in the eighties. These were frank efforts to agree on price schedules for a given locality.[55]

More direct methods of stabilizing the situation were adopted in many cities. Sometimes competing interests organized consolidated delivery companies.[56] In some cities several companies united to form a single corporation. Shortly after 1900 seven of the principal concerns of St. Louis merged to establish the Polar Wave Ice and Fuel Company. Combina-

[51] C. D. Wingfield, in paper read at the Annual Convention of the Southern Ice Exchange, 1897, *Ice and Refrigeration*, xii (1897), 183-184.

[52] Norvell, *loc.cit.*, 369.

[53] W. J. Rushton, "Early Days of the Manufacture of Ice," *Ice and Refrigeration*, li (1916), 152; *Ice and Refrigeration*, ci (1941), 10.

[54] *Ice and Refrigeration*, li (1916), 197.

[55] *Ice Trade Journal*, viii (Sept. 1884), 2; xi (Nov. 1887), 5; Editorial, xii (Feb. 1889), 4.

[56] C. D. Wingfield, *loc.cit.*, 183-184.

tions were not always confined to a single city. In 1910 the Atlantic Ice and Coal Corporation appeared. Its purpose was to provide centralized ownership and management for a number of ice-making plants in Georgia, Tennessee, Alabama, and other states.[57]

The motive behind at least one combination was not stabilization of the industry as such, but rather the accumulation of speculative profits. This was the American Ice Company, incorporated under New York laws in March 1899. An over-capitalized creation of a notorious promoter and speculator, Charles W. Morse, it controlled the greater part of the ice supply of New York City. In May 1900 it doubled its prices there, causing considerable suffering among the poor. Prosecutions, begun at the instigation of the press, revealed that Morse had sold 4,200 shares of stock to Mayor Van Wyck at fifty cents on the dollar and then had lent him money to pay for them. Other political leaders and office-holders, mostly members of Tammany, owned large blocks of stock. Two share-holding dock commissioners saw to it that Morse received exceptionally favorable docking privileges. These revelations caused such resentment that the company lowered its prices, but the prosecutions came to naught. After further manipulation Morse withdrew from the business with a profit estimated at $12,000,000.[58]

The public, though it obtained ice cheaply enough during price wars, was convinced that it usually was being gouged. A basic reason for this widespread belief was failure to understand the economics of the industry. Consumers saw that ice made by nature or manufactured from water costing as little as eight cents a thousand gallons was sold for thirty to sixty cents a hundred pounds, and they concluded that tremendous

[57] *Ice and Refrigeration*, xxiv (1903), 108; xxxviii (1910), 112.

[58] O. Wilson, "The Admiral of the Atlantic Coast: The Career of Charles W. Morse," *The World's Work*, xiii (1907), 8719; M. Lerner, "Charles Wyman Morse," *Dictionary of American Biography*, xiii, 239-240. For contemporary accounts see "The Ice Trust Outrage," *Gunton's Magazine*, xviii (1900), 515-519; "The Ice Trust Exactions," *The Outlook*, lxv (May 19, 1900), 144; lxv (June 9, 1900), 328-329; and "The Ice Trust in Politics," *The Outlook*, lxv (June 16, 1900), 376.

profits were made.[59] It is not surprising that combinations in the industry were interpreted by the press and the public as "ice trusts," as ruthless attempts to rob the poor.[60]

Some combinations undoubtedly were able to control prices, at least for a limited period, but on the whole high charges were difficult to maintain for long. Since the capital now necessary to enter the business was comparatively small, high rates in any locality served usually only to attract competition.[61]

Other factors were more important in keeping prices up. Distribution costs were high, due not only to the bulk and perishability of ice, but also to wasteful methods and to duplication of effort by competing companies. The seasonal nature of the industry itself made for heavy costs, as did the maintenance of sufficient capacity to meet peak requirements which might be of only a few days' duration.[62] The main cause of high prices in Massachusetts, concluded a firm of economists and engineers who investigated the situation there in 1913, was not trusts, but the unrestricted operation of small, inefficient firms. In order to stay in business, these concerns had to sell at high prices which provided larger, more efficient enterprises with an excuse for greater charges than were necessary for their profitable operation.[63] An important reason, of course, for short periods of especially high prices was an excess of demand over supply, a situation that sometimes existed after warm winters in the North, where cities still depended on natural ice for a considerable part of their needs.

Public indignation against the industry occasionally flared into action. In a wave of feeling against the prevailing prices numerous attempts were made in 1906 and 1907 to prosecute

[59] C. S. Campbell, "The Ice Man in Relation to the Public," *Ice and Refrigeration*, XLVIII (1915), 127.

[60] For accounts of press attacks on the ice industry see editorials in *Ice and Refrigeration*, XIII (1897), 107; XXXI (1906), 230; and XLI (1911), 50.

[61] C. D. Wingfield, *loc.cit.*, 183-184; Special contributor in *Ice and Refrigeration*, XII (1897), 16-17.

[62] Campbell, *loc.cit.*, 127.

[63] Dwelley-Nichols Engineering Company, *op.cit.*, 269. Natural ice was still the dominant factor in Massachusetts.

dealers on anti-trust charges.[64] Little came of most of these efforts, but in 1914 the Supreme Court of Missouri fined the Polar Wave Ice and Fuel Company of St. Louis $50,000 for violating the state anti-trust law.[65] About 1906 many cities made plans to set up municipal plants, but they usually were stopped by the courts, which generally held that cities did not have the power to enter the business. In 1913 only one municipal plant was in actual operation, but there was a growing interest in them, for rates were higher than before, and several states had passed home-rule acts for all cities or enabling acts for specific municipalities.[66] Municipal plants had been established in a few small cities by 1918,[67] but they never became important generally in the United States. Action against the industry sometimes took extra-legal form. In 1911 there was rioting at ice plants and stoning of wagons in several cities.[68]

This public dissatisfaction with the industry was significant. It helped create among consumers a receptive mood for a substitute for ice refrigeration.

[64] S. Beeson, "Prosecuting the Ice Men," *The World To-Day*, XI (1906), 939-941; *Ice and Refrigeration*, XXXIII (1907), 202.

[65] *Ice and Refrigeration*, XLVI (1914), 282.

[66] J. W. Wentworth, *A Report on Municipal and Government Ice Plants in the United States and Other Countries* (New York, 1913), 1.

[67] *Ice and Refrigeration*, LV (1918), 104.

[68] Editorial, *ibid.*, XLI (1911), 50.

CHAPTER VIII

REFRIGERATED TRANSPORTATION
1890-1917

THE refrigerator cars of the early nineties left much to be desired. Their dimensions, upon which the adequacy of air circulation in part depended, were determined by guess rather than by scientific methods, and the temperatures that could be maintained varied from forty-five to sixty degrees. Thirty-two to thirty-four feet long, they ranged in weight from 46,000 to 52,000 pounds, but had a capacity of only twenty tons.[1]

The excessive weight and relatively small loading space of early cars were corrected in models that were developed in the twentieth century. At the same time that the cars were lengthened by several feet and their capacities increased to thirty tons, their weight was reduced to 38,000 pounds. Better design and insulation made possible lower temperatures.[2]

Two main types were in use—the beef or brine-tank car and the ventilator refrigerator car. The beef type, designed to provide the temperatures near freezing needed to preserve fresh meat, was equipped at each end with two or four galvanized-iron tanks filled with crushed ice and salt. Air was cooled as it circulated over the cold tank surfaces. Insulation consisted of layers of hair felt separated by dead air space.[3]

The ventilator refrigerator car, which received its name from the fact that it could be used for either ventilated or refrigerated shipments, found employment in the transportation of a great many products, the most important of which were fruits and vegetables. It differed from the beef car mainly in the construction of its ice compartments. Air was cooled not

---

[1] E. F. Carry, "American Practice in Refrigerator Car Construction," *Premier Congrès International du Froid*, III (Paris [1908?]), 637.

[2] *ibid.*, 637; J. M. Culp, "Report No. 1, Question XVI, Perishable Goods," International Railway Congress Association, *Proceedings, Eighth Session, 1910*, III (Brussels, 1912), xvi/6, xvi/9.

[3] Culp, *loc.cit.*, xvi/12-xvi/13; Federal Trade Commission, *Food Investigation. Report on Private Car Lines* (Washington, 1920), 31.

by contact with metal surfaces, but by passing through the ice mass itself. The ice was placed at first in tanks of galvanized iron, one at each end, which were provided with openings, upper and lower, to permit the entrance and exit of the circulating air.[4] Later the tanks were superseded by ice bunkers made of wooden slats so constructed that the ends of the car formed one side of the bunkers. The bunkers were separated by slatted bulkheads from the rest of the car. Block rather than crushed ice generally was used in them.[5] Salt was sometimes added to obtain lower temperatures in the transportation of perishables, but its use with fruit and vegetable shipments did not become common until the nineteen-twenties.[6]

In some cases refrigerator cars were insulated, but had no bunkers. The ice they carried was packed around the containers of the perishables that were being transported. It was possible to ship milk, beer, fish, and dressed poultry in this fashion. Refrigerator cars of this type were in most cases merely made-over old standard models.[7]

The ventilator refrigerator type was far from perfect.[8] This was brought out by studies which United States Department of Agriculture investigators made of the transportation of fruits and dressed poultry. Products that were loaded at

[4] *Ice and Refrigeration*, xxvii (1904), 205.

[5] M. E. Pennington, "Development of a Standard Refrigerator Car," *A.S.R.E. Journal*, vi (1919-1920), 9; P. K. Blinn, *Development of the Rockyford Cantaloupe Industry* (Colorado Agricultural College, Agricultural Experiment Station, *Bulletin 108*) (Fort Collins, 1906), 17; Federal Trade Commission, *op.cit.*, 31-32; See *Ice and Refrigeration*, xxvii (1904), 208, for a description of the Bohn Siphon System car, in which S-shaped metal plates, slats, and mesh formed the inner wall of the bunker. The function of the plates, which became very cold, was to improve circulation by deflecting warm air down into the ice chamber, where it fell until it came in contact with a sloping metal plate which caused it to flow out along the floor of the car.

[6] M. E. Pennington, "Low Temperature in Transit," *Ice and Refrigeration*, lxvii (1924), 324.

[7] Federal Trade Commission, *op.cit.*, 32; G. C. White, "Improved Transportation Service for Perishable Products," *Proceedings of The Second Pan American Scientific Congress*, iii (Washington, 1917), 402.

[8] Beef cars, almost all of which were owned by the packers, were more satisfactory.

the center and top of the car were found to be particularly subject to decay, an evil that indicated poor air circulation.[9] By 1917 some excellent equipment had been built, but not in sufficient quantity to transport any sizable part of the nation's vast traffic in perishables. A large number of refrigerator cars were still in service which were scarcely worthy of the name. Extensive replacement or rebuilding was needed.[10]

Investigations that were prerequisite to any general improvement were made by bureaus of the Department of Agriculture. In 1908 studies on the preservation of eggs and dressed poultry in transit were initiated by the Food Research Laboratory of the Bureau of Chemistry. In 1916 these were supplemented by investigations on the shipment of fruits and vegetables begun by the Bureau of Plant Industry. With cooperation from the railroads accurate records were made of the performance of different types of equipment under varying conditions of actual operation. These tests showed that better circulation with its benefits of more rapid cooling, lower temperatures in transit, and more uniform temperatures throughout the car could be achieved if floor racks and a wire-basket bunker with an insulated bulkhead were adopted.[11] Another conclusion was that improved insulation and construction would make possible important economies in the use of ice.[12]

Attempts to contrive cars cooled by mechanical refrigeration dated from the eighteen-eighties. An experimental Chi-

[9] Pennington, "Development of a Standard Refrigerator Car," *loc.cit.*, 2-3.

[10] W. A. Sherman, *Merchandising Fruits and Vegetables* (Chicago, 1928), 172-173, 301-302; M. Cooper, *Practical Cold Storage*, 2nd edn. (Chicago, 1914), 515, 518-519.

[11] By raising the contents of the car a few inches off the floor, the racks prevented them from obstructing circulation. The wire-basket bunker permitted freer contact between air and ice, while the insulated bulkhead made possible the use of greater salt concentrations without danger of freezing products loaded near the bulkhead.

[12] Pennington, "Development of a Standard Refrigerator Car," *loc. cit.*, 3-11; U. S. Department of Agriculture, Bureau of Plant Industry, *Report of the Chief of the Bureau of Plant Industry* (Washington, 1917), 26-28.

cago-to-Florida shipment of beef from Armour and Company was made in 1888 in a car cooled by chloride-of-ethyl-compression machinery.[13] Several patents were issued over the years for units based on different principles, but although a few cars were built for demonstration purposes, none was adopted for regular service.[14]

The reason for this, charged one early experimenter, C. C. Palmer,[15] was the opposition of railroads and others who had heavy investments in ice plants, harvesting and storing facilities, icing stations, and existing cars.[16] This charge contained elements of truth. But the fundamental reason for the reluctance of railroads and others to bring mechanical refrigeration into general use was the problem of economic feasibility. Cars were assigned to refrigerated shipments only part of the time. An example of the problem was the transportation of the citrus crop from southern California. Of the loads transported, not more than about one-third were refrigerated, for almost one-half of the fruit moved under ventilation alone, and a very high percentage of the westbound traffic required no ice in the bunkers.[17] Under such conditions the installation of expensive machinery would have raised materially the cost of providing refrigerated service.

The refrigerator car was almost exclusively an American development. Few countries were faced with the long rail hauls and the climatic conditions that made it a necessity in the United States. In 1910 there were only about 1,085 such cars in continental Europe, while two years earlier there had been about 85,000 equipped with bunkers in the United States in addition to about 50,000 with insulation but no

[13] "A New Refrigerator Car," *The Railroad Gazette*, xx (March 2, 1888), 143-144.

[14] *Ice and Refrigeration*, xxvi (1904), 299; lii (1917), 107.

[15] The 1888 Chicago-to-Florida shipment of beef mentioned above was made in a Palmer car.

[16] *Transactions of the American Society of Refrigerating Engineers*, ix (1913), 151-152.

[17] J.-S. Leeds, "Refrigeration of Citrus Fruits in Transit from California," *Premier Congrès International du Froid*, iii (Paris [1908?]), 611.

bunkers.[18] In 1913 more than half of all refrigerator cars in Europe were in Russia, where the factors of distance and climate were most similar to those of North America.[19]

Marine refrigerating machinery was developed in connection with the transoceanic meat trade.[20] Charles Tellier, the versatile French inventor, attempted in 1868 to import meat from Uruguay to France in a ship refrigerated by an ammonia-compression system, but failed because of an accident to the machinery. In 1876 a French company experimented by sending meat to Buenos Aires in a steamship cooled by three Tellier methyl-ether units; in the following year the return voyage was made with a cargo of Argentine meat, part of which did not arrive in good condition. A Marseilles firm sponsored more successful voyages in 1877 and 1878, but these experiments, which failed to enlist the interest of the French public and the support of French capital, were not repeated. Thomas S. Mort of Sydney, Australia, was another pioneer in the development of mechanical refrigeration on shipboard. About 1876 he made an unsuccessful attempt to send meat to England in a sailing vessel equipped with an ammonia-compression machine.[21]

Cold-air machines were favored by Henry and James Bell, J. J. Coleman, Andrew McIlwraith, and W. S. Davidson, the Britons who pioneered in the transportation to England of meat in mechanically refrigerated ships. Such equipment was in use by 1883 on vessels carrying meat from both American

[18] E. F. McPike, "The Refrigerator Car—Retrospective and Prospective," *Transactions of the American Society of Refrigerating Engineers*, IX (1913), 128.

[19] G. L. Eir, in paper read before fifteenth general assembly of the Association of Railway Officials, referred to in *Ice and Refrigeration*, XLV (1913), 300.

[20] For an account of early attempts to ship beef under mechanical refrigeration from Texas to New Orleans and New York see W. R. Woolrich, "Mechanical Refrigeration—Its American Birthright," *Refrigerating Engineering*, LIII (1947), 248.

[21] J. T. Critchell and J. Raymond, *A History of the Frozen Meat Trade*, 2nd edn. (London, 1912), 18-19, 21, 28-30; C. Tellier, *Histoire d'une Invention Moderne, le Frigorifique* (Paris [1910]), 258. Mort was assisted by E. D. Nicolle, a French engineer.

continents and from Australia and New Zealand. It predominated in marine refrigeration throughout the eighties.[22]

But these machines had little potentiality. They were low in efficiency and expensive in their consumption of power. After 1890 they gradually were superseded by more efficient systems and came to be installed only in the relatively few situations where economy of power was of secondary importance.[23]

Machines of the vapor-compression type began to displace cold-air units during the nineties. First carbon-dioxide systems were introduced and then ammonia. Carbon-dioxide units were safer to employ in engine rooms and other enclosed spaces. Ammonia systems were preferred where economy of power was of primary importance and where leakage of ammonia was less likely to be disastrous.[24] These vapor-compression systems were, of course, great improvements over the early units of their type first tried on shipboard in the seventies.

The methods of using the low temperatures produced to cool cargo space varied. In ships equipped with open-cycle air systems, cold air was circulated rapidly by the operation of the machine, which drew in air, cooled it, and returned it to the provision chamber. When vapor-compression machinery was employed, rapid circulation was obtained by forcing the air to be cooled over nested coils, through ducts to the space to be refrigerated, and then back to the coils once more. When natural circulation would suffice, cooling was accomplished by pumping brine through pipes distributed in the cargo space. In some installations the refrigerant was evaporated in the pipes.[25]

[22] Critchell and Raymond, *op.cit.*, 24-26, 30-31, 39-41, 340. Y. F. Rennie, *The Argentine Republic* (New York, 1945), 146.

[23] *ibid.*, 340; J. A. Ewing, *The Mechanical Production of Cold*, 2nd edn. (Cambridge, 1921), 45, 122-123; "Recent Improvements in the Refrigerating Industry," *Scientific American*, CIX (Oct. 4, 1913), 261.

[24] Ewing, *op.cit.*, 122-123; J. Creen, "Marine Refrigeration," *Ice and Refrigeration*, XLII (1912), 95.

[25] W. G. Sickel, "Refrigeration on Ocean Steamships," *Premier Congrès International du Froid*, III (Paris [1908?]), 755; Critchell and Raymond, *op.cit.*, 340.

Great Britain, dependent on supplies of perishable foods from overseas for her very existence, led the world in the development and manufacture of refrigerating machinery for marine use. American contributions in the field were slight. As late as 1935 only about ten per cent of all refrigerated cargo ships were equipped with machines built by manufacturers in the United States.[26] One American firm, the United Fruit Company, based its extensive banana trade on marine refrigeration, but it relied mainly in the period before 1917 on ships and equipment of British design.[27]

[26] "A Survey of Marine Refrigerating Machinery," *Refrigerating Engineering*, xxxi (1936), 168-169.

[27] C. F. Greeves-Carpenter, "Refrigeration Has Built the Banana Industry," *Refrigerating Engineering*, xxvii (1934), 66-68; H. J. Ward, "Sea-Transport of Bananas by Refrigeration," *Premier Congrès International du Froid*, iii (Paris [1908?]), 780.

CHAPTER IX

COLD STORAGE: ITS CONFLICT WITH
PUBLIC OPINION, 1890-1917

IN cold-storage warehouses refrigerating machinery was installed extensively during the eighteen-nineties. Soon it dominated the field, superseding ice in the large city warehouses. Ice, however, continued to be used in some small storages.[1]

After 1890 the mechanically refrigerated cold-storage warehouse was developed rapidly. By 1917 it had become an efficient instrument for food preservation. Progress was achieved in methods of controlling the temperature and quality of the air and in the construction of the building itself.

Distribution of refrigeration throughout the space to be cooled was accomplished by varied methods. For some purposes chilled brine was pumped through pipes suspended in the rooms; for others brine was dispensed with and the refrigerant itself permitted to evaporate in the pipes. In some cases it was desirable to concentrate the pipes in a single compartment and to distribute refrigeration by forcing air over them and then through the storage space.[2] These methods made possible the maintenance and accurate regulation of a wide range of temperatures.[3]

[1] F. G. Urner, quoted by G. K. Holmes, *Cold Storage and Prices* (U. S. Department of Agriculture, Bureau of Statistics, *Bulletin 101*) (Washington, 1913), 8; F. A. Horne, "Development of the Cold Storage Industry," *Ice and Refrigeration*, LI (1916), 170; D. A. Willey, "The Diversity of Uses for Cold Storage," *Scientific American Supplement*, LV (Jan. 24, 1903), 22628; M. E. Pennington, "Relation of Cold Storage to the Food Supply and the Consumer," American Academy of Political and Social Science, *Annals*, XLVIII (1913), 155; M. Cooper, "Cold Storage By Means of Ice," *Transactions of the American Society of Refrigerating Engineers*, III (1907), 53.

[2] *Ice and Refrigeration*, IX (1895), 388-389; J. E. Siebel, *Compend of Mechanical Refrigeration . . .* , 8th edn. (Chicago, 1911), 195-196. For methods of arranging inlet and outlet ducts in the storage rooms see M. Cooper, "Air Circulation in Cold Stores," *Ice and Refrigeration*, XXI (1901), 60-64.

[3] Pennington, *loc.cit.*, 155.

A measure of control over the quality of air in the storage was achieved. Relative humidity was regulated as early as 1895 by varying the temperature of the pipe surfaces by which the air was cooled.[4] The lower the temperature of these surfaces, the greater the amount of atmospheric moisture that condensed upon them and the drier became the air. This was a method of removing moisture; a satisfactory way of adding it was not developed.[5] Purity was regulated by admitting fresh air and by removing foul.[6]

The warehouse building itself underwent changes which increased its efficiency. The old, brick, wood-lined construction was replaced by concrete, a development which permitted greater cleanliness. As the superior insulating value of minute cellular structure was learned, compressed corkboard and mineral wool were used instead of boards, sawdust, shavings, cardboard, and hollow tile.[7]

The industry grew rapidly between 1890 and 1917. The general introduction of mechanical refrigeration in the nineties proved a tremendous stimulus to its expansion.[8] By 1904 there were in the United States 620 public cold-storage warehouses—establishments which held goods for customers—with a combined capacity of about 102,500,000 cubic feet. Ten years later these concerns had increased to 898 in number and to 200,000,000 cubic feet in capacity.[9] In 1916 it was estimated that there were at least 1,000 public cold-storage warehouses with an aggregate space of 250,000,000 cubic feet. It was also estimated that an additional 200,000,000 cubic feet of refrigerated space existed in breweries, packing plants,

[4] *Ice and Refrigeration*, IX (1895), 388-389.

[5] T. W. Heitz, *The Cold Storage of Eggs and Poultry* (U. S. Department of Agriculture, *Circular No. 73*) (Washington, 1929), 15.

[6] Siebel, *op.cit.*, 198; Cooper, "Air Circulation in Cold Stores," *loc.cit.*, XX (1901), 223; *Ice and Refrigeration*, XXX (1906), 46.

[7] Pennington, *loc.cit.*, 155-156; J. E. Starr, "Refrigeration Twenty-Five Years Ago," *Ice and Refrigeration*, LI (1916), 144-145; J. F. Nickerson, "The Development of Refrigeration in the United States," *Ice and Refrigeration*, XLIX (1915), 174.

[8] Heitz, *op.cit.*, 4; Horne, *loc.cit.*, 170; F. G. Urner, quoted by Holmes, *op.cit.*, 8.

[9] *1915 Ice and Refrigeration Blue Book* (Chicago, 1915), 15, 17.

creameries, dairies, and produce- and fruit-storage houses.[10] A survey made just before the United States entered the World War disclosed that the country had a total of about 475,000,-000 cubic feet for use in the process of wholesale marketing. Of this about 125,000,000 was maintained at freezing temperatures.[11]

The warehouses were first established in the large consuming centers such as New York and Boston and in cities adjacent to the nation's producing centers. Later, small storages were built in the growing areas themselves, but the largest proportion of cold-storage space continued to be located in the cities.[12] Small producing-area warehouses were advantageous in that they made possible prompt refrigeration and enabled the farmer to market his products when and where conditions were most favorable. But these benefits were largely offset by the high initial and operating costs of small storages.[13]

The principal concentrations of space were in the north central and middle Atlantic states. In 1920, the first year for which complete statistics became available, over thirty-three per cent of the total refrigerated space in the United States was located in the east north central states, over twenty-five per cent in the west north central, and nineteen per cent in the middle Atlantic. A little less than six per cent was to be found in New England and about the same amount in the Pacific states. Only about ten per cent was located in the South.[14]

[10] S. S. Vandervaart, "Commercial Uses of Refrigerating Machinery," *Ice and Refrigeration*, LI (1916), 179.

[11] I. C. Franklin, "The Effect of the War Upon the Cold Storage Industry," *A.S.R.E. Journal*, VI (1919-1920), 237-238.

[12] Horne, *loc.cit.*, 171; I. C. Franklin, "The Service of Cold Storage in the Conservation of Foodstuffs," U. S. Department of Agriculture, *Yearbook, 1917* (Washington, 1918), 365.

[13] Franklin, "The Service of Cold Storage in the Conservation of Foodstuffs," *loc.cit.*, 365; W. A. Taylor, "The Influence of Refrigeration on the Fruit Industry," U. S. Department of Agriculture, *Yearbook, 1900* (Washington, 1901), 573; M. Cooper, *Practical Cold Storage*, 2nd edn. (Chicago, 1914), 455.

[14] E. A. Duddy, *The Cold-Storage Industry in the United States* (Chicago, 1929), 87. East north central states included Wisconsin, Illinois, Michigan, Indiana, and Ohio. West north central states were North and South Dakota, Nebraska, Kansas, Minnesota, Iowa, and

Cold Storage: Conflict, 1890-1917

The main reason for the rapid expansion of cold storage after 1890 was the need created by the growth of large urban centers. The vast city populations, far removed from adequate growing areas, demanded throughout the year perishable foods the production of which was subject to wide seasonal fluctuations. If these foods were to be provided in a condition closely approaching freshness, refrigerated preservation on a large scale was essential.[15] Fortunately, this need became acute at the same time that mechanical refrigeration became thoroughly practical. Had it not been for the refrigerating machine, which could produce cheap, dependable, and controllable low temperatures at any time in any clime, strict limits might have been set to the progress of urbanization.

The United States held a position of world leadership in the application of mechanical refrigeration to cold storage. The application of it in Europe was small in scale except in Great Britain, where dependence on imports of large quantities of perishable foods made extensive facilities a necessity.[16]

A flood of adverse criticism accompanied the extension of this new technique of food preservation. This criticism, which found expression in the press and in legislative halls, consisted of two charges: first, that cold storage was a menace to health; and second, that it was responsible for increases in the prices of foodstuffs.[17]

The health-menace charge emerged early in the nineties and

Missouri. Middle Atlantic states were New York, Pennsylvania, and New Jersey. The South included Oklahoma, Texas, Arkansas, Louisiana, Kentucky, Tennessee, Mississippi, Alabama, Georgia, Florida, West Virginia, Maryland, Delaware, Virginia, and North and South Carolina.

[15] Pennington, *loc.cit.*, 159; Heitz, *op.cit.*, 4; F. A. Horne, "Cold Storage Warehousing in the United States," *Ice and Refrigeration*, LXVII (1924), 148.

[16] For analyses of cold-storage facilities outside the United States in the nineteen-twenties see U. S. Department of Commerce, Bureau of Foreign and Domestic Commerce, *Trade Information Bulletin No. 209* (South America), *No. 229* (Mexico, Central America and West Indies), *No. 280* (Australia and New Zealand), *No. 330* (United Kingdom), *No. 338* (Canada), and *No. 388* (Continental Europe).

[17] H. A. Haring, "Cold Storage Regulation," *Ice and Refrigeration*, LXVIII (1925), 419.

continued to be pressed vigorously well into the second decade of the twentieth century. It was alleged that foods were held so long in storage that they were neither wholesome nor palatable when they reached the market.[18]

The allegation that cold storage was dangerous to health had some justification in view of the practices that prevailed in the early days of refrigerated warehouses. The science of preservation by low temperatures was not generally understood. Data were lacking on such vital factors as proper temperature, air circulation, and the length of time that various commodities should be held. Eggs were not candled to make certain that only sound ones were stored, nor was attention given to their containers and to methods of packing, both important elements in satisfactory holding.[19] With the passage of time warehousing methods improved, but some malpractices lingered on. Occasionally food was held too long, and correct methods of handling were not employed.[20] A particularly bad practice still followed by some warehouse customers as late as 1917 was to return goods to cold storage after they had been on display in the market at high temperatures and had, of course, deteriorated.[21]

The actual danger from cold-storage foods, despite the existence of some faulty handling, was slight. Contrary to popular opinion, refrigerated warehouses generally gave foods a clean, sanitary environment, much better than the butcher's icebox and the household refrigerator. Comparatively little storage food was condemned as unfit by health authorities,

[18] *ibid.*, LXVIII (1925), 419; Franklin, "The Service of Cold Storage in the Conservation of Foodstuffs," *loc.cit.*, 366.

[19] Horne, "Cold Storage Warehousing in the United States," *loc.cit.*, 151; Franklin, "The Service of Cold Storage in the Conservation of Foodstuffs," *loc.cit.*, 366; Heitz, *op.cit.*, 50.

[20] W. C. Mullendore, *History of the United States Food Administration* (Stanford University, 1941), 250; Commonwealth of Massachusetts, *Report of the Commission to Investigate the Subject of the Cold Storage of Food and of Food Products Kept in Cold Storage* (Boston, 1912), 192.

[21] M. E. Pennington, *Studies of Poultry from the Farm to the Consumer* (U. S. Department of Agriculture, Bureau of Chemistry, *Circular No. 64*) (Washington, 1910), 29-30; Franklin, "The Service of Cold Storage in the Conservation of Foodstuffs," *loc.cit.*, 367.

though enormous quantities of so-called "fresh" produce met such a fate each year.[22] Studies by the Food Research Laboratory of the United States Department of Agriculture revealed that meats, poultry, fish, butter, eggs, and some other products, if received in good condition and properly stored, could be held for from nine to twelve months without an appreciable loss of flavor. They could be kept even longer than this, it was learned, without losses in food value and wholesomeness.[23] That storage periods usually were less than a year was revealed by another Department of Agriculture investigation.[24] No convincing proof was advanced that cold-storage foods as such were a menace to public health. In 1911, in testimony before the Massachusetts Commission to Investigate the Subject of the Cold Storage of Food, Professor William T. Sedgwick, a leading American epidemiologist, declared that he did not know of "any well-authenticated, carefully investigated case of food poisoning, or other ill effects, due distinctly to cold storage."[25]

The charge that cold storage increased food prices became common about 1909 and was repeated intermittently for about a decade and a half. The main substance of this accusation was the claim that speculators used the refrigerated warehouse to withhold perishable foods from the market until prices soared, and huge profits could be reaped.[26]

This complaint no doubt had its origin in the fact that dealing in foods preserved in cold-storage warehouses was speculative in nature, for it involved buying perishable goods during periods of surplus production when prices were low and selling them during periods of scant supply when prices were high. Operations of this nature had long been carried on in

[22] Pennington, "Relation of Cold Storage to the Food Supply and the Consumer," *loc.cit.*, 156, 158.

[23] Franklin, "The Service of Cold Storage in the Conservation of Foodstuffs," *loc.cit.*, 368-369.

[24] G. K. Holmes, *Cold-Storage Business Features. Reports of Warehouses* (U. S. Department of Agriculture, Bureau of Statistics, *Bulletin 93*) (Washington, 1913), 37.

[25] *Report of the Commission to Investigate Cold Storage*, 272.

[26] Haring, *loc.cit.*, LXVIII (1925), 419-420; Editorial, *Ice and Refrigeration*, XLV (1913), 365.

such commodities as cotton and wheat; the refrigerated warehouse made possible their extension to perishable products such as eggs and butter. Speculation performed a valuable economic function. By increasing demand in seasons of surplus and by augmenting supply during seasons of scarcity, it tended to equalize demand and supply throughout the year and to prevent extreme fluctuations in prices. But this function was abused if goods were held in an attempt to manipulate the market for profit. Such attempts, though infrequent, were sometimes undertaken by dealers in food held in cold storage.[27]

Strict limits to speculative holdings were set by the operation of economic pressures. The costs of storing foods—storage charges, insurance, interest, shrinkage, probable depreciation if they were held too long—increased with each passing month. Products had to be removed before these mounting costs eliminated the chance of making a profit. Another limit was the hazard involved in carrying storage stocks so long that they had to compete with the products of the next period of surplus production. So disadvantageous was their position in such competition that warehousemen were wont to call in loans made to dealers on security of such stocks or to ask for additional margins. This in itself was a check on long holdings.[28] That these restraints were not theoretical alone

[27] Massachusetts, *Report of the Commission to Investigate Cold Storage*, 91-93; Section on "Economic Results of Cold Storage" in "Report of the Secretary," U. S. Department of Agriculture, *Yearbook . . . , 1911* (Washington, 1912), 31; Franklin, "The Service of Cold Storage in the Conservation of Foodstuffs," *loc.cit.*, 366-367; J. H. Frederick, *Public Warehousing: Its Organization, Economic Services and Legal Aspects* (New York, 1940), 115; F. M. Shoemaker, "Is the Public Attitude toward Refrigerated Products Changing, and What Is Necessary to More Favorable Conditions?" *Proceedings of the Twenty-Fifth Annual Meeting of the American Warehousemen's Association . . .* (Pittsburgh, 1916), 233-234.

[28] Loans on stocks held in cold storage also had harmful aspects. Warehousemen, in competing for business, tended to grant loans on too liberal a basis. This practice encouraged speculators to widen the scope of their operations and contributed to market instability. Massachusetts, *Report of the Commission to Investigate Cold Storage*, 195. For an example of the effect of these loans see M. L. Larkin, "The

was indicated by a study published by the United States Department of Agriculture in 1913 which showed that most food products placed in cold storage were held there less than one year.[29]

A safeguard against the manipulation of the market by speculators was the fact that cold-storage stocks were owned by a vast number of wholesale dealers. In general, public warehousemen did not own the goods they stored. This being the case, any effective efforts at manipulation required that a great number of wholesalers act in combination. Manifestly, such concerted action was difficult to achieve.[30]

The effect of cold storage on food prices was given detailed consideration in a Department of Agriculture study begun in 1911 in an effort to evaluate the charges being leveled at the new instrument of food preservation. Cold storage, it was concluded after a careful analysis of prices for selected commodities from 1880 to 1911, had contributed in general, though not in all cases, toward a uniformity of prices throughout the year. Whether or not it played a large part in causing prices to move to higher or lower levels was not determined, for the study recognized that price trends were the product of many factors. No justification, however, was found for the contention of warehousemen that cold storage had reduced prices.[31]

The reasons for the charges directed against the cold-storage warehouse were to be found in the widespread prejudice against this new method of preservation and in the general anxiety over the increasing cost of living.

The popular feeling against cold storage was due in large

Butter Market," *The Journal of Political Economy*, xx (1912), 267-275.

[29] Franklin, "The Service of Cold Storage in the Conservation of Foodstuffs," *loc.cit.*, 367-368; Holmes, *Cold-Storage Business Features*, 37, 43-44.

[30] Holmes, *Cold-Storage Business Features*, 7; B. H. Hibbard, *Marketing Agricultural Products* (New York, 1930), 97.

[31] G. K. Holmes, *Cold Storage and Prices* (U. S. Department of Agriculture, Bureau of Statistics, *Bulletin 101* (Washington, 1913), 7, 24, 69, 72.

part to conditions that prevailed in the early days of refrigerated warehousing, when unfit products were stored and improper methods were employed. The poor quality of foods held under such conditions convinced many that refrigerated preservation itself was an evil.[32] This conviction was reinforced by the concern over pure food that came to the fore in the first decade of the twentieth century, a concern which tended to render any form of commercial food preservation suspect.[33]

An explanation even more fundamental for the prejudice against cold storage lay in the average city-dweller's ignorance of the problems of food supply that had been raised by the growth of metropolitan communities. Defining freshness as it was defined on the farm or in the small town in terms of chronological rather than physiological age, he demanded fresh products at seasons of the year when they could not be supplied. To satisfy this demand, it was common for wholesalers and retailers to sell the better grades of cold-storage products as fresh at high prices. The poorer grades from the warehouses and sometimes even the poorer grades of fresh products were sold as storage goods at lower prices. This only confirmed the consumer in his bias, though many violently opposed to cold-storage foods ate them in the belief they were fresh. One of the evil consequences of this practice was that in order to deceive the consumer, retailers allowed products that were stored in frozen condition—fish and poultry, for example—to thaw before sale. When thawed, these products deteriorated more rapidly than when frozen.[34]

[32] Horne, "Cold Storage Warehousing in the United States," *loc.cit.*, 151; Heitz, *op.cit.*, 50.

[33] Haring, *loc.cit.*, LXVIII (1925), 421.

[34] Pennington, "Relation of Cold Storage to the Food Supply and the Consumer," *loc.cit.*, 154-155, 161-162; Editorial, *Ice and Refrigeration*, XXXIX (1910), 213; C. L. Hunt, "Cold Storage Foods and Confidence," *Good Housekeeping Magazine*, LII (1911), 343-344. The term "fresh" is used here in the sense that the average consumer understood it. Elsewhere in this work the term usually has been employed in the sense of a commodity having its original qualities unimpaired. Actually, a perishable product that has been stored properly in a refrigerated warehouse can be called fresh with greater justifica

The concern over the rising cost of living, which reached an acute stage about 1909, was the basis for much of the criticism directed against cold storage. In the search for a reason for the greater cost of food, a vocal segment of the public came to believe that the refrigerated warehouse was largely responsible.[35] It was a comparatively new factor in the economy; that many should jump to the conclusion that it was at fault was not surprising.

This very human tendency to blame the new and strange may have been stimulated by politicians with ulterior motives. So, at least, thought the Committee on Cold Storage of the American Warehousemen's Association, which in a 1913 report charged that beginning in 1910 the Republicans had blamed cold storage for the high cost of living in an effort to save the high tariff. The Democrats had continued the complaint in 1913, said the Committee, "in spite of the low tariff."[36] Substance was lent to this accusation by the fact that in February 1910 the Senate established a committee, headed by Henry Cabot Lodge, to make an exhaustive investigation into the cost of living and its increase since 1900. In less than two months the group made a preliminary report which recommended that a limit be put on the time goods could be kept in cold storage. It believed that this would "tend to an equalization of prices and in some cases to a reduction." A bill providing for a time limit was introduced and referred to the Committee on Manufactures, which held extensive hearings. The Lodge committee made a more detailed

---

tion than many a product only a few days old that has not been kept at low temperatures.

[35] Commonwealth of Massachusetts, *Report of the Commission on the Cost of Living* (Boston, 1910), 175; R. H. Switzler, "Cold Storage Legislation—State and Federal," *Transactions of the American Society of Refrigerating Engineers*, VII (1911), 85; "The Benefits of Cold Storage," *The Independent*, LXVIII (April 14, 1910), 821. For a review of the various interpretations of rising food prices in this period see J. D. Black, "Professor Schultz and C.E.D. on Agricultural Policy in 1945," *Journal of Farm Economics*, XXVIII (1946), 677-678.

[36] "Report of the Committee on Cold Storage," *Proceedings of the Twenty-Third Annual Meeting of the American Warehousemen's Association* . . . (Pittsburgh, 1914), 108-113.

report in February 1911, listing cold storage among many factors contributing to the advance in prices.[37] The minority members, while they had concurred in the recommendation for a cold-storage time limit, accused the Republicans of restricting the committee to the period since 1900 so that no comparison could be drawn between the effect of one tariff and another on prices. The Republicans "seem to think," declared the minority, "that, like grand larceny, the tariff has a statute of limitations of ten years, and in its case they plead that statute in bar of any conviction under an indictment that it has been an accessory to the higher cost of living."[38]

Legislation regulating the cold-storage industry was the natural outcome of the allegations made against it. Some early attempts at regulation were made by municipalities, but these failed, and city authorities came to recognize that this was a matter for the states or the federal government.[39] It was in the states that the main story of cold-storage legislation was to be found. By the end of 1911 five states had laws[40] and by 1915 eleven states. The first statutes were unreasonable and worked a hardship on the industry, but after 1911 they improved.[41] A significant milestone was the adoption in 1914 of a sound draft cold-storage act by the National Conference of Commissioners on Uniform State Laws.[42] In 1916 a fairminded and progressive warehouseman pronounced the existing laws and proposals as "for the most part reasonable and satisfactory."[43]

[37] "Recommendations Touching the Use of Cold Storage," *Senate Report*, 61 Cong., 2 sess., no. 522; "Investigation Relative to Wages and Prices of Commodities," *Senate Document*, 61 Cong., 3 sess., no. 847, I, 5-7, 11-12, 124-125.

[38] "Investigation Relative to Wages and Prices of Commodities," *loc.cit.*, I, 148-149, 159.

[39] Switzler, *loc.cit.*, 86.

[40] For a summary and copies of these laws see Massachusetts, *Report of the Commission to Investigate Cold Storage*, 104-105, 208-223.

[41] Haring, *loc.cit.*, LXVIII (1925), 420-421.

[42] A copy of the draft act is printed in *Ice and Refrigeration*, XLVII (1914), 181.

[43] Horne, "Development of the Cold Storage Industry," *loc.cit.*, 171.

State laws, though they varied in details, followed a common pattern. Provision usually was made for licensing warehouses on the basis of sanitary inspections by state food commissioners or boards of health. Most laws required that packages of food be marked with both the date of entry into storage and the date of withdrawal and directed retailers to display placards to inform the consumer that he was purchasing cold-storage food. These labeling provisions, which were often drafted in great detail, proved difficult to enforce. Restrictions were set to the time food could be held in storage. Designed as safeguards against unfitness and against speculation, time limits at first were so strict that they interfered with the main function of storage, the preservation of foods for use in seasons of scant production. They were revised, however, and in most places fixed at twelve months. Provisions intended to prevent the return to storage of goods once withdrawn impeded in some cases legitimate inter-warehouse transfers, but this defect gradually was corrected. Requirements of monthly or quarterly reports of holdings were common features of state legislation.[44]

A federal law was not enacted. But from 1910 on proposals for such legislation were almost always before Congress. One of the most important of these arose from the recommendations of the Lodge committee on the increased cost of living. In April 1910 Senator Lodge introduced a bill which embodied his group's recommendations by proposing to prohibit the sale or transportation in interstate or foreign commerce of food stored more than one year. This bill was referred to the Committee on Manufactures, which after hearings reported on the measure in March 1911. The committee struck out everything but the enacting clause and substituted a more extreme proposal calling for a time limit of from three

[44] Several of the state laws are printed in U. S. Public Health Service, *Public Health Reports*, XXVII-XXXII (Washington, 1912-1918). A compilation of state laws relating to cold storage is attached as an appendix to U. S. Congress, House, Committee on Agriculture, *Cold Storage Legislation. Hearings . . . . August 11-26, 1919*, 66 Cong., 2 sess. (Washington, 1919). Haring, *loc.cit.*, LXVIII (1925), 491-493; LXIX (1925), 41-44.

to seven months on various products. This substitute was referred back to the Committee on Manufactures, where after
further hearings in 1911 it died.[45] A similar fate was met by
other bills introduced both in the Senate and in the House,
where Representative McKellar of Tennessee was a particularly vigorous advocate of regulation. That these bills did
not make greater progress in Congress was due largely to the
effective testimony presented against arbitrary time limits by
representatives of the United States Department of Agriculture and of the cold-storage industry.[46]

A decline in public prejudice was evident after 1911. It
was reflected most clearly, perhaps, in the trend toward more
reasonable state legislation. In 1915 the changing attitude of
the public was considered in part responsible for the increased demand for storage space.[47] In 1916 the Committee
on Cold Storage of the American Public Health Association
noted that the bias against the new method of food preservation was breaking down, though at the same time it reported
continued existence of the feeling that cold storage was a device for cornering food and raising prices.[48]

One of the first of several factors that contributed to a more
intelligent attitude was the report published in 1912 by the
Massachusetts Commission to Investigate the Subject of the

[45] "Report of Committee and Hearings Held before the Senate Committee on Manufactures Relative to Foods Held in Cold Storage," *Senate Report*, 61 Cong., 3 sess., no. 1272, iii-iv; U. S. Congress, Senate,
Committee on Manufactures, *Foods Held in Cold Storage. Hearings . . .
Sixty-Second Congress . . .* (Washington, 1911), passim; F. A. Horne,
"Effectiveness of Existing Laws Regulating Cold Storage Warehouses
and Products," *Proceedings Third International Congress of Refrigeration*, iii [Chicago, 1914?], 573.

[46] "Report of Committee and Hearings Held before the Senate Committee on Manufactures Relative to Foods Held in Cold Storage,"
loc.cit., passim; U. S. Congress, Senate, Committee on Manufactures,
op.cit., passim. For an account of the work of the industry in present-
ing testimony to a congressional committee see "Report of the Joint
Committee Representing Cold Storage Warehousemen and Affiliated
Industries," *Proceedings of the Twenty-Fifth Annual Meeting of the
American Warehousemen's Association* (Pittsburgh, 1916), 189-191.

[47] Editorial, *Ice and Refrigeration*, xlix (1915), 250.

[48] "Fifth Report of the Committee on Cold Storage," *The American
Journal of Public Health*, vii (1917), 306.

Cold Storage of Food which concluded that its effect on both health and prices was beneficial. Noting, however, that the industry had a public character, it called for legislation including a twelve-month time limit and labeling to identify cold-storage goods and to inform the consumer how long they had been held.[49] This temperate report had a sobering effect on the attitude toward the industry and served as a handbook on regulation throughout the country.[50]

Another factor that improved the attitude of the public toward cold storage was the work of the United States Department of Agriculture. The research of Department scientists and their testimony before Congressional committees did not corroborate the popular belief that refrigerated storage was a menace to public health. Nor did the investigations into the economic aspects of cold storage substantiate the charge that it caused prices to increase. The Department's system of voluntary reports by warehousemen, which was begun in 1914 with one commodity and expanded in 1916 and 1917, contributed to better understanding. These monthly reports, which included statistics on the amount of foods in storage and on receipts and deliveries, gave needed publicity to the economic phases of refrigerated warehousing.[51]

Among the forces that came to the defense of cold storage was the American Public Health Association, whose committee on the subject recognized the importance of refrigerated warehouses in conserving food resources and in making available sufficient wholesome supplies. Though handicapped by lack of funds, the committee forwarded its reports to legislators introducing regulatory measures and to the chairmen of the committees to which these bills were referred.[52]

[49] *Report of the Commission to Investigate Cold Storage*, 196-199.

[50] Haring, *loc.cit.*, LXVIII (1925), 420. Haring refers to the *Report of the Commission on the Cost of Living*, but he obviously has confused this with the *Report of the Commission to Investigate Cold Storage*.

[51] W. Broxton, "Cold Storage Reporting," *Proceedings of the Fiftieth Annual Meeting of the American Warehousemen's Association* . . . (Chicago, 1941), 280-284.

[52] "Report of the Committee on Cold Storage," *The American Journal of Public Health*, VI (1916), 1120-1121; "Fifth Report of the Committee on Cold Storage," *loc.cit.*, 315.

The industry was active in its own defense. Its efforts were confined principally to marshaling testimony for presentation to legislative committees. Before 1917, however, progressive warehousemen were calling for a program of education directed at the consumer himself.[53]

By 1917 the cold-storage warehouse, though its troubles were not over, had weathered a severe storm of criticism. This criticism was beneficial rather than injurious in its effects, for it resulted in a more general understanding of the new method of food preservation and the important role it played in the American economy.[54]

[53] Shoemaker, *loc.cit.*, 236-237.
[54] Horne, "Development of the Cold Storage Industry," *loc.cit.*, 171.

CHAPTER X

REFRIGERATION AND THE PRODUCTION AND
DISTRIBUTION OF FOOD, 1890-1917

REFRIGERATION was a vital factor in the production and distribution of food in the quarter-century between 1890 and the World War. Nowhere was it more important than in the meat-packing industry. This industry, transformed by the agency of natural ice in the eighties, relied increasingly thereafter on the newer methods of producing low temperatures.

Though in packing plants the refrigerating machine was not accepted generally until about 1890, its domination of the field soon was complete. By 1914 the machinery installed in American packing houses, almost all of which was of the ammonia-compression type, had a refrigerating capacity of well over ninety thousand tons a day.[1]

A large amount of refrigeration was essential in the industry. After slaughter the carcasses were placed in a chill room, where they were cooled from a little less than blood heat to about thirty-five degrees Fahrenheit. When this was accomplished, they were removed to other refrigerated rooms for storage pending shipment. If the meat were not to be sold fresh, it was kept at low temperatures in other rooms while curing progressed. The advantage of the preliminary chill room was that it prevented the increases in temperature and humidity that would occur in storage and curing chambers if warm carcasses were placed in them immediately after butchering. Some meat—a relatively small amount—was frozen at the packing plant. By permitting storage for longer periods than was possible at temperatures above freezing, this practice, not feasible in the days of ice refrigeration, helped packers to adjust their business to fluctuations in supply and demand.[2]

[1] *1915 Ice and Refrigeration Bluebook* (Chicago, 1915), 41; R. Hoagland, C. N. McBryde, and W. C. Powick, *Changes in Fresh Beef During Cold Storage Above Freezing* (U. S. Department of Agriculture, *Bulletin No. 433*) (Washington, 1917), 3.

[2] Hoagland, McBryde, and Powick, *op.cit.*, 2; D. A. Willey, "The

Improved methods of refrigerating chill rooms were evolved as packers sought more effective ways of removing heat from freshly killed animals. This task, the most difficult phase of packing-plant refrigeration, was accomplished at first by the closed-coil system, which employed brine or evaporator coils banked in overhead bunkers constructed so as to permit natural air circulation. Though this was a vast improvement over cooling by ice, the rate of heat transfer from the air to the coils was limited by the superficial area of the pipe used. Another disadvantage was the necessity of removing at frequent intervals the large amounts of moisture that condensed in the form of frost on the surface of the coils. As a result of efforts to overcome these limitations the sheet-brine system was developed in 1900. This involved hanging in the bunkers numerous sheets of muslin over which cold brine was allowed to flow. By increasing the area available for the transfer of heat, this device reduced the time required for chilling carcasses. Furthermore, it speeded the absorption of moisture rising from the meat and eliminated the necessity of defrosting. Later even greater efficiency was made possible by the development of the brine-spray system, in which the sheets were replaced by nozzles which produced a spray of finely divided particles of chilled brine. In 1917 all three systems were still being used, but the trend was toward brine-spray installations.[3]

Meat was by no means the only product of the packing plant that required low temperatures. Refrigeration was essential in the preparation of lard, oleomargarine, and a variety of other products for which coolness was less obviously neces-

---

Diversity of Uses for Cold Storage," *Scientific American Supplement,* LV (Jan. 24, 1903), 22628; J. J. Cosgrove, *Sanitary Refrigeration and Ice Making* (Pittsburgh, 1914), 248; H. D. Tefft, "Refrigeration Functions In a Packing Plant," *Refrigerating Engineering,* XXXVI (1938), 168.

[3] Hoagland, McBryde, and Powick, *op.cit.,* 3; A. Cushman, "The Packing Plant and Its Equipment," *The Packing Plant* (Chicago, 1924), 115-118; H. P. Henschien, *Packing House and Cold Storage Construction* ... (Chicago, 1915), 59-68; R. F. Wheaton, "Mass Production of Meat Needs Thousands of Tons of Cooling," *Refrigerating Engineering,* XXV (1933), 317.

sary, such as fertilizer, glue, and hides. As in the methods of chilling meat improved techniques made possible more efficient operation. The lard-chilling roll, for example, cut to one-third the time required to cool lard and provided a means of controlling accurately the consistency of the product.[4]

The effect of refrigeration on the industry after 1890 was much the same as in the period before, when dependence had been mainly on natural ice. The trends that had been evident in the eighties were broadened and intensified, but their course was not altered.

The trend toward geographic centralization, which in the case of slaughter for the fresh-meat trade depended chiefly on the refrigerator car, became after 1890 even more pronounced than before. The extent to which packing ceased to be a local industry was indicated by United States Department of Agriculture estimates that in 1916 over sixty-nine per cent of the cattle slaughtered in the entire country were butchered in plants engaged in interstate trade. Of the total number of these cattle eighty-one per cent were killed in twelve packing centers, the most important of which were Chicago, Kansas City, Omaha, and St. Louis. These twelve also accounted for over sixty-five per cent of the calves handled in interstate concerns, for over seventy-eight per cent of the sheep and lambs, and for fifty-eight per cent of the hogs.[5]

The trend toward concentration of control was so strong after 1890 that it became a major public problem. By 1916 the Big Five packers—the Armour, Swift, Morris, Wilson, and Cudahy interests—butchered over eighty-two per cent of the cattle slaughtered by interstate firms as well as over seventy-six per cent of the calves, eighty-six per cent of the sheep, and sixty-one per cent of the swine.[6]

Foods other than meat seemed likely to become the do-

[4] Cushman, *loc.cit.*, 120-121; S. S. Vandervaart, "Commercial Uses of Refrigerating Machinery," *Ice and Refrigeration*, LI (1916), 179-180.

[5] Federal Trade Commission, *Food Investigation. Report . . . on the Meat-Packing Industry* (Washington, 1918-1920), I, 117-118, 121.

[6] *ibid.*, I, 106.

minion of the large packers. By 1917 the Big Five were engaged so extensively in the distribution of meat substitutes, of canned fruits and vegetables, and of staple groceries and vegetables that the Federal Trade Commission feared packer domination of all important foods.[7]

The monopolistic position of the big packers in the meat industry and their increasingly strong situation in the distribution of other foods was due in large part to their ownership of the expensive equipment—refrigerator cars, branch houses, and other cold-storage facilities—which was essential for the distribution of perishable foods on a large scale. Ownership brought with it important competitive advantages that were not available to those who did not have equipment of their own.[8] The privately owned refrigerator car was the most important feature of the distribution system of the packing industry and the one that was more frequently a target for the complaints made against the Big Five.

Packer ownership of refrigerator cars began in the seventies, when Gustavus F. Swift and others tried to persuade the railroads to provide the equipment they needed in order to develop the fresh-meat trade to the East. Rebuffed in these efforts, the packers were forced to construct their own cars.[9] These the railroads agreed to haul, paying a rental fee that came to be based on mileage. With the passage of the years the railroads and railroad-owned companies built and operated large numbers of ventilator refrigerator cars, but they built no units of the beef type that were necessary for satisfactory year-around transportation of fresh meat. At the end of 1917 the Big Five had over ninety-one per cent of the beef cars in the United States. Independent packing companies held a little less than seven per cent, while other private-car companies had the few that remained. Though there were no patents

[7] *ibid.*, I, 35-36, 390.
[8] *ibid.*, I, 31, 41-42, 104-105, et passim.
[9] L. D. H. Weld, "Private Freight Cars and American Railways" (Faculty of Political Science of Columbia University, eds., *Studies in History, Economics and Public Law*, xxxi, Number 1) (New York, 1908), 17, 78; J. O. Armour, *The Packers, the Private Car Lines, and the People* (Philadelphia, 1906), 23.

preventing independents from building equipment, economic obstacles proved an effective bar. Large amounts of capital were needed to acquire beef cars, and the cars could not be operated as efficiently by a small packer as by his big competitor who had great fleets of them at his command.[10]

Complaints were made even before 1890 that ownership of beef refrigerator cars by the packers was a powerful factor in enabling them to achieve a monopolistic position in the meat industry. For many years these complaints centered around the mileage payments made by the railroads for the use of packer-owned equipment.

Mileage payments, it was claimed, were so large that they constituted rebates on the freight rates paid by their packer owners, giving them a crushing advantage over competitors who did not receive them. These claims received substantial backing from the Interstate Commerce Commission. Hearings held in Washington in 1889 did not yield conclusive evidence that rebates had been paid, but the Commission did conclude that mileage rates were large enough to prove a material advantage to a shipper owning cars over one who did not. Regulation of these payments, the commissioners held, should be a public function.[11] By 1902 the Commission had come to the belief that the mileage payments were substantially rebates and that they discriminated against shippers who had no cars. It pointed to common reports that the packers threatened railroads with loss of traffic if they refused to make the payments demanded.[12] In 1904 mileage rates were characterized as "extremely profitable,"[13] a finding that was supported by the

[10] Federal Trade Commission, *Food Investigation. Report on Private Car Lines* (Washington, 1920), 81, 83; M. W. Watkins, Section on "Social Aspects" under "Meat Packing and Slaughtering," *Encyclopaedia of the Social Sciences*, x, 260.

[11] *Third Annual Report . . . December 1, 1889* (Washington, 1889), 15-18, 108. A summary of the investigations of the private-car line question is included in Federal Trade Commission, *Report on Private Car Lines*, 64-68.

[12] *Sixteenth Annual Report . . . December 15, 1902* (Washington, 1902), 80-82.

[13] *Eighteenth Annual Report . . . December 19, 1904* (Washington, 1904), 16.

estimates of private-car earnings published a few months later by the Commissioner of Corporations.[14]

Between 1902 and 1906 the abuse of mileage payments was criticized with particular intensity. It figured prominently in hearings held by the Interstate Commerce Commission, by the Bureau of Corporations, and by the committees on interstate commerce in both houses of Congress.[15] The part these payments were alleged to have played in the formation of the beef trust was brought vividly to the attention of the public by Charles Edward Russell in "The Greatest Trust in the World," a series of articles published in *Everybody's* and in book form.[16] Ray Stannard Baker penned a similar attack for *McClure's*.[17] The packers were defended before the public by J. Ogden Armour, who in his book, *The Packers, the Car Lines, and the People,* denied flatly that the refrigerator cars owned by his firm were used either directly or indirectly to obtain rebates, discriminations, or concessions.[18]

The complaints against rebates in the form of mileage payments declined after 1906. Though the Interstate Commerce Commission did not receive full authority to regulate mileage rates until 1917, the Hepburn Act did give it power to determine the maximum amount that would be reasonable for the railroads to pay. Profits from the operation of cars declined materially. In 1917 the refrigerator cars of the Big Five and their affiliated companies yielded an average net return on investment that was estimated by the Federal Trade Commis-

[14] U. S. Department of Commerce and Labor, Bureau of Corporations, *Report of the Commissioner of Corporations on the Beef Industry, March 3, 1905* (Washington, 1905), xxxv-xxxvi.

[15] "Hearings Before the Committee on Interstate and Foreign Commerce of the House of Representatives on Bills to Amend the Interstate Commerce Act," *House Document*, 58 Cong., 3 sess., no. 422; "Regulation of Railway Rates: Hearings Before the Committee on Interstate Commerce, Senate of the United States . . . ," *Senate Document*, 59 Cong., 1 sess., no. 243.

[16] *The Greatest Trust in the World* (New York, 1905), 11-13, 28-31, et passim.

[17] "The Railroads on Trial," *McClure's Magazine*, xxvi (1905-1906), 318-331.

[18] See especially pp. 38-39.

sion to be only about four and a half per cent. This was not regarded as excessive.[19]

However, other complaints against big-packer ownership came to the fore after 1906. In these the independent packers were joined by distributors of non-meat food products who were encountering increasing competition from the large packers. Both claimed that ownership of refrigerator cars gave the Big Five undue advantages. A thorough examination was accorded their allegations in the course of the investigation of the meat-packing industry begun by the Federal Trade Commission in 1917.

Important advantages over other packers quite apart from mileage payments, the Commission found, accrued to the Big Five from their ownership of most of the country's beef-type cars. Possession of these units, which were equipped with the brine tanks necessary for maintaining especially low temperatures, gave Armour and the others the assurance of always being able to ship fresh meat. Unless independent packers were large enough to acquire their own cars, it was practically impossible for them to send fresh meat out of the area in which they slaughtered it, for ventilator refrigerator cars, the only type the railroads could furnish them, were unsatisfactory for this purpose. The only alternatives left to small independents were to confine themselves to curing pork products or to slaughtering for local consumption. Even here they faced competition from the Big Five, whose peddler cars and branch houses enabled them to vie with the local slaughterer in his own domain. If an independent packer could afford his own cars, he was still at a disadvantage, for his powerful rivals, whose great freight tonnage gave them leverage in dealing with the railroads, could see that their cars were promptly moved and promptly returned, that they were carefully handled, and that they were used only for transporting their own commodities. Among further advantages of the Big Five was their ownership of stations for icing refrigerator cars. These not only enabled them to ice their own shipments at

[19] *Report on Private Car Lines*, 19, 68-69, 81.

cost and those of others at a profit, but also gave them opportunity to obtain information on their competitors' business and to discriminate against them by providing the icing service improperly.[20]

The Big Five, concluded the Commission, also derived from their refrigerator cars important advantages over the wholesalers with whom they competed in the distribution of non-meat food products. One of these lay in the use of extra space in their peddler cars for less-than-carload shipments of groceries and other merchandise to outlying districts. Since these cars received the special handling given to the meat traffic of the big packers, the non-meat products loaded in them benefited by a service that was not available to the goods of other distributors. Thus the packers were able to make prompt and regular deliveries and to sell at points their competitors were not able to reach as frequently, if at all. In shipping to their branch houses they were able to count on a similar speed and dependability of service that others could not obtain. Underlying these advantages were mixing rules and minimum weight regulations that permitted them to use their cars profitably for transporting a variety of commodities.[21]

Drastic measures for preventing the inequalities arising from large-packer ownership of refrigerator cars were recommended by the Commission in 1919. It suggested two possible courses: either that all refrigerator cars and equipment necessary for their proper use be owned and operated as a government monopoly, or that the cars and equipment be owned and operated by the railroads under government license.[22] Neither of these alternatives, however, was destined for acceptance.

Fruit and vegetable production in the United States expanded enormously in the twenty-five years after 1890. Several factors contributed to this. For one thing, a large market for such products was available in the swollen urban com-

[20] *ibid.*, 18-20, 82-84, 99-101; *Report on the Meat-Packing Industry,* I, 40-41.
[21] *Report on the Meat-Packing Industry,* IV, 55-84.
[22] *Report on Private Car Lines,* 14-15.

munities of the North and East. For another, these markets had been connected with remote agricultural districts by efficient rail transportation. But neither of these factors could have been decisive without refrigeration. The ice-making machine, the refrigerator car, and the cold-storage warehouse were key developments in the growth of the fruit and vegetable industry.[23]

The refrigerator car could not be used effectively in transporting the fruits and vegetables grown in the warmer sections of the United States until factories capable of manufacturing cheap and abundant quantities of ice had been established at shipping points. About 1880 refrigerator cars came to be used regularly and extensively by the meat-packers, whose shipments originated in the natural-ice belt. But it was not until about 1890, when ice-manufacturing plants had been erected throughout the South, that the refrigerator car found general employment in the transportation of that section's fruits and vegetables.[24] From California the first refrigerated shipments of fruits on a large scale were made about the same time as from the South, but natural ice obtained in the Sierras was used rather than the manufactured product.[25] The growers of the central valley of California continued to depend to a

[23] E. A. Duddy and D. A. Revzan, "The Physical Distribution of Fruits and Vegetables" (The Journal of Business of the University of Chicago, *Studies in Business Administration*, VII, No. 2) (Chicago, 1937), 14.

[24] W. A. Sherman, *Merchandising Fruits and Vegetables* (Chicago, 1928), 33-34; Weld, *loc.cit.*, 27, 52-53; A. M. Soule, "Vegetables, Fruit and Nursery Products and Truck Farming in the South," *The South in the Building of the Nation*, VI (Richmond, 1909), 130; U. S. Department of the Interior, Census Office, *Twelfth Census, 1900* (Washington, 1901-1902), VI, 305. Even after 1890 the manufactured-ice supply of the South occasionally proved inadequate. In 1895, for example, the Armour Refrigerator Car Line was forced to ship natural ice from Cincinnati to Macon, Georgia, for use in icing its fruit cars. At the time Macon had three ice factories. *Ice Trade Journal*, XIX (Sept. 1895), 5.

[25] Weld, *loc.cit.*, 47; *Ice and Refrigeration*, I (1891), 215; II (1892), 23; M. C. B., "The California Fruit-supply in New York," *Garden and Forest*, VI (1893), 432; Letter from Sacramento, Calif., June 16, 1895, in *Ice Trade Journal*, XVIII (July 1895), 1; Letter from San Francisco, Calif., Feb. 17, 1897, in *Ice Trade Journal*, XX (March 1897), 1.

large extent on natural ice well into the twentieth century, and in the nineteen-forties the Pacific Fruit Express was still harvesting ice in the Rockies.[26] In southern California citrus-growers, however, were faced with conditions more like those existing in the southeastern and Gulf states. They had to depend on the ice factory, though as late as 1906 some natural ice from the mountains of northern California was hauled to the southern part of the state for use in refrigerator cars.[27] Numerous plants were erected during the nineties in the areas in which citrus shipments originated.[28]

Improvements in the technique of using the refrigerator car in the fruit and vegetable traffic were mainly the work of organizations that specialized in operating them and of the United States Department of Agriculture.

Specialized refrigerator-car organizations, which were either independent car companies, railroad-controlled concerns, or separate departments of the railroads, developed successful methods of coping with the complexities of the trade in perishables. The independent companies led the way, while the others, whose formation was inspired by their success, were quick to follow. In order to provide proper refrigeration service for the cars at the shipping point and along the route to their destination, these organizations made elaborate arrangements to assure the availability of adequate ice supplies and developed efficient methods of icing and re-icing. Frame platforms the same height as the tops of the cars were constructed at key points. Along either side of these structures solid trains of refrigerator cars could be serviced in a short time by laborers who had merely to push the ice from the platform through the roof hatches into the bunkers. To assure rapid handling, managers received advance telegraphic notice of the arrival of a train and of the number of cars to be serv-

[26] W. C. Phillips, "Plants and Car Icing Facilities of Pacific Fruit Express," *Ice and Refrigeration*, LXIII (1922), 289; "Icemen's Shivers," *Business Week*, Nov. 29, 1941, 52.

[27] Armour, *op.cit.*, 79.

[28] For some of these see *Ice and Refrigeration*, II (1892), 118, 362; III (1892), 41; Letter from P. C. B., San Francisco, Calif., Nov. 17, 1898, in *Ice Trade Journal*, XXII (Dec. 1898), 2.

iced. Inspectors were maintained to supervise operations. Concerned only with the traffic in perishables, these specialized organizations could see that the shipments entrusted to them were not delayed.[29]

The United States Department of Agriculture contributed to improvements in the technique of using the refrigerator car through its investigations of the storage and transportation of perishables. These studies were instigated by William A. Taylor, a Department official who had been in charge of the continuous exhibition of American fresh fruits at the Paris Exposition in 1900. The difficulties he had experienced in shipping and storing different fruits convinced him that scientific study of the problems involved was needed. In 1901 an appropriation for such work was obtained from Congress, and G. Harold Powell, horticulturist at the Delaware Agricultural Experiment Station, was called to take charge. The work, which in various forms became a continuous activity, soon revealed that two important factors in satisfactory preservation of fruit during transportation were careful handling and adequate cooling before loading in refrigerator cars.[30]

The importance of careful handling was shown by a Bureau of Plant Industry investigation directed by Powell into the reasons for the decay of oranges while in transit from California. This inquiry, undertaken at the request of growers, revealed that decay was being caused by fungi which entered

[29] J.-S. Leeds, "Refrigeration of Citrus Fruits in Transit from California," *Premier Congrès International du Froid*, III (Paris [1908?]), 603-611; Armour, *op.cit.*, 79-83, 259-261; G. C. White, "Improved Transportation Service for Perishable Products," *Proceedings of the Second Pan American Scientific Congress*, III (Washington, 1917), 408; J. H. Bawden, "Distribution and Sale of Perishable Crop Products," L. H. Bailey, ed., *Cyclopedia of American Agriculture*, IV (New York, 1909), 245; C. M. Secrist, "Facilities for Replenishing Ice for Refrigeration in Transit," *Railway Age Gazette*, LV (Sept. 26, 1913), 568; Phillips, *loc.cit.*, 291; Federal Trade Commission, *Report on Private Car Lines*, 236.

[30] A. V. Stubenrauch, "The Fruit Storage and Transportation Investigations of the U. S. Department of Agriculture," *Ice and Refrigeration*, LI (1916), 187; S. J. Dennis, "The Precooling of Fruit in the United States," IInd International Congress of Refrigeration, Vienna, 1910, *English Edition of the Reports and Proceedings* (Vienna, 1911), 464-465.

the fruit through abrasions in the skin. The result of rough treatment during picking and in the packing house, these abrasions could be avoided by better handling.[31] Powell's findings were corroborated by subsequent investigations.[32] This work resulted in materially reduced losses during transportation, especially in orange shipments from California, where the California Fruit Growers Exchange was able to effect improvements in methods of handling fruit in packing houses.[33]

The value of cooling fruit before placing it in refrigerator cars was discovered early in the course of the investigations. At the time the studies were begun, the prevailing practice was to load the fruit while still warm. Many hours elapsed before enough of the natural heat of the fruit could be removed by the ice in the bunkers to check ripening processes and the growth of destructive fungi. As a consequence the fruit kept poorly, especially when carried in the upper and middle part of the car.[34] In 1904, while trying to reduce the losses in peach shipments from Georgia, Powell tried cooling before loading—precooling, as the practice came to be known. Since very satisfactory results were obtained, similar experiments were made with the orange, peach, and grape crops of California, with the pears and berries of the Pacific Northwest, with the strawberries of Louisiana, and with the oranges,

[31] G. H. Powell, assisted by A. V. Stubenrauch et al., *The Decay of Oranges in Transit from California* (U. S. Department of Agriculture, Bureau of Plant Industry, *Bulletin No. 123* (Washington, 1908), 70-71.

[32] L. S. Tenney, assisted by G. W. Hosford and H. M. White, *The Decay of Florida Oranges while in Transit and on the Market* (U. S. Department of Agriculture, Bureau of Plant Industry, *Circular No. 19*) (Washington, 1908); A. V. Stubenrauch, H. J. Ramsey, and L. S. Tenney, assisted by A. W. McKay et al., *Factors Governing the Successful Shipment of Oranges from Florida* (U. S. Department of Agriculture, *Bulletin No. 63*) (Washington, 1914); A. W. McKay, G. L. Fisher, and A. E. Nelson, *The Handling and Transportation of Cantaloupes* (U. S. Department of Agriculture, *Farmers' Bulletin 1145*) (Washington, 1921).

[33] R. M. MacCurdy, *The History of the California Fruit Growers Exchange* (Los Angeles, 1925), 52-53.

[34] W. A. Taylor, "Influence of Precooling on Transportation of Fruit," L. H. Bailey, ed., *Cyclopedia of American Agriculture*, IV (New York, 1909), 248-249.

lettuce, and celery of Florida.[35] Precooling, it was learned, helped to cut down losses from deterioration. It made unnecessary the practice of picking and shipping such fruits as peaches and plums while still green. By making possible closer loading, it increased the carrying capacity of the cars, and by enabling growers to ship grapes, cherries, and other highly perishable fruits for greater distances, it extended their market area.[36]

Precooling was used most extensively in the citrus industry of southern California, where by 1910 several plants for accomplishing it were in operation. These were of two types— warehouse precooling and car precooling. In the warehouse type the fruit was chilled in refrigerated rooms before being loaded, a method that made possible thorough and economical cooling. Since the warehouses were located near the groves, the fruit could be chilled promptly after picking. Moreover, the grower was enabled to hold his fruit for a while and thus to have some control over the time it was marketed. A disadvantage of these plants was their high initial cost, which made it profitable to build them only at large producing points. In the car type, which was owned by the railroads, the fruit was loaded and then cooled rapidly by forcing cold air from a central refrigerating plant through the cars. This system was effective for handling the shipments of growers to whom warehouse plants were not available, but it was not precooling in the strict sense of the word, and it did not chill the fruit thoroughly. Since it was practical to install the expensive equipment required only at the principal shipping points, the time that elapsed between picking and cooling was greater

[35] A. V. Stubenrauch and S. J. Dennis, "The Precooling of Fruit," U. S. Department of Agriculture, *Yearbook* . . . , *1910* (Washington, 1911), 439-442; Stubenrauch, *loc.cit.*, 188; H. J. Ramsey, *Handling and Shipping Citrus Fruits in the Gulf States* (U. S. Department of Agriculture, *Farmers' Bulletin 696*) (Washington, 1915), 26; H. J. Ramsey and E. L. Markell, *The Handling and Precooling of Florida Lettuce and Celery* (U. S. Department of Agriculture, *Bulletin No. 601*) (Washington, 1917), 27-28; Powell, *op.cit.*, 71.

[36] Stubenrauch and Dennis, *loc.cit.*, 448; J. F. Nickerson, "The Development of Refrigeration in the United States," *Ice and Refrigeration*, XLIX (1915), 174.

than when the warehouse system was employed. A further disadvantage was that it did not permit as heavy loading.[37]

The adoption of precooling was delayed by a long controversy between citrus-shippers of California and the railroads. This dispute began over the rates charged by the carriers for servicing cars of fruit that had been precooled by the shippers in their own warehouses. The issue finally was settled by the Supreme Court of the United States, which in 1914 upheld an Interstate Commerce Commission ruling in favor of the citrus men.[38] In spite of this difficulty, in 1916 it was estimated that ten per cent more salable oranges were being brought from California to the East than before precooling.[39]

The most striking effect of the refrigerator car on fruit and vegetable production was the impetus it gave to regional specialization. By making it possible to transport the most perishable of crops for hundreds and even thousands of miles, it stimulated the growing of fruits and vegetables in areas particularly well-adapted by climate and soil, regardless of how far they lay from the principal markets.[40] The refrigerator car, of course, was not the sole reason for this development. Commercial canners located in the producing areas were the market for substantial amounts of some crops. High-speed ventilated shipments continued to be used for the transportation of some commodities. But without the refrigerator car regional specialization in the fruit and vegetable industry would have been limited.[41]

[37] Stubenrauch and Dennis, *loc.cit.*, 437-447; Dennis, *loc.cit.*, 476, 481; B. W. Redfearn, "Methods of Pre-cooling Perishable Goods at Loading Stations," *Railway Age Gazette*, LV (Sept. 26, 1913), 568-569; Powell, *op.cit.*, 55-57; M. Cooper, *Practical Cold Storage*, 2nd edn. (Chicago, 1914), 511-513.

[38] Atchison, Topeka and Santa Fe Railway Company v. United States, 232 U. S. 199; Redfearn, *loc.cit.*, 569.

[39] V. W. Killick, "Pre-Cooling California Oranges to Save Millions of Dollars Annually," *Scientific American*, CXV (Oct. 28, 1916), 387.

[40] Fruit and vegetable growing, which before the days of rapid transportation and refrigeration had been carried on in the areas surrounding the cities almost as a single operation, tended to separate into two distinct industries. W. A. Sherman, "Fruit and Vegetable Industry," *Encyclopaedia of the Social Sciences*, VI, 510.

[41] McKay et al., "Marketing Fruits and Vegetables," U. S. Department of Agriculture, *Agriculture Yearbook, 1925* (Washington,

Under the influence of the car, numerous areas far from the markets of the North and East specialized in growing deciduous fruits. One of the most important of these areas was Georgia, famous for its peaches. In 1889 only 150 carloads were shipped from its orchards, but in 1898 over eleven times that many were dispatched and in 1905 about 5,000 carloads.[42] California, the width of the continent from its market, concentrated on several fruits—grapes, peaches, pears, plums, and apples. Carloadings of these increased from over 4,500 in 1895 to between 8,000 and 10,000 in 1905, over ninety-five per cent of which were refrigerated.[43] Among other specialized areas was the Pacific Northwest, where Washington and Oregon developed vast apple, pear, and cherry orchards and grew large quantities of raspberries.

Citrus-growing areas developed rapidly. In 1886 California shipped 1,000 carloads of oranges, protected only by ventilation. With the introduction of the refrigerator car, the traffic in oranges increased rapidly. Between 1890 and 1895 from 4,000 to 7,000 carloads a year were forwarded and between 1900 and 1907 from 25,000 to 32,000. A large part of the movement continued to depend on ventilation alone, but by the season of 1904-1905 fifty-one per cent of the citrus traffic of the Santa Fe Refrigerator Despatch Company moved under refrigeration.[44] In the citrus districts of Florida, the refrigerator car was not so important a factor. For many years only a very small portion of the oranges from Florida were refrigerated, since the crop moved from September to January, when little warm weather was encountered. In the second decade of the twentieth century, however, refrigerator cars

---

1926), 683, believe that the refrigerator car "has had more influence in commercial-fruit and vegetable growing than any other single factor."

[42] Weld, *loc.cit.*, 52-53; G. H. Powell, "The Handling of Fruit for Transportation," U. S. Department of Agriculture, *Yearbook* . . . , *1905* (Washington, 1906), 350.

[43] Powell, "The Handling of Fruit for Transportation," *loc.cit.*, 350; *Twelfth Census, 1900*, vi, 305.

[44] Powell, *The Decay of Oranges in Transit from California*, 10-11; Weld, *loc.cit.*, 50.

came to be used for a considerable portion of the fruit sent North in September and October.[45]

Areas that specialized in growing truck crops for distant markets also were stimulated. The increasing importance of the refrigerator car was indicated by the fact that in 1900 the trucking district of North Carolina, nearer to market than many areas, depended on it for over eighty per cent of its shipments.[46] The Carolinas, the Gulf states, Arkansas, Missouri, and Tennessee entered into strawberry production on a large scale. Crystal Springs, Mississippi, became the center of a great tomato-growing industry, while Rocky Ford, Colorado, became a similar focal point for cantaloupe production when in the nineties the refrigerator car made eastern markets accessible. By 1920 more than 21,000 cars of cantaloupes were shipped annually in the United States. Of these about four-fifths originated at points in Colorado, New Mexico, Arizona, Nevada, and California—six to twelve days from market. Without refrigerated transportation this would have been impossible. Many areas came to specialize in vegetable production, especially in the south Atlantic and Gulf states and in California. Refrigerator cars were not needed for all vegetables, but were important for successful marketing of lettuce, celery, cauliflower, cabbage, asparagus, and others of the more perishable crops.[47]

By 1917 a remarkably full development of areas particularly well-suited by climate and soil to the production of fruits and vegetables had taken place. Other specialized areas were destined to appear, but in the main they were only supplemental to those already in existence.[48]

The established fruit and vegetable areas that lay close to

[45] Leeds, *loc.cit.*, 602; Ramsey, *op.cit.*, 24.
[46] *Twelfth Census, 1900,* VI, 305.
[47] Weld, *loc.cit.*, 56, 60-63; P. A. Bruce, *The Rise of the New South*, (G. C. Lee, ed., *The History of North America*, XVII) (Philadelphia, 1905), 64-68; McKay, Fisher, and Nelson, *op.cit.*, 3; L. C. Corbett et al., "Fruit and Vegetable Production," U. S. Department of Agriculture, *Agriculture Yearbook, 1925*, 420-421; Sherman, "Fruit and Vegetable Industry," *loc.cit.*, 509.
[48] An excellent account of regional specialization in the fruit and vegetable industry as of 1925 is Corbett et al., *loc.cit.*, 151-452.

the large eastern markets felt the competition of distant specialized centers. In some cases the time that locally grown produce dominated the market was shortened at both ends of the seasons by products shipped in from afar. A few home-grown crops did not have the market to themselves even at the height of their seasons. Local areas were able to survive under these conditions, but at the price of making adjustments and readjustments, particularly in the type of products grown.[49]

Serious problems accompanied the growth of the fruit and vegetable industry under the influence of the refrigerator car. One of these centered on the privately owned car, which was the source of as much contention as it was in the meat-packing industry. Others arose from the fact that refrigerated transportation barred the grower from personal contact with the market in which he sold his produce. Still another resulted from the inadequacy of city market facilities, which were ill-adapted to the needs of the long-distance traffic in perishables.

The privately owned car was a factor from the very beginning of refrigerated transportation on a large scale. In 1888 F. A. Thomas and Son, Chicago commission merchants, obtained from Carlton B. Hutchins, a Detroit inventor, the use of fifty-five cars. These they employed to provide a through refrigerated service from shipping point to destination that bore a relation to the railroads similar in principle to that of the passenger service furnished by the Pullman sleeping cars. After operating the cars east of the Mississippi, the Thomases sent them to California to bring back deciduous fruits, an enterprise that proved successful and soon was organized as the California Fruit Transportation Company.[50] The prosperity of this firm was so pronounced that it attracted competitors, among whom was P. D. Armour, who about 1890 began acquiring cars adapted to the fruit traffic. A fierce struggle

---

[49] W. P. Hedden, *How Great Cities Are Fed* (Boston, 1929), xiii-xiv; Sherman, *Merchandising Fruits and Vegetables*, 36-37, 456-457.

[50] Russell, *op.cit.*, 34-37; W. A. Taylor, "The Influence of Refrigeration on the Fruit Industry," U. S. Department of Agriculture, *Yearbook, 1900* (Washington, 1901), 575-576; Weld, *loc.cit.*, 18-19.

ensued in California, marked by rate cutting and rebating, which ended with Armour attaining a position of supremacy.[51] His operations extended to other parts of the country as well, particularly the Southeast. In 1900 several railroads—the Santa Fe was most active—were furnishing shippers along their lines with refrigerator cars, either directly or through companies they controlled.[52] Nevertheless, in 1904 the Armour Car Lines Company, in the opinion of the Interstate Commerce Commission, possessed a practical monopoly of the movement of fruit in large quantities in most sections of the United States. No other company, it believed, could handle the various fruit crops of Michigan or the peach harvest of Georgia.[53]

The principal objection to the privately owned refrigerator car in the fruit and vegetable industry lay in the charges made for icing under the exclusive contract. This was an agreement upon which the Armour Car Lines came to insist as a prerequisite to permitting its cars to go on the line of any railroad for use in moving produce originating there. According to the terms of the contract, the railroad agreed to allow the Armour Company alone to provide refrigerator service for a particular kind of traffic, usually fruit, and to pay the company a fixed mileage fee for the rental of its cars. Armour agreed to furnish whatever cars were needed and to provide for their icing. Charges for icing were fixed by the car company and paid by the shipper. Where these contracts were in force, service improved, it was generally agreed, but icing charges increased. In 1904 the Interstate Commerce Commission was of the opinion that under exclusive contracts the cost had advanced from 50 to 150 per cent and that in most cases the charges were "utterly unreasonable."[54]

[51] Weld, *loc.cit.*, 19; Russell, *op.cit.*, 37-40; Baker, *loc.cit.*, 399-400; Armour, *op.cit.*, 25-26.

[52] *Twelfth Census, 1900*, vi, 305; Weld, *loc.cit.*, 27-30.

[53] *Eighteenth Annual Report, 1904*, 14; Baker, *loc.cit.*, 400. Although other packing interests had refrigerator cars in the fruit trade, Armour was by far the most extensively engaged. Russell, *op.cit.*, 43-44.

[54] Interstate Commerce Commission, *Eighteenth Annual Report, 1904*, 14-15; Weld, *loc.cit.*, 94-132.

Agitation against the operations of packer-owned refrigerator cars in the fruit and vegetable trade coincided in the first years of the twentieth century with the agitation against their abuses in the meat-packing industry itself. Both commission men and growers protested against unfair icing charges, while C. E. Russell and R. S. Baker dealt at length with the issue in their indictments of the beef trust. In hearings conducted in 1904 by the Interstate Commerce Commission and in 1905 by the congressional committees on interstate commerce, exclusive contracts and icing charges received their share of attention.[55]

After 1906 complaints against the private-car system in the fruit and vegetable industry subsided. Two factors were responsible. First, the Hepburn Act of 1906 brought refrigeration or icing charges directly under the supervision of the Interstate Commerce Commission. Exclusive contracts continued to exist, but the railroads were made responsible for the icing charges. A shipper who considered a rate unreasonable could bring it to the attention of the Commission. Second, western producers became less dependent on cars of private companies as several western railroads followed the example of the Santa Fe and formed subsidiary car companies of their own. Armour cars, operated under the name of the Fruit Growers Express, were restricted for the most part to the Southeast.[56]

The produce-grower who participated in the long-distance trade that the refrigerator car did so much to make possible had no direct personal contact with his market. As a result he had no reliable information on market conditions. Most of what he knew came from his commission merchant, not a disinterested source. Since he was separated from many of his competitors by great distances, there was no way of dis-

[55] Baker, *loc.cit.*, 398-411; Russell, *op.cit.*, passim; Weld, *loc.cit.*, 101. For a defense of the operations of private cars in the fruit and vegetable industry see Armour, *op.cit.*, passim, especially 29-35, 70-71, 75-77, 226.

[56] Federal Trade Commission, *Report on Private Car Lines*, 68, 163, 165.

covering by his own observation what they were doing. No official estimates of acreage or production existed to guide him. Furthermore, he seldom had personal contact with the commission merchant or buyer who received his goods, and there were no generally recognized standards of quality to serve as a common language. These conditions combined to create widespread hardship and confusion. Glutted markets were common and very difficult for a grower to avoid. Misunderstandings were chronic between growers and receivers, who generally were blamed for all low prices. The reigning confusion proved an invitation to dishonesty, not only on the part of commission men and buyers, but of producers as well.[57]

The first steps toward improvement of these conditions were being taken in 1917, but much remained to be done. Market-information work was begun by the United States Department of Agriculture in 1913 with the creation of the Office of Markets, which after two years of research and survey started active service to the fruit and vegetable industry. Many cooperative marketing organizations had been tried, but the only one outstandingly successful was the California Fruit Growers Exchange. How much was needed to be done to put the trade on a stable basis was indicated by the fact that at the organization of the Food Administration during the World War there was no fruit or vegetable for which definite standards were accepted throughout the United States.[58]

City market facilities were ill-adapted to the needs of the long-distance traffic. Markets had been laid out when produce had been brought to the city in boats or wagons from surrounding farms. Often they were not favorably located in relation to railroad terminals, a situation that made for congestion and inconvenience when large quantities of perishables

[57] Sherman, *Merchandising Fruits and Vegetables*, 38-40, 45-49; R. L. Spangler, *Standardization and Inspection of Fresh Fruits and Vegetables* (U. S. Department of Agriculture, *Miscellaneous Publication No. 604*) (Washington, 1946), 3-4.

[58] T. S. Harding, *Some Landmarks in the History of the Department of Agriculture* (U. S. Department of Agriculture, *Agricultural History Series No. 2*) (Washington, 1942), 64; Sherman, *Merchandising Fruits and Vegetables*, 148, 177.

began to arrive by rail. Little was done to adapt markets to the new conditions.[59]

The refrigerator ship was a comparatively minor factor in the growth of the industry in the United States. It was not used to any significant extent for transporting perishables in domestic trade,[60] but it did open foreign markets to some American fruit-growers. Apples and oranges were the fruits exported in largest quantities. Though they did not require refrigeration during ocean transportation under all conditions, it was the refrigerator ship that made it possible for exports of green and ripe apples to increase from just over 135,000 barrels in 1891 to over 1,700,000 in 1917 and for exports of oranges to increase over sixteen times in value between 1900 and 1917.[61] The refrigerator ship, of course, was a vital feature of the banana trade carried on by the United Fruit Company.[62]

The cold-storage warehouse was used for the preservation of fruits and vegetables in ways that varied with each product. Apples, which kept better under refrigeration than other fruits, were stored most extensively. They could be placed in cold storage in the fall or winter and not withdrawn until the following spring or summer.[63] At the peak of the 1915

[59] Hedden, *op.cit.*, 42-44; Sherman, *Merchandising Fruits and Vegetables*, 173.

[60] White, *loc.cit.*, 413-414.

[61] U. S. Treasury Department, Bureau of Statistics, *Statistical Abstract of the United States, 1900* (Washington, 1901), 194; U. S. Department of Commerce, Bureau of Foreign and Domestic Commerce, *Statistical Abstract of the United States, 1917* (Washington, 1918), 456; W. G. Sickel, "Refrigeration on Ocean Steamships," *Premier Congrès International du Froid*, III (Paris [1908?]), 757-758, 764-765. American fruits were exported under refrigeration in the middle of the nineteenth century, when it was the practice to ship them along with cargoes of natural ice bound for tropic ports. These early fruit exports, however, never attained large proportions. "Ice: and the Ice Trade," *Hunt's Merchants' Magazine*, XXXIII (1855), 176; L. Weatherell, "The Ice Trade," U. S. Department of Agriculture, *Report of the Commissioner, 1863* (Washington, 1863), 440-441; Taylor, *loc.cit.*, 578.

[62] Sickel, *loc.cit.*, 765-766.

[63] Letter from W. G. Campbell to H. W. Wiley, May 4, 1910, in "Report of Committee and Hearings Held before the Senate Committee

storage period over 4,200,000 barrels were being held in re-frigerated warehouses in the United States.[64] Pears were an important storage fruit, but only the hardier varieties could be kept as long as apples.[65] Oranges were not held for long periods, for they did not keep well and could be obtained in city markets direct from the groves at all seasons of the year.[66] Quick-ripening fruits—peaches, summer pears, grapes, and small fruits—were not adapted to long-term storage. Peaches could be kept one or two months at the most, delicate varieties of pears a little longer, grapes under certain conditions three or four months,[67] but most small fruits only a few days.[68] Vegetables were not preserved in cold storage as extensively as fruits. Some, such as onions and cauliflower, were kept for six to nine months. Potatoes usually were not stored, although exceptions to this rule were northern-grown seed potatoes held for second-crop planting in the South. Celery was some-times stored from one to three months, but most of the delicate vegetables, if warehoused at all, were kept only a few days.[69]

Improvements in refrigerated warehousing were to an im-

on Manufactures Relative to Foods Held in Cold Storage," *Senate Report*, 61 Cong., 3 sess., no. 1272, p. 172.

[64] U. S. Department of Agriculture, Bureau of Agricultural Economics, *Cold-Storage Holdings, Year Ended December 31, 1933 . . .* (U. S. Department of Agriculture, *Statistical Bulletin No. 48*) (Washington, 1934), 8.

[65] G. H. Powell and S. H. Fulton, *Cold Storage, with Special Reference to the Pear and Peach* (U. S. Department of Agriculture, Bureau of Plant Industry, *Bulletin No. 40*) (Washington, 1903), 11. Letter from Campbell to Wiley, May 4, 1910, *loc.cit.*, 172.

[66] McKay et al., *loc.cit.*, 673; Ramsey, *op.cit.*, 27-28.

[67] Letter from Campbell to Wiley, May 4, 1910, *loc.cit.*, 172.

[68] S. H. Fulton, *The Cold Storage of Small Fruits* (U. S. Department of Agriculture, Bureau of Plant Industry, *Bulletin No. 108*) (Washington, 1907), 7-8.

[69] Letter from Campbell to Wiley, May 4, 1910, *loc.cit.*, 172; W. Stuart, *Potato Storage and Storage Houses* (U. S. Department of Agriculture, *Farmers' Bulletin 847*) (Washington, 1917), 21; P. M. Kiely, "Cold Storage for Fruits, Etc.," *Ice and Refrigeration*, XX (1901), 37; M. B. C., "Cold Storage," *Garden and Forest*, VII (1894), 352. For a list indicating the wide variety of products placed in cold storage see statement of Charles H. Utley in "Report of Committee and Hearings Held before the Senate Committee on Manufactures Relative to Foods Held in Cold Storage," *loc.cit.*, 136-137.

portant extent the work of the federal Department of Agriculture. Practical warehousemen developed better methods out of their experience, to be sure, but the assistance of the Department was essential, for it had greater scientific resources and was in a superior position to study the steps, so necessary to proper preservation, that had to be taken long before the product reached the warehouse.

The cold-storage investigations of the Department began in 1901 and were continued along with the closely related studies of preservation during transportation. Apples were the subject of the first inquiries, which were headed by Powell. Soon these were extended to pears, peaches, grapes, small fruits, and eventually, to celery. They showed clearly that there was much more than mere low temperature to successful storage. After demonstrating that such seemingly unrelated factors as soil and climate affected keeping qualities, the investigations emphasized the importance of picking at the correct time and storing at the optimum temperature, the necessity for careful handling and prompt refrigeration, and the value of proper wrapping and packaging.[70]

A new technique in the storage of fruits and vegetables—freezing preservation—made its appearance about 1905 in the eastern part of the United States. It was used for holding small fruits longer than was possible at temperatures above freezing so that they could be employed by manufacturers of jellies, jams, ice cream, and pies when fresh fruits were out of season. Frozen fruits were more satisfactory for these purposes than those canned. The freezing process was simple:

[70] Stubenrauch, *loc.cit.*, 187-188; G. H. Powell and S. H. Fulton, *The Apple in Cold Storage* (U. S. Department of Agriculture, Bureau of Plant Industry, *Bulletin No. 48*) (Washington, 1903), 63-64; H. J. Ramsey et al., *The Handling and Storage of Apples in the Pacific Northwest* (U. S. Department of Agriculture, *Bulletin No. 587*) (Washington, 1917), 31; Powell and Fulton, *Cold Storage, with Special Reference to the Pear and Peach*, 21-22, 24-26; Fulton, *op.cit.*, 22-23; A. V. Stubenrauch and C. W. Mann, *Factors Governing the Successful Storage of California Table Grapes* (U. S. Department of Agriculture, *Bulletin No. 35*) (Washington, 1913), 30-31; H. C. Thompson, *Celery Storage Experiments* (U. S. Department of Agriculture, *Bulletin No. 579*) (Washington, 1917), 24-26.

fruits merely were packed with sugar in large barrels, which then were placed in rooms maintained at low temperatures. About 1910 cold packing, as freezing came to be known to the trade, was extended to the Pacific Northwest, but its progress was slow until after the World War.[71] The freezing of vegetables had been tried by 1917, but was in a stage of development even more experimental.[72]

The most important consequence of the cold-storage warehouse was to lengthen the time during which fruits and vegetables could be sold. Extended marketing periods not only made possible an increase in total demand, but also tended to prevent the demoralizing low prices that were inevitable when an entire crop had to be disposed of within, perhaps, a few days or weeks. By increasing demand and stabilizing prices, cold storage served as an incentive to growers. The production of pears, for example, increased from a little over three million bushels in 1890 to well over fourteen million in 1919. Part of this growth, of course, was due to the effects of the refrigerator car, of the canning industry, and of the nation's enlarging population, but cold storage was an important contributing factor.[73] The production of apples, the fruit stored in

[71] D. K. Tressler and C. P. Evers, *The Freezing Preservation of Foods*, 2nd edn. (New York, 1947), 276, 383; H. Carlton, *The Frozen Food Industry* (Knoxville, 1941), 3-4; H. C. Diehl et al., *The Frozen-Pack Method of Preserving Berries in the Pacific Northwest* (U. S. Department of Agriculture, *Technical Bulletin No. 148*) (Washington, 1930), 1; D. M. Taylor, "Cold Packing of Fruits in the Pacific Northwest," *Ice and Refrigeration*, LXXXI (1931), 337.

[72] *Ice and Refrigeration*, LII (1917), 272; LIII (1917), 141; F. W. Knowles, "How Foods Are Frozen in the Northwest," *Food Industries*, XII (April 1940), 54-55.

[73] G. H. Powell, "Relation of Cold Storage to Commercial Orcharding," *Ice and Refrigeration*, XXV (1903), 240-241; Powell and Fulton, *The Apple in Cold Storage*, 9-10; Powell and Fulton, *Cold Storage, with Special Reference to the Pear and Peach*, 11; H. C. Wallace, "Storage and Distribution of Perishable Foods," *Ice and Refrigeration*, LXII (1922), 479; J. H. Frederick, *Public Warehousing* (New York, 1940), 114-115; U. S. Department of Commerce and Labor, Bureau of Statistics, *Statistical Abstract of the United States, 1911* (Washington, 1912), 158; U. S. Department of Commerce, Bureau of Foreign and Domestic Commerce, *Statistical Abstract of the United States, 1921* (Washington, 1922), 164.

greatest quantity, also increased, though there was a downward trend after 1909, partly because they were displaced to some extent by other fruits that had been made available by refrigerated transportation and storage.[74]

Among other results was greater production of choicer, more delicate varieties of fruit. Before cold storage, the hardier varieties of apples such as the Roxbury and Ben Davis were dominant in spite of their relatively poor quality, for they kept fairly well without refrigeration. But with the coming of the cold-storage warehouse, the more delicate and better quality types such as the McIntosh and Delicious grew in favor.[75]

The dairy industry was one of man's activities most dependent on refrigeration. The distribution of fresh milk required low temperatures to check the multiplication of bacteria, which proceeded most rapidly when the milk was between seventy and one hundred degrees Fahrenheit.[76] So important were low temperatures in retarding spoilage that many of the larger cities prohibited the entry of milk that was warmer than sixty degrees.[77]

On the farm the cooling began as soon as possible. The most satisfactory method was to allow the milk to flow over cold metal surfaces, but also it could be chilled by placing it in cans set in tanks of cold water. The refrigerant used most commonly with each of these methods was well water, for refrigerating machines were too expensive for the ordinary

[74] H. P. Gould and F. Andrews, *Apples: Production Estimates and Important Commercial Districts and Varieties* (U. S. Department of Agriculture, *Bulletin No. 485*) (Washington, 1917), 45-47; E. Rauchenstein, *Economic Aspects of the Apple Industry* (University of California Agricultural Experiment Station, *Bulletin 445*) (Berkeley, 1927), 5, 11-13.

[75] E. L. Overholser, "History of Fruit Storage and Refrigeration in the United States," *Better Fruit*, xxix (Aug. 1934), 8; Taylor, "The Influence of Refrigeration on the Fruit Industry," *loc.cit.*, 572; J. C. Folger and S. M. Thompson, *The Commercial Apple Industry of North America* (New York, 1921), 334.

[76] J. T. Bowen, *The Application of Refrigeration to the Handling of Milk* (U. S. Department of Agriculture, *Bulletin No. 98*) (Washington, 1914), 11.

[77] *1915 Ice and Refrigeration Blue Book*, 634.

farmer,[78] and ice was not always available. On southern farms ice was not employed to any great extent even after the appearance of ice factories, since to haul it from plant to farm was a difficult and expensive task that in many cases was entirely impractical. In some parts of New England natural ice was used extensively, but in most of the sections where it could be had for the cutting few bothered to obtain it. In an effort to correct this situation, the United States Department of Agriculture pointed out repeatedly the advantages northern dairy farmers could earn by harvesting an ice crop each winter.[79]

From farm to city milk was protected against temperature rises by procedures that varied greatly both in form and effectiveness. During the wagon haul to the railroad, which brought to large cities the bulk of their supply, milk was kept cool by insulated jackets fitted around the cans. Often even this simple device was not employed. At the railroad the cans sometimes merely were placed on a platform or baggage truck to await the train. More satisfactory care was provided by the receiving stations, operated for the most part by city dealers, that were constructed at many shipping points. These differed greatly in equipment, some even having bottling facilities, but all made provision—refrigerating machinery in the larger stations—for keeping the milk cool while awaiting shipment. During rail transport, as elsewhere, methods varied. Sometimes the cans were hauled in baggage cars, which served well enough if the milk were already cool, the cans jacketed, and the run not over four hours. A better technique involved

---

[78] Bowen, *op.cit.*, 65-66.
[79] O. Erf, "Refrigeration of Dairy Products," L. H. Bailey, ed., *Cyclopedia of American Agriculture*, 2nd edn., III (New York, 1910), 232; J. T. Bowen and G. M. Lambert, *Ice Houses and the Use of Ice on the Dairy Farm* (U. S. Department of Agriculture, *Farmers' Bulletin 623*) (Washington, 1915), 1, 23-24; L. C. Corbett, *Ice Houses* (U. S. Department of Agriculture, *Farmers' Bulletin 475*) (Washington, 1911), passim; J. A. Gamble, *Cooling Milk and Cream on the Farm* (U. S. Department of Agriculture, *Farmers' Bulletin 976*) (Washington, 1918), 2; U. S. Department of Agriculture, Bureau of Animal Husbandry, *Ice: Do You Have It on Your Farm?* (Washington, 1918).

the use of refrigerator cars, which were of two kinds. In one, either an old standard refrigerator with bunkers removed or a baggage-type car with little or no insulation, the cans were placed on the floor and surrounded by crushed ice. In the other, which was fitted with ice bunkers or brine tanks, the cans were refrigerated by circulation of air. This was by far the better equipment for long hauls.[80]

In the city the milk was received at bottling plants, where it was cooled to temperatures of from forty-five to fifty degrees. When it was pasteurized, care was taken to chill it rapidly after the heating, for otherwise the process in most cases would have been worse than useless. The milk, when sufficiently cold, was bottled and placed in a storage room to await delivery.[81] Ice continued to be employed in small plants, but large ones used refrigerating machinery, which made possible more rapid cooling, more accurate temperature control, and in addition, greater cleanliness and convenience.[82]

Refrigeration enabled dairy farmers to sell their fluid milk in distant cities. Limits to the distance their product could be shipped were set by its extreme perishability, but early in the twentieth century cities drew their supplies from farms as far as four hundred miles away.[83]

The butter industry in all its phases was dependent on low temperatures. In the creamery, which was gradually displacing the farm in butter production, refrigeration was employed to chill the cream after pasteurization, to maintain a constant temperature in the vats where it was ripened, to provide the exact temperatures required in churning, to cool the water

[80] Bowen, *op.cit.*, 69-74; M. J. Rosenau, *The Milk Question* (Boston, 1912), 287; Federal Trade Commission, *Report on Private Car Lines*, 32.

[81] Erf, *loc.cit.*, 245; J. A. Ruddick, "The Refrigeration of Dairy Products," *Ice and Refrigeration*, xxxviii (1910), 97; Bowen, *op.cit.*, 74.

[82] H. N. Parker, *City Milk Supply* (New York, 1917), 338; Bowen, *op.cit.*, 74-78; F. Fernald, "Ice-Making and Machine Refrigeration," *The Popular Science Monthly*, xxxix (1891), 29.

[83] G. K. Holmes, *Systems of Marketing Farm Products and Demand for Such Products at Trade Centers* (U. S. Department of Agriculture, *Report No. 98*) (Washington, 1913), 123.

used in working the butter, and to cool the storage space where the finished product was held pending shipment. Mechanical refrigeration was introduced in the nineties and by 1915 was in general use in larger creameries, where in most parts of the country it proved more economical than ice and even in the extreme North no more expensive. Small plants, however, continued to rely on ice. An advantage of the machine as important as its economy was the accurate control of temperature it afforded, a control that made possible a superior product.[84]

Shortly after its manufacture butter was hauled in refrigerator cars to city markets, where it was either directed immediately into retail channels or held by wholesalers in cold storage for future sale. The amount stored assumed large proportions. In 1915, the first year for which federal statistics on holdings were available, over a hundred million pounds were in storage on the first of September, the peak of the season that year.[85] The establishment of mechanically refrigerated warehouses was a great boon to butter-storers, for it enabled them to provide the exceptionally low temperatures needed—ten degrees below zero indicated a Department of Agriculture investigation[86]—if their product were to be held in good condition for several months.

Refrigeration was a powerful influence in determining the character of the butter industry in the United States. The refrigerator car continued to make possible production for eastern markets in areas far to the West, where costs were so low that dairy farmers could pay transportation charges and still undersell their eastern competitors. In the twentieth century the butter output of New England and the middle Atlantic

[84] Bowen, *op.cit.*, 78-82; *Ice and Refrigeration*, ix (1895), 105; xi (1896), 315; xxxviii (1910), 178; *1915 Ice and Refrigeration Blue Book*, 49; Erf, *loc.cit.*, 245.

[85] U. S. Department of Agriculture, Bureau of Agricultural Economics, *op.cit.*, 10.

[86] C. E. Gray, "Investigations in the Manufacture and Storage of Butter. I.—The Keeping Qualities of Butter Made Under Different Conditions and Stored at Different Temperatures," U. S. Department of Agriculture, Bureau of Animal Industry, *Bulletin No. 84* (Washington, 1906), 22.

states declined so sharply that it became a comparatively negligible factor in the country's supply, while the production of other sections increased. In 1919 the east north central states produced over twenty-five per cent of the total output, while the west north central states were responsible for almost thirty-three per cent.[87] The cold-storage warehouse, by enabling the market to absorb seasonal surpluses, tended to stabilize prices and to encourage production.[88]

The cheese industry also depended on low temperatures, but its requirements differed greatly from those of butter. Cheese, once made, went through a curing or ripening process in which temperature was a factor that determined the quality of the product and the speed at which curing progressed. The optimum temperature varied with the type of cheese, the requirements of the hard varieties being less rigorous than those of the soft, though cheddar, a hard type, suffered definite loss of quality if exposed to temperatures over sixty degrees Fahrenheit. When curing was attempted at the factory, as was sometimes the case, the necessary cool environment was provided either by ice, which was favored in smaller establishments, or by refrigerating machines. In most instances curing was not carried on at the factory, but instead at large city cold-storage warehouses. For protection during shipment to the city, reliance was placed on the refrigerator car.[89]

Refrigeration was an important factor in the development of the cheese industry. By providing a means of controlling the curing process, it made possible not only uniform high quality, but also more efficient and profitable marketing operations. Moreover, by affording the dairymen of Wisconsin,

[87] O. F. Hunziker, *The Butter Industry* (LaGrange, Illinois, 1920), 28; *Statistical Abstract of the United States, 1921*, 158-159.

[88] Ruddick, *loc.cit.*, 97; E. Wiest, "The Butter Industry in the United States" (Faculty of Political Science of Columbia University, eds., *Studies in History, Economics and Public Law*, LXIX, Number 2) (New York, 1916), 179.

[89] *1915 Ice and Refrigeration Blue Book*, 49; J. A. Ruddick, "The Making of Cheddar Cheese," L. H. Bailey, ed., *Cyclopedia of American Agriculture*, 2nd edn., III (New York, 1910), 213-214; C. Thom, "Soft Cheese in America," *Cyclopedia of American Agriculture*, 2nd edn., III, 220-226; Holmes, *op.cit.*, 38-42.

Michigan, and Minnesota a means of protecting their product during transportation, it permitted them to sell in eastern markets, to exploit fully their advantages in cheese production.

The segment of the dairy industry most obviously dependent on refrigeration was ice-cream manufacturing. Factories designed to supply ice cream on a wholesale basis, which dated back at least to the eighteen-fifties, first employed as a refrigerant mixtures of ice and salt. About 1890 mechanical refrigeration was introduced, but it made slow progress until about 1910, when equipment adapted to the needs of the smaller manufacturers was available. In 1909 only a little more than one hundred factories had refrigerating machinery, but four years later almost six hundred establishments had installed it. It gave an impetus to the industry by eliminating the unreliability, the inconvenience, and the dampness that accompanied the use of ice and by making possible the manufacture of a better and more uniform product.[90]

Another food industry in which temperature control was essential was the distribution of eggs, for the speed at which the incubation of fertile and the deterioration of infertile eggs proceeded varied directly with the temperature.[91]

In spite of the importance of low temperature in handling eggs, practically no refrigeration was used during the early stages of their marketing. Farmers generally treated them as though they were imperishable, keeping them in the pantry or kitchen cupboard until a sufficient number had been gathered to make a trip to town worthwhile. Even if the importance of keeping eggs cool were recognized, refrigeration was almost never available, and the cellar used as a substitute was likely to be damp, a condition favorable to the development of mold. The farmer sold his eggs to the village storekeeper, from whom he received payment in goods. The mer-

[90] F. B. Fulmer, "Manufacture of Ice Cream," *Ice and Refrigeration*, LXVIII (1925), 429; *1915 Ice and Refrigeration Blue Book*, 636; *1928 Ice and Refrigeration Blue Book and Buyers' Guide* . . . (New York, 1928), 76-77; *Ice and Refrigeration*, XXXII (1907), 281.

[91] M. E. Pennington, "Relation of Cold Storage to the Food Supply and the Consumer," American Academy of Political and Social Science, *Annals*, XLVIII (1913), 157-158.

chant, who engaged in the egg business only because he felt it necessary in order to keep the farm trade, held eggs until he had enough to send to a shipper. During the time they were in his hands—a period that might be two days or two weeks—they were not handled much more carefully than on the farm.[92]

After the eggs reached the shipper they usually received better attention.[93] When packed, they were forwarded in refrigerator cars to the cities, where they entered retail distribution channels[94] immediately or were stored in great mechanically refrigerated warehouses, where methods and equipment improved with the years. The quantities stored assumed large proportions. On August 1, 1917, for example, almost seven million cases, thirty dozen to the case, were reported in storage.[95]

The United States Department of Agriculture was actively interested in improving egg-handling methods. In 1913 it began work with a demonstration car, a refrigerated egg-packing establishment on wheels that was sent on long trips through the Midwest, where its staff showed both farmers and dealers how proper care and grading would yield greater profits. Since fertile eggs deteriorated more rapidly than infertile, farmers were encouraged to remove roosters from their flocks after the hatching season. Farmers were shown the importance of gathering eggs promptly and of keeping them in the coolest place available. Shippers were urged to chill them to below fifty degrees before loading in refrigerator cars.[96]

Not all could be shipped in the shell to city markets. Considerable numbers of cracked, small, dirty, and other second-grade eggs reached major concentrating centers in the producing sections in wholesome condition, but unable to with-

[92] Holmes, *op.cit.*, 48.

[93] *ibid.*, 48.

[94] The care given eggs in many retail stores was no better than they received on the farm.

[95] U. S. Department of Agriculture, Bureau of Agricultural Economics, *op.cit.*, 14.

[96] M. E. Pennington, H. C. Pierce, and H. L. Shrader, "The Egg and Poultry Demonstration Car Work in Reducing Our $50,000,000 Waste in Eggs," U. S. Department of Agriculture, *Yearbook . . . , 1914* (Washington, 1915), 364-365, 369-376, 378.

stand any longer the rigors of marketing. The practice of salvaging these by freezing them began about 1900. The whites and yolks were removed from the shell, frozen either separately or together, and sold to bakers and confectioners. The egg-freezing industry, however, failed to grow rapidly, for its product was inferior, principally because of a general failure to take necessary sanitation measures and the too-common practice of freezing eggs that were unfit.[97]

Proper techniques for commercial freezing were developed by the Department of Agriculture in an effort to reform the industry. This work was directed by the Chief of the Food Research Laboratory of the Bureau of Chemistry, Mary E. Pennington, a capable scientist who along with G. Harold Powell deserved much of the credit for the great improvements made in applying refrigeration to food handling during the first two decades of the twentieth century.[98] Working in co-operation with commercial establishments in the Midwest, government scientists learned that a quality product depended on careful grading of the eggs and on the most scrupulous cleanliness. To assure the attainment of these conditions, they devised elaborate equipment and outlined in detail procedures to be followed.[99] The egg-freezing industry did not become large in this period, but a firm foundation was laid for its future expansion.

Both the geography and the economics of the egg industry felt the impact of refrigeration. The refrigerator car enabled the states of the Corn Belt to develop their egg-producing

[97] M. E. Pennington, M. K. Jenkins, and W. A. Stocking, assisted by S. H. Ross et al., *A Study of the Preparation of Frozen and Dried Eggs in the Producing Section* (U. S. Department of Agriculture, *Bulletin No. 224*) (Washington, 1916), 4-5; J. H. Radabaugh, "Economic Aspects of the Frozen-Egg Industry in the United States," *Ice and Refrigeration*, XCVII (1939), 265.

[98] For an interesting sketch see B. Heggie, "Ice Woman," *The New Yorker*, XVII (Sept. 6, 1941), 23-26, 29-30.

[99] M. E. Pennington, *Practical Suggestions for the Preparation of Frozen and Dried Eggs* (U. S. Department of Agriculture, Bureau of Chemistry, *Circular No. 98*) (Washington, 1912), 5-6, 9-12; Pennington, Jenkins, and Stocking, *op.cit.*, 20-21; M. K. Jenkins, *The Installation and Equipment of an Egg-Breaking Plant* (U. S. Department of Agriculture, *Bulletin No. 663*) (Washington, 1918), 24-25.

potential to the fullest, while the cold-storage warehouse helped adjust demand to seasonal fluctuations in supply.[100]

Few foods required more refrigeration than dressed poultry. As soon as a bird was killed and picked, its animal heat had to be removed. Methods of doing this differed. In some cases the bird was chilled by immersion in cold running water to which ice sometimes was added. This was satisfactory when only a few days elapsed between killing and eating, but a better method, particularly when poultry was killed for distant markets, was to chill by circulation of air in mechanically refrigerated rooms. When a bird was thoroughly cold, it was ready for shipment. At first it was customary to pack poultry with ice in barrels or boxes. This was adequate if the distance to market were not great, but it eventually was recognized that dressed poultry kept better during long hauls if packed dry in boxes and shipped in refrigerator cars, where low temperatures were maintained by circulating air.[101] If poultry were to be put in cold storage, as was the case with substantial quantities, it was necessary to freeze it solid. When this was done properly, a bird could be stored for twelve months and still be in better condition than one kept only five and a half days at the average temperature of the domestic icebox.[102] Freezing, which involved merely placing the bird in a very cold room, was accomplished either at a producing-area packing plant or at a city warehouse.[103]

[100] T. W. Heitz, *The Cold Storage of Eggs and Poultry* (U. S. Department of Agriculture, *Circular No. 7* ` (Washington, 1929), 2; M. A. Jull et al., "The Poultry Industry," U. S. Department of Agriculture, *Yearbook, 1924* (Washington, 1925), 386.

[101] M. E. Pennington, "The Handling of Dressed Poultry a Thousand Miles from the Market," U. S. Department of Agriculture, *Yearbook . . . , 1912* (Washington, 1913), 288-292; M. E. Pennington, *Studies of Poultry from the Farm to the Consumer* (U. S. Department of Agriculture, Bureau of Chemistry, *Circular No. 64*) (Washington, 1910), 14-24; D. J. Lambert, "Preparing and Marketing Poultry Products, and the Care of Eggs," L. H. Bailey, ed., *Cyclopedia of American Agriculture*, 2nd edn., iii (New York, 1910), 545-546.

[102] Pennington, "Relation of Cold Storage to the Food Supply and the Consumer," *loc.cit.*, 157.

[103] Pennington, *Studies of Poultry from the Farm to the Consumer*, 29-33.

Improved methods of handling dressed poultry were the objectives of Department of Agriculture studies directed by Miss Pennington during the decade before 1917. The behavior of poultry under refrigeration, the investigations showed, depended not on low temperatures alone, but on a complex of factors that included techniques used in killing and picking the fowl and in applying refrigeration. One result of this work was the development of a method of killing that assured the thorough bleeding necessary for satisfactory refrigerated preservation, while another was the discovery that poultry decomposed more rapidly when wholly or partially eviscerated than when left undrawn.[104]

The dressed-poultry industry was influenced by refrigeration in much the same manner as other perishable-food enterprises.[105] A special advantage of cold storage to the trade was that it enabled growers to supply the demand for particular types of poultry at any season of the year. Broiler chickens, for example, could be killed in the summer and fall, frozen, stored, and sold during the winter and spring, when few live birds of proper size were available.[106]

The fish trade, one of the first large users of refrigeration, employed it on an increasing scale after 1890. As soon as fish were taken from the water, refrigeration was brought into play, for crushed ice preserved catches until they could be landed. Not until the nineteen-twenties were refrigerating machines installed on American vessels to assist in this work.[107]

[104] *ibid.*, 39-42; M. E. Pennington and H. M. P. Betts, *How to Kill and Bleed Market Poultry* (U. S. Department of Agriculture, Bureau of Chemistry, *Circular No. 61*) (Washington, 1910), 6, 15; M. E. Pennington, with E. Witmer and H. C. Pierce, *The Comparative Rate of Decomposition in Drawn and Undrawn Market Poultry* (U. S. Department of Agriculture, Bureau of Chemistry, *Circular No. 70*) (Washington, 1911), 17.

[105] *Twelfth Census, 1900*, v, ccxxvi; Jull et al., *loc.cit.*, 386.

[106] Heitz, *op.cit.*, 49-50.

[107] R. H. Fiedler, "The Factory Ship—Its Significance to Our World Trade and Commerce," The American Fisheries Society, *Transactions, Sixty-sixth Annual Meeting* . . . (Washington, 1937), 430. In 1898 two vessels were fitted with refrigerating machines at Gloucester and sent to Newfoundland to freeze herring and squid for use later as bait.

If fish were to be sold within about a month after being caught, they were repacked in crushed ice on landing to protect them from deterioration. Properly packed thus in barrels, casks, boxes, or bins, they could be shipped by express or freight to all parts of the country.[108] But if fish were to be kept longer than a few weeks, it was necessary to freeze them in order to retard decomposition.

Freezing was first accomplished through the agency of ice-and-salt mixtures, but in 1892 a refrigerating machine was adapted to the task at Sandusky, Ohio. Before long mechanical refrigeration had supplanted ice and salt in other lake ports and along both the Atlantic and Pacific coasts. Methods varied, but in the best of these establishments fish were "sharp" frozen by placing them in pans on shelves constructed of pipes in which cold brine was circulated or liquid refrigerant allowed to evaporate. It was considered best to freeze fish whole, but many that were split and dressed were so preserved. When thoroughly frozen, they were sprinkled with cold water, a process that facilitated their removal from the pans and covered them with a thick coating of ice that served as protection from desiccation during storage.[109]

Large quantities of many kinds of fish came to be preserved by freezing. At the peak of the 1917 storage season over seventy million pounds of all types were reported on hand in the nation's warehouses.[110] Nevertheless, the industry had not developed fully, mainly because defects in techniques resulted in a product of low quality.

Refrigeration was a great boon to American fishermen. Without it they would have been confined to supplying fresh fish to a narrow strip along the coast and to whatever market they could find for their product in its less palatable dried,

---

C. H. Stevenson, "The Preservation of Fishery Products for Food," U. S. Fish Commission, *Bulletin*, xviii (1898), 388.

[108] H. E. Williams, *Protection of Food Products from Injurious Temperatures* (U. S. Department of Agriculture, *Farmers' Bulletin No. 125*) (Washington, 1901), 12.

[109] Stevenson, *loc.cit.*, 373, 377-382.

[110] U. S. Department of Agriculture, Bureau of Agricultural Economics, *op.cit.*, 41.

salted, smoked, or canned forms. Under its influence an extensive industry grew not only along the New England coast, where cities were close at hand, but along the shores of the South Atlantic, Gulf, and Pacific states, hundreds and thousands of miles from the major centers of population. The fisheries of the Pacific Northwest were able to extend their markets to the eastern United States and beyond the Atlantic to European cities as well.[111] Without refrigeration fisheries would have been unable to adjust the supply of their product to fluctuations in demand, which sectarian customs made more intense and frequent than in the case of other perishable commodities.[112]

The diet of American city-dwellers benefited by the revolution which refrigeration worked in the production and distribution of food. After 1890 the nation's growing urban populations were enabled to draw upon the resources of the entire country for most of their fresh foods. The period during which they could obtain seasonal products was extended—with some perishables for a few weeks or months—with others until the yearly production cycle began once more. Accompanying this mitigation of the evils of glut and scarcity was a trend toward greater uniformity of prices throughout the year, a tendency that enabled more family budgets to include fresh foods for a greater part of the time. Under these conditions city diet became more nutritious and appetizing and city life more wholesome and alluring.

There were limits to what refrigeration accomplished. Large numbers of urbanites still ate insufficient quantities of fresh foods—partly because of poverty, partly because of ignorance,[113] and partly, in the case of the immigrant population, because of adherence to the food habits of the country

[111] Stevenson, *loc.cit.*, 358, 368, 383-384; "Growth of the Fish Refrigeration Industry on the Pacific Coast," *Pacific Fisherman*, xxvi (Aug. 1928), 12. Refrigeration was used not only in the fresh and frozen-fish trade, but also in the distribution of mild-cured products.

[112] H. F. Taylor, *Refrigeration of Fish* (U. S. Department of Commerce, Bureau of Fisheries, *Document No. 1016*) (Washington, 1927), 503-504.

[113] R. O. Cummings, *The American and His Food* (Chicago, 1940), 118-121, 230.

of their origin. Moreover, the inhabitants of smaller cities did not benefit as much from the refrigerator car and the cold-storage warehouse as did those of great metropolises, while residents of small rural towns were affected even less.[114] The farmer's diet had not yet been changed significantly by refrigeration, but it should be remembered that nutrition was not so critical a problem in rural as in metropolitan areas. Although major deficiencies prevailed in parts of the South, farm people tended to fare somewhat better than their city cousins, for they consumed larger amounts of meat and livestock products.

[114] W. R. Woolrich, "Railroad Service and Frozen Foods," *Refrigerating Engineering*, XXXVIII (1939), 277.

CHAPTER XI

THE DIVERSE APPLICATIONS OF
REFRIGERATION, 1890-1918

EVEN before the advent of the machine the techniques of refrigeration were not restricted to problems of food preservation. As early as the eighteen-fifties manufacturers of lard oil used ice in processing their product during the warm summer months.[1] The Atlantic cable owed much to refrigeration, for the coils stowed on the *Great Eastern* were packed in ice to prevent heat from melting or softening the insulation.[2] But the application of low temperatures to a wide variety of tasks waited upon the machine, which surpassed ice in dependability, flexibility, and convenience.

A factor which helped to make possible greater diversity was the development of air conditioning, the art of purifying air and controlling its temperature, humidity, and circulation within an enclosure.[3] Refrigeration could play an important role in accomplishing these objectives; it could be used not only for cooling, but for dehumidification as well, although this employment sometimes was unnecessary if an abundant supply of cold water were available. To achieve dehumidification, advantage was taken of the fact that the amount of water vapor a given volume of air could hold was directly proportional to its temperature. Air was passed over cold

---

[1] F. Tudor and T. T. Sawyer, "Ice Trade," Boston Board of Trade, *Third Annual Report* (Boston, 1857), 81; J. C. Schooley, *A Process of Obtaining a Dry Cold Current of Air From Ice* (Cincinnati, 1855), 16. For a discussion of the importance of lard oil see C. W. Towne and E. N. Wentworth, *Pigs, From Cave to Corn Belt* (Norman, 1950), 162-163.

[2] "Refrigeration," *American Artisan*, x n. s. (Feb. 16, 1870), 105.

[3] For definitions of air conditioning see J. A. Moyer and R. U. Fittz, *Air Conditioning*, 2nd edn. (New York, 1938), 1; W. H. Carrier and F. L. Busey, "Air-Conditioning Apparatus; Principles Governing its Application and Operation," American Society of Mechanical Engineers, *Transactions*, xxxiii (1911), 1055; J. C. Fistere, "An Introduction to Air Conditioning," *The Architectural Forum*, lvii (1932), 167-174.

surfaces or through a cold spray, and any moisture above what could be held at the low temperature was condensed. When this air was heated, or simply discharged into a room, it gained in capacity to absorb water and consequently decreased in relative humidity.[4]

Air conditioning began to make significant progress early in the twentieth century.[5] In 1902 Willis H. Carrier, an engineer with the Buffalo Forge Company, devised for a lithographing plant a system to dehumidify air and control its moisture content. Not satisfied with this, he invented in 1904 the central-station spray apparatus for conditioning air, and in 1906 he developed dew-point control, a method of regulating relative humidity by altering at the apparatus the temperature at which moisture in the air begins to condense. These two inventions were fundamental to the advance of the art. In 1907 Carrier air-conditioned a silk mill in Wayland, New York, with a system that still was in operation in 1950.[6] Another pioneer was Colonel Stewart W. Cramer, a Charlotte, North Carolina, mill-owner and operator, who first suggested

[4] For simplified explanations see O. W. Ott, *Essential Features of Comfort Air Conditioning in Non-Technical Language*, 2nd edn. (Los Angeles, 1934), 25-26; W. H. Stangle, *An Air-Conditioning Primer; the A-B-C of Air Conditioning* (New York, 1940), 148; L. K. Wright, *The Next Great Industry; Opportunities in Refrigeration and Air Conditioning* (New York, 1939), 176-177, 182; Carrier Corporation, *You are an Engine Air-Cooled* (n. p., n. d.), 5-6. Dehumidification could also be accomplished through the use of absorbent substances. Stangle, *op.cit.*, 148-150; J. R. Allen, J. H. Walker, and J. W. James, *Heating and Air Conditioning*, 6th edn. (New York, 1946), 494-497. Schooley, the Cincinnati experimenter with refrigerators and cold-storage warehouses, understood clearly the relation between the temperature of the air and its water-vapor content. Schooley, *op.cit.*, 4-5.

[5] As early as 1891 Eastman was using refrigeration to dehydrate air for drying out films and emulsions. J. E. Starr, "Refrigeration Twenty-Five Years Ago," *Ice and Refrigeration*, LI (1916), 144.

[6] "Weathermakers: Carrier Corp.," *Fortune*, XVII (April 1938), 91; W. H. Carrier, "The Control of Humidity and Temperature as Applied to Manufacturing Processes and Human Comfort," *Heating, Piping and Air Conditioning*, I (1929), 537. The writer wishes to acknowledge the very helpful information on early air-conditioning installations that he has received from Margaret Ingels, Engineering Editor of the Carrier Corporation, Syracuse, N.Y., who is working on a biography of Dr. Carrier.

the term "air conditioning."[7] A landmark was passed in 1911, when Carrier formulated the theoretical basis of the new technique in a paper he presented at the annual meeting of the American Society of Mechanical Engineers.[8]

Industry was the first principal market for air conditioning.[9] The curing of tobacco, the handling of deliquescent salts in pharmaceutical plants, the fabrication of precision instruments—all benefited from its adoption. The infant motion-picture industry employed air-conditioning machinery to control studio temperatures and to dry developed films, and the Ford Motor Company installed it in its Woodward Avenue plant. Even mining was affected, for refrigerating machines were used to chill water for systems that supplied cool, dry air to men working the lower levels of very deep mines, where increased human efficiency justified the investment.[10] Firms engaged in textile manufacture found the new technique particularly useful, for it permitted the production without seasonal irregularities of goods that were uniform in quality. It should be remembered, however, that textile-plant systems at this time usually were designed only to add moisture to the air and did not include refrigerating machinery.[11]

Air conditioning primarily for human comfort lagged far behind the achievements in the industrial field. This was true

[7] A. E. Stacey, Jr., "Development of Heating, Ventilating and Air-Conditioning Equipment," American Management Association, *Lighting and Air Conditioning for the Modern Plant* (New York, 1940), 18; W. B. Henderson, "Air Conditioning as an Industry in the United States," *Ice and Refrigeration*, XCI (1936), 121.

[8] W. H. Carrier, "Rational Psychrometric Formulae; Their Relation to the Problems of Meteorology and of Air Conditioning," American Society of Mechanical Engineers, *Transactions*, XXXIII (1911), 1005-1039. A companion piece was Carrier and Busey, *loc.cit.* For the significance of these papers see "Weathermakers: Carrier Corp.," *loc.cit.*, 124, and R. E. Cherne, "Developments in Refrigeration as Applied to Air Conditioning," *Ice and Refrigeration*, CI (1941), 27.

[9] Cherne, *loc.cit.*, 29.

[10] Stacey, *loc.cit.*, 18-19; S. S. Vandervaart, "Commercial Uses of Refrigerating Machinery," *Ice and Refrigeration*, LI (1916), 181-183.

[11] Carrier, "Rational Psychrometric Formulae," *loc.cit.*, 1005; Henderson, *loc.cit.*, 121; P. L. Davidson and J. deB. Shepard, "Air Conditioning in Textile Mills," *Refrigerating Engineering*, LVIII (1950), 155.

despite the fact that mankind had long dreamed of making life indoors during warm weather more tolerable.[12] A number of Americans had been stirred by such dreams. Gorrie's device for the prevention of malaria involved air cooling,[13] while John C. Schooley claimed that his cold-storage system would "ventilate any public hall or assembly room during an external temperature of 95°, and produce a pure refreshing atmosphere free from any dampness flowing through the apartment at a temperature of 60° to 70°, if required, at a cost but trifling in comparison with the results obtained."[14] In 1867 Daniel E. Somes, a patent lawyer who had served a term in the House of Representatives, proposed that the Capitol be supplied with fresh air that had been cooled by forcing it over pipes filled with cold water.[15] A few Englishmen did more than dream. As early as 1836 the Houses of Parliament were provided with a rudimentary form of air conditioning,[16] and in 1887 the English Linde Company used mechanical refrigeration to cool a rajah's palace in India.[17]

Several air-cooling systems designed mainly for comfort were installed in the United States after 1890. In 1895 the library of a St. Louis home was cooled by a De La Vergne machine,[18] and by 1910 ice was being used for railroad coaches and Pullman cars.[19] Most activity was in the public-

[12] For illustrations of primitive ventilating devices see D. L. Fiske, "The Origins of Air Conditioning," *Refrigerating Engineering*, XXVII (1934), 122-126.

[13] For a similar device of 1903 see "Air-Cooling Apparatus," *Scientific American*, LXXXVIII (May 16, 1903), 372.

[14] Schooley, *op.cit.*, 17.

[15] D. E. Somes, *Mr. D. E. Somes' Plan for Ventilating, Cooling and Heating the Capitol, and for Purifying and Moistening the Air* ([Washington?] 1867).

[16] W. H. Carrier, R. E. Cherne, and W. A. Grant, *Modern Air Conditioning, Heating and Ventilating* (New York and Chicago, 1940), 2.

[17] Cherne, *loc.cit.*, 27.

[18] *Ice and Refrigeration*, IX (1895), 104. There were those who already envisioned extensive use of refrigeration to cool living quarters. See F. L. Oswald, "Summer Refrigeration," *The North American Review*, CXLV (1887), 261-266.

[19] E. F. McPike, "Transportation of Perishable Freight in America: Present Practice and Desiderata," IInd International Congress of Re-

building field. Refrigerated coils were relied upon in 1902 in the New York Stock Exchange,[20] and mechanical refrigeration was employed to make more comfortable the main rotunda at the Louisiana Purchase Exposition and the public rooms of Chicago's Auditorium Hotel.[21] Here and there a hospital room was equipped with an air-cooling system.[22]

That ordinary cold storage was not limited in its usefulness to food preservation was demonstrated clearly before the first World War. A case in point was its use to check the ravages of moths. Just as low temperatures inhibited the forces that caused food decay, they curbed the activity of the insects that destroyed fur and woollen garments. The first storage house to take commercial advantage of this fact may have been the Terminal Warehouse Company of New York, which about 1890 began to develop a fur-storage business. Soon refrigerated warehouses throughout the country were holding furs and valuable woollen goods during the summer months. The growth of this business was accompanied by experiments both in industry and in the Division of Entomology of the United States Department of Agriculture which indicated the best temperatures and proved that sudden changes were necessary if moths were actually to be killed.[23]

Storage at low temperatures was a boon to nurserymen and florists. It permitted them to lay up bulbs and to retard their natural period of germination; it enabled them to meet seasonal

---

frigeration, Vienna, 1910, *English Edition of Reports and Proceedings* (Vienna, 1911), 1015; *Ice and Refrigeration*, xxxiii (1907), 44.

[20] The New York Stock Exchange system was designed by Alfred R. Wolff, a consulting engineer of New York City. It operated successfully for twenty years. Letter, M. Ingels, Carrier Corporation, Syracuse, N.Y., to writer, May 22, 1952.

[21] *Ice and Refrigeration*, xxxiii (1907), 126-128.

[22] Vandervaart, *loc.cit.*, 184.

[23] W. C. Reid, "Rise and Development of Cold Storage of Furs and Fabrics," *Ice and Refrigeration*, li (1916), 172-173; L. O. Howard, "Some Temperature Effects on Household Insects," *Proceedings of the Eighth Annual Meeting of the Association of Economic Entomologists* (U. S. Department of Agriculture, Division of Entomology, *Bulletin No. 6—New Series*) (Washington, 1896), 13, 17; E. A. Back, *Clothes Moths and Their Control* (U. S. Department of Agriculture, *Farmers' Bulletin No. 1353*) (Washington, 1923), 18.

demands by delaying the blossoming of plants; it afforded a means of holding cut flowers.[24] Low-temperature storage was even brought into play to keep living tissue for surgical use and to preserve human bodies in morgues.[25]

In the metal trades mechanically produced cold was employed to assist in the tempering of cutlery, plowshares, and tools. The complete control it afforded at critical stages in the manufacturing process made not only for uniform quality but for economy as well.[26] But the most interesting use of refrigeration in the industry was to remove moisture from the air delivered to blast furnaces. For many years it was known that moisture removal would help, but no practical means of accomplishing this existed before the development of powerful refrigerating machinery. In 1894 James Gayley, manager of the Edgar Thomson Works of the Carnegie Steel Company, received his first patent for a method of condensing water vapor from the blast. The first successful demonstration of this technique took place in 1904 at the Isabella furnaces of the United States Steel Corporation in Etna, Pennsylvania; before the World War a number of other installations were made in American plants. The dry blast made possible greater iron production with smaller quantities of coke and limestone, but after about 1916 installations ceased. Despite its advantages, it was unprofitable commercially, for the initial investment was prohibitive, the power demand excessive, and the maintenance and replacement costs high.[27]

[24] Vandervaart, *loc.cit.*, 185; M. B. C., "Cold Storage," *Garden and Forest*, VII (1894), 353.

[25] Vandervaart, *loc.cit.*, 184-185. The Cook County, Illinois, Morgue early in the nineties became the first in the nation to be so equipped. The bodies were kept in cases refrigerated by brine pipes. *Ice and Refrigeration*, IV (1893), 13.

[26] Vandervaart, *loc.cit.*, 181.

[27] R. VD. Dunne, "Dry Blast and Production for War," *Refrigerating Engineering*, XLIV (1942), 19; L. E. Riddle, "Reminiscences of the First Application of Dry Blast," *Blast Furnace and Steel Plant*, XXVIII (1940), 464-470; J. H. Hart, "The Uses of Mechanical Refrigeration in Metallurgical Practice," *The Engineering Magazine*, XXXVI (1908-1909), 777; S. S. Van der Vaart, "Growth and Present Status of the Refrigerating Industry in the United States," *Premier Congrès International du Froid*, III (Paris [1908?]), 347; B. Walter,

Many other manufacturing industries enlisted the aid of refrigeration. Textile mills used it in the processes of mercerizing, bleaching, and dyeing. Oil refineries found it essential, as did the manufacturers of such diverse products as paper, drugs, soap, glue, shoe polish, perfume, celluloid, and photographic materials.[28] In the nineties Carl Linde used the refrigerating machine to make possible the commercial production of liquid air.[29]

The food-processing industries owed much to the refrigerating machine. Sugar mills were equipped with units that cooled storage rooms and assisted in the refining process.[30] Confectioneries, particularly those which produced chocolate candy, required its services, as did bakeries, yeast manufactories, and establishments that processed tea.[31] In the dairy field it was essential not only for milk, butter, cheese, and ice cream, but in the manufacture of condensed milk and milk powder. In the preparation of the condensed product refrigeration cooled the concentrated milk after its excess water had been removed by a heating method. In the case of the powder it froze the milk; the resulting ice crystals were then removed by a centrifuge, and the remaining pasty matter was dried out by heating.[32]

Brewers continued to depend on refrigeration for almost

"Refrigeration Applied to Air Supply for Blast Furnaces," *Ice and Refrigeration*, xxxv (1908), 340; E. T. Murphy, "The Varied Uses of Refrigeration," *Ice and Refrigeration*, lx (1921), 382.

[28] Vandervaart, "Commercial Uses of Refrigerating Machinery," *loc.cit.*, 180-186.

[29] J. E. Siebel, *Compend of Mechanical Refrigeration*, 8th edn. (Chicago, 1911), 182-183; J. H. Awberry, "Carl von Linde: A Pioneer of 'Deep' Refrigeration," *Nature*, cxlix (June 6, 1942), 630; J. C. Goosman, "History of Refrigeration," *Ice and Refrigeration*, lxviii (1925), 479. For the uses of liquid air, which included the manufacture of nitrates, see Siebel, *op.cit.*, 182, 187-188, and "Science Remaking Everyday Life: II. Cold Almost as Useful as Heat," *The World's Work*, xlv (1922-1923), 165-166.

[30] Vandervaart, "Commercial Uses of Refrigerating Machinery," *loc.cit.*, 184; Siebel, *op.cit.*, 259.

[31] Vandervaart, "Commercial Uses of Refrigerating Machinery," *loc.cit.*, 181, 184, 186; *Ice and Refrigeration*, xxi (1901), 19.

[32] Vandervaart, "Commercial Uses of Refrigerating Machinery," *loc.cit.*, 180; *Ice and Refrigeration*, xxv (1903), 206.

every step from the storage of raw materials to the aging of the finished brew. Of the 1,345 breweries listed in the internal-revenue returns for 1915, all but about 95 were fitted out with refrigerating machinery, mainly of the ammonia-compression type. A number of changes were made in the techniques of using this equipment, one of which was the replacement of brine-circulation systems by direct-expansion installations. Responsible for this modification was the increasing reliability of pipe fittings, which made it safe to place ammonia-evaporator coils directly in the refrigerated rooms.[33] Distillers and wine-makers, the industrial cousins of the brewers, also found machine-made cold increasingly useful.[34]

Among the more dramatic developments of the years under consideration was the application of refrigeration to a stubborn civil-engineering problem. In many places the presence of excessive water seepage or of silt or quicksand made excavation impossible, thus blocking the working of valuable mineral deposits and the construction of tunnels. A solution to this vexing problem was worked out in 1883 by a German mining engineer, F. H. Poetsch, who originated a method of freezing the troublesome earth by means of pipes sunk in the ground through which refrigerated brine was pumped. By 1891 the Poetsch process had made it possible to sink through quicksand the shaft of the Chapin Mine at Iron Mountain, Michigan.[35]

In the first World War refrigeration played a part that demonstrated how important it had become to our economy.

---

[33] Vandervaart, "Commercial Uses of Refrigerating Machinery," *loc.cit.*, 179; G. J. Patitz, "Progress in Brewery Refrigeration," *Ice and Refrigeration*, LI (1916), 168-169; J. A. Ewing, *The Mechanical Production of Cold*, 2nd edn. (Cambridge, 1921), 114; F. P. Siebel, Sr., "Brewery Refrigeration. History of Progress During the Last Twenty-Five Years," *Ice and Refrigeration*, CI (1941), 55.

[34] J. E. Siebel, "Refrigeration and the Fermenting (Brewing) Industry in the United States," *Premier Congrès International du Froid*, III (Paris [1908?]), 77; Vandervaart, "Commercial Uses of Refrigerating Machinery," *loc.cit.*, 179.

[35] Vandervaart, "Commercial Uses of Refrigerating Machinery," *loc.cit.*, 182; D. E. Moran, "Refrigeration and Engineering," *Ice and Refrigeration*, I (1891), 264.

In munitions factories it provided the strict control of temperature and humidity necessary.[36] In our fighting ships carbondioxide machines were installed to keep magazines well below the temperatures at which high explosives become unstable.[37]

But the most important assignment of refrigeration was to help food win the war. Wartime exigencies required the refrigerated warehouses of the nation not only to fulfill their normal function, but in addition to supply vast quantities of frozen beef to the American Expeditionary Forces and the armies of the Allies. Indeed, more than half of the meat sent to our army was frozen. A serious strain was placed upon the freezing capacity of American warehouses by the magnitude of this work.[38]

Emergency regulation was imposed on the cold-storage industry by the United States Food Administration. The background of this move lay in the long controversy over the refrigerated warehouse, but the immediate motivation was fear that the abnormal demand for both food and warehouse space might lead to speculation in the former and to exorbitant rates for the latter.[39] After conferring with leading warehousemen, the Food Administration proclaimed rules in the fall of 1917 that called for licensing, that prohibited public warehousemen from owning or dealing in storage goods, that forbade them to make loans of more than seventy per cent of the value of the foods held, that required certain commodities to be labeled "cold storage," and that specified that no goods unfit for human consumption should be received. A schedule of rates was to be filed by the individual warehousemen; later, maximum rates were established for the more important commodities. Reporting of quantities of certain foods in storage,

[36] Vandervaart, "Commercial Uses of Refrigerating Machinery," *loc.cit.*, 182-183; Cherne, *loc.cit.*, 28; *Ice and Refrigeration*, LVII (1919), 92.

[37] *Ice and Refrigeration*, LI (1916), 177.

[38] I. C. Franklin, "The Effect of the War Upon the Cold Storage Industry," *A.S.R.E. Journal*, VI (1919-1920), 239-243; F. A. Horne, "Cold Storage Warehousing in the United States," *Ice and Refrigeration*, LXVII (1924), 150-151.

[39] Franklin, *loc.cit.*, 243-244; *Ice and Refrigeration*, LIII (1917), 39-40.

begun on a voluntary basis by the Department of Agriculture, the Food Administration made obligatory.[40] Influential elements in the industry approved heartily of most of the regulations because of their stabilizing effect, and at the annual meeting of the American Warehousemen's Association just after the war a call was raised for consideration of what should be done to replace the regulatory power of the Food Administration when its authority should expire.[41]

[40] U. S. Food Administration, "Special Rules and Regulations Governing Licensees Engaged in Business as Cold-Storage Warehousemen, Including Official Interpretations No. xxiv," *Ice and Refrigeration*, LV (1918), 88-89; Franklin, *loc.cit.*, 244, 248-249; F. A. Horne, "Address Before Convention of New York State Cold Storage Association," *Ice and Refrigeration*, LV (1918), 87; W. C. Mullendore, *History of the United States Food Administration, 1917-1919* (Stanford University, California, 1941), 250-252.

[41] Horne, "Address Before Convention of New York State Cold Storage Association," *loc.cit.*, 87; A. M. Read, "Preparing for Peace," *Proceedings of the Twenty-Eighth Annual Meeting of the American Warehousemen's Association* . . . (Pittsburgh, 1919), 367-369.

CHAPTER XII

THE TECHNICAL ADVANCE OF
REFRIGERATION, 1917-1950

THE progress made after 1917 in the techniques of refrigeration was complex in nature and based on sound technical knowledge. Great numbers of individuals and organizations made their contributions. So interdependent was the field that innovations often were of key importance far outside the application for which they were designed. Gone was the day of rule-of-thumb methods when refrigeration was more an art than a science. Engineering schools began to furnish the industry with men trained in its special requirements.[1] More emphasis was placed on research,[2] and better facilities for the exchange of information were provided by technical and trade associations, which covered every phase of the refrigerating industries.[3] Perhaps the most important of these was the American Society of Refrigerating Engineers, which disseminated technical data. Among its many services was the financial assistance it gave the United States Bureau of Standards in compiling complete, reliable tables on the properties of ammonia.[4] Although association on an international scale was resumed in 1919, such activity left much to be desired, and by 1949 a call was going up for some new basis of international cooperation.[5]

[1] For surveys of refrigeration training facilities see J. F. Nickerson, "Instruction in Refrigeration Engineering in the United States," *Ice and Refrigeration*, LXVIII (1925), 164-167, and American Society of Refrigerating Engineers, Committee on Education, *Educational Guide to Refrigeration and Air Conditioning Training Facilities in the U. S. (A.S.R.E. Misc. Publ. No. 6)*. Published as supplement to January 1946 issue of *Refrigerating Engineering*.

[2] "A Review of Sources and Research in Refrigeration," *Refrigerating Engineering*, XXV (1933), 34-35.

[3] For a classified list of these associations see D. L. Fiske, "The Refrigerating Industries," *Refrigerating Engineering*, XXV (1933), 308.

[4] J. E. Starr, "Refrigeration," *Mechanical Engineering*, LII (1930), 353-358; H. Sloan, "Refrigeration Then and Now," *Refrigerating Engineering*, XXXIX (1940), 9.

[5] *Ice and Refrigeration*, LVIII (1920), 129; W. R. Woolrich, "Ob-

The compression machine, more supreme than ever where refrigeration-tonnage requirements were large, underwent major refinements in the hands of its manufacturers.

The compressor, the heart of a refrigerating system, was improved in efficiency. Multi-cylinder units with high rotative speeds were developed which featured decided advantages over the ponderous giants of the early days. Their increased capacity, coupled as it was with reduced size and weight, lessened not only first costs but also the amount of floor space and headroom that had to be devoted to machinery, while their quiet and steady operation permitted mounting on upper floors.[6]

The multi-stage compressor, designed to function in two or more steps with provision for cooling the compressed gas between stages, was a development of significance. Two-stage units had been built in the United States at an early date. It was recognized that although their initial cost was high, they required less power per ton of refrigeration than the standard type and were subject to less mechanical strain. Their virtual abandonment about 1895 was due primarily to the fact that most applications—ice manufacturing and brewing, for example—did not need the very low temperatures for which two-stage machines were particularly adapted. Even when low-temperature installations were under consideration, the two-stage compressor usually was bypassed in favor of an ammonia-absorption system. But about 1916 such equipment began once more to be produced.[7] The steady increase in its employment that followed was due partly to improvements in the machinery itself and partly to the growing volume of low-

servations on Refrigeration in Great Britain and Western Europe," *Refrigerating Engineering*, LVII (1949), 698, 700.

[6] L. W. Morse, "Progress in Mechanical Refrigeration During the Last Twenty-Five Years," *Ice and Refrigeration*, CI (1941), 16. A typical ammonia compressor of 1885 was twenty feet in height. Its 1941 counterpart of practically the same capacity was only about four feet high. *Ice and Refrigeration*, CI (1941), 14.

[7] H. Sloan, "Low Temperatures by Means of Multi-Stage and Other Compression Systems," *Refrigerating Engineering*, XLV (1943), 419-420; Sloan, "Refrigeration Then and Now," *loc.cit.*, 10.

temperature requirements and the declining popularity of absorption systems.[8]

A radical innovation in design was the centrifugal compressor, brought out with a specially adapted refrigerant in 1921 by Willis H. Carrier and subsequently improved by the Carrier Corporation and two of its competitors, the York Ice Machinery Corporation and the Worthington Pump and Machinery Company.[9] Of the same basic design as centrifugal pumps and fans and compressing the refrigerant by centrifugal force, the new machine had certain distinct advantages. Although its initial cost was rather high, it was generally less expensive in large installations than conventional reciprocating machines which used refrigerants other than ammonia. It was particularly economical when relatively high-pressure steam was available from some other operation to serve as motive power.[10]

Related closely to improvements in compressors was the development of new refrigerants. The most important step in this field was the introduction of the Freon refrigerants, a series of fluorine derivatives of hydrocarbons. Their advantages were significant; they were noninflammable, nonexplosive, nontoxic, and noncorrosive. Since the pressures required in their use were less than needed for ammonia, they made possible lighter construction of machinery, while their low boiling points made them particularly desirable for low-tem-

[8] For a discussion of the mechanical features of these compressors see J. A. Moyer and R. U. Fittz, *Refrigeration, Including Household Automatic Refrigerating Machines* (New York, 1928), 54-55, and W. R. Woolrich, *Handbook of Refrigerating Engineering* (New York, 1938), 75-76.

[9] W. H. Carrier, "How Air Conditioning Has Advanced Refrigeration," *Mechanical Engineering*, LXV (1943), 333; V. R. H. Greene, note in *Ice and Refrigeration*, CI (1941), 368. The best historical survey is W. A. Grant, "A History of the Centrifugal Refrigeration Machine," *Refrigerating Engineering*, XLIII (1942), 82-86, 120, 122. Carrier was looking for safe, efficient, and flexible equipment for air-conditioning purposes.

[10] Carrier, *loc.cit.*, 334; A. A. Berestneff, "A New Development in Absorption Refrigeration," *Refrigerating Engineering*, LVII (1949), 553.

perature work.[11] These new refrigerants widened the utility of the various compressors. Their nonpoisonous character permitted the reciprocating type to be used in air conditioning, where safety was a prime consideration, and the employment of one of the series, Freon-11, in the centrifugal type made it possible for such units to produce very low temperatures.[12]

The progress made in compressors and refrigerants was paralleled elsewhere. Many kinds of condensers were used, but the most important innovations were the shell-and-tube and the evaporative models. The former, employed widely where ample water was available and where water contained a high concentration of minerals, consisted of tubes set in a cylindrical steel shell. The compressed gas entered the shell; its heat was removed by water circulating through the tubes. The evaporative condenser, used where water was scarce and expensive, featured closely spaced piping banked in an enclosure. The hot gas was pumped into the pipes; its heat was removed by the evaporation of a thin film of water which was sprayed over the metal surfaces. To assure adequate evaporation, air was circulated through the enclosure.[13] Both condensers and evaporators were rendered more efficient by the evolution of the finned coil, which augmented the surface available for heat transfer by means of its thin metal fins attached at right angles to the tubing.[14] Evaporator operation

[11] J. S. Beamensderfer, "The Common Refrigerants," *Industrial and Engineering Chemistry*, xxvii (1935), 1029; R. J. Thompson, "Freon, a Refrigerant," *Industrial and Engineering Chemistry*, xxiv (1932), 623; D. K. Tressler and C. P. Evers, *The Freezing Preservation of Foods*, 2nd edn. (New York, 1947), 31; "Chemical Industry," *Fortune*, xvi (Dec. 1937), 83-84; B. Sparkes, *Zero Storage in Your Home* (New York, 1944), 69. For a survey of the entire refrigerant field see "What the Refrigerants Have Contributed: The Extent of Use, History and Sources of Common Media," *Refrigerating Engineering*, xxviii (1934), 305-312.

[12] Carrier, *loc.cit.*, 334.

[13] J. A. Moyer and R. U. Fittz, *Air Conditioning*, 2nd edn. (New York, 1938), 129-137; Morse, *loc.cit.*, 16-17; E. Brandt, "Refrigerated Warehouses—1949 Plants and Methods," *Ice and Refrigeration*, cxvi (Jan. 1949), 34.

[14] S. C. Moncher, *Commercial Refrigeration and Comfort Cooling* . . . (Chicago, 1940), 21-22.

was made automatic by the development of the thermostatic expansion valve.[15]

The ammonia-absorption machine after 1916 lost ground rapidly to the compression system. A number of factors contributed to its decline. Among them was the concentration of manufacturers and power companies on promoting electric-motor-driven compressors. This tended to block improvements in absorption machinery. Other factors were the application of improved multi-stage compressors to low-temperature work, where the absorption system had been most effective, the materially higher initial cost of an absorption plant, the shift to the manufacture of ice from undistilled water, and the hazards of using ammonia as a refrigerant in air-conditioning projects.[16] Beginning in the thirties there was a limited revival of absorption refrigeration. Important in explaining this were improvements in design, largely the work of Europeans, and the desire to find a profitable means of utilizing cheap steam.[17]

Absorption machines based on refrigerants other than ammonia were the subject of considerable activity. In the late twenties a system for cooling refrigerator cars was designed that depended on the ability of silica gel, or silicon dioxide, to absorb large quantities of sulphur dioxide, which was used as the refrigerant.[18] Other absorption machines were brought out for air-conditioning work. One of these used methylene chloride as the refrigerant and dimethyl ether of tri-ethylene glycol as the absorbent, while another was based on the combination of water and lithium-bromide salt. The most promising development was a machine built by the Carrier Corporation that featured a water refrigerant and a lithium-bromide absorbent. Intended for air conditioning, it was expected to find employment in many places where the cost of steam was low in comparison with that of electric power.[19]

[15] Morse, *loc.cit.*, 18-19.

[16] *ibid.*, 17; Moyer and Fittz, *Refrigeration*, 29-30; Berestneff, *loc.cit.*, 553.

[17] Morse, *loc.cit.*, 17-18; S. Ruppright, "The Absorption System Comes Back," *Refrigerating Engineering*, xxxiv (1937), 93-95, 124.

[18] Berestneff, *loc.cit.*, 553; *Ice and Refrigeration*, lxxiv (1928), 217, 219.

[19] Berestneff, *loc.cit.*, 553-556, 609.

One new technique was steam-jet refrigeration, a device for cooling water by the effect of evaporation from its own surface. The idea of cooling water by such means was old; indeed, it was the basis of William Cullen's ice-making machine of 1755. But early in the twentieth century Maurice LeBlanc of France used steam to speed the evaporative process, and about the time of the first World War his system was introduced into the United States. In these units steam at high pressure was ejected from nozzles over a tank of water. The steam entrained particles of water vapor, removed them, and thus maintained the partial vacuum necessary for high-speed evaporation. The steam-jet system found its chief employment in air conditioning, to which the safety of water and the low initial cost of the machine commended it. The relative costs of electricity, steam, and water in any given installation determined whether or not it was used.[20]

The cold-air machine, which was obsolescent as early as the eighteen-nineties, was revived. After 1945 a French concern installed a modern, improved unit on a new passenger-cargo liner, where its safety, the absence of a refrigerant-leakage problem, and the elimination of a considerable amount of piping were considered important advantages.[21] In the United States improved variants of the old machines were used for air conditioning military and commercial passenger aircraft. For this purpose they had the assets of small volume, light weight, absence of danger from refrigerant leaks, and the fact that cabin-pressurization equipment could supply compressed air to the refrigeration unit at the low altitudes where

[20] J. C. Bertsch, "Steam Jet Refrigeration," *Ice and Refrigeration*, XCII (1937), 315-316; J. R. Allen, J. H. Walker, and J. W. James, *Heating and Air Conditioning*, 6th edn. (New York, 1946), 482-485; J. R. Dalzell and C. L. Hubbard, *Air Conditioning, Heating and Ventilating . . .* (Chicago, 1940), 359-363; W. H. Stangle, *An Air Conditioning Primer* (New York, 1940), 154-155. A related system used a centrifugal pump to maintain the partial vacuum. Woolrich, *Handbook of Refrigerating Engineering*, 309-310.

[21] L. Denis, "Shipboard Air Conditioning," *Refrigerating Engineering*, LVII (1949), 134-136.

pressurization was unnecessary and the need for cooling the greatest.[22]

No development was more important than the perfection of the small domestic machine, the household refrigerator. Here, as in the case of the large machines, the compression system held sway. But the pioneering was not done by the old-line manufacturers. Perhaps they saw no point in jeopardizing their business of equipping ice plants.[23]

The introduction of the household machine was the work of a great number of concerns,[24] but among the most important in technical contributions were the Kelvinator Corporation and the General Electric Company. Kelvinator was the result of an association that began in 1914 between Edmund J. Copeland and Arnold H. Goss, two Detroiters with wide experience in the automobile industry. In February 1918 they sold their first refrigerator.[25] Although Kelvinator was formed for the express purpose of developing and promoting a domestic unit, such activity was of course only a side line with General Electric, which in 1911 had begun to manufacture

[22] P. C. Scofield, "Air Cycle Refrigeration," *Refrigerating Engineering*, LVII (1949), 558-560.

[23] G. F. Taubeneck, "Refrigeration Has 'Come of Age'," *Rand McNally Bankers Monthly*, LIV (1937), 264.

[24] For an idea of the great number of early manufacturers see *Ice and Refrigeration*, LXV (1923), 3; E. W. Lloyd, "Electric Household Refrigeration," *Ice and Refrigeration*, LXX (1926), 656; G. Muffly, "Twenty-five Years of Household Electric Refrigeration Development," *Ice and Refrigeration*, CI (1941), 39; H. B. Hull, *Household Refrigeration; A Complete Treatise on the Principles, Types, Construction, and Operation of Both Ice and Mechanically Cooled Domestic Refrigerators, and the Use of Ice and Refrigeration in the Home*, 3rd edn. (Chicago, 1927), 187-298. Considerable information is to be found in the Special Supplement to *Electric Refrigeration News*, reprinted from the issues of March 27 and April 10, 1929, and in the Industry Pioneer Number, *Air Conditioning and Refrigeration News*, XIX (Oct. 7, 1936).

[25] J. W. Beckman, "Pioneer Explains Barriers Facing Early Manufacturers," *Electric Refrigeration News*, VI (July 6, 1932), 9-11; W. H. Long, "Goss Relates Story of Kelvinator's Early History," *Air Conditioning & Refrigeration News*, XXV (Oct. 26, 1938), 11; *Air Conditioning & Refrigeration News*, XXV (Oct. 26, 1938), 1-2, 11. The first Kelvinator is on exhibit at the National Museum, Washington, D. C.

the Audiffren machine.[26] But this experience led to a long series of more practical models that put the company in the forefront of the domestic-refrigerator field.[27]

Certain basic technical advances marked the evolution of the household machine. Foremost among these was dependable automatic control, for a machine would have been worse than useless had manual regulation been necessary. One of the first practical automatic controls was the thermostatic switch developed by Copeland for the first Kelvinators.[28]

The hermetically sealed unit, which featured both motor and compressor enclosed in a gas-tight case, was a major step forward. At first compressors were driven with power transferred by belt from an electric motor. This arrangement was unsatisfactory, since noise, refrigerant and oil leaks, and belt failure were inevitable.[29] The sealed unit, which eliminated the belt, was introduced in 1925 by General Electric, whose experience with the Audiffren machine had suggested its advantages.[30]

Essential to the success of the household machine was the self-contained refrigerator. In the early days of the industry refrigerating systems—compressor, condenser, and evaporator—were built separately and installed in the cabinet, usually an ice refrigerator, of the purchaser. Sometimes the motor and compressor were located away from the cabinet, frequently in the basement. Separate construction of machines did not work out well, for the cabinets to which they were connected were of different sizes that often were ill-adapted to the capabilities of a given unit. Besides, the old wooden ice-

[26] *Supra*, 101.

[27] "What the Refrigerating Machine Companies Have Contributed," *Refrigerating Engineering*, xxviii (1934), 304.

[28] Muffly, *loc.cit.*, 39-40, 42; J. F. Wostrel and J. G. Praetz, *Household Electric Refrigeration, Including Gas Absorption System* (New York, 1938), 121-122; Beckman, *loc.cit.*, 9-10; Long, *loc.cit.*, 11.

[29] American Society of Refrigerating Engineers, *The Refrigerating Data Book*, 5th edn. (New York, 1943), 345.

[30] *Ice and Refrigeration*, lxix (1925), 265; "What the Refrigerating Machine Companies Have Contributed," *loc.cit.*, 304; Hull, *op.cit.*, 227-229; Muffly, *loc.cit.*, 40; "The Nudes Have It," *Fortune*, xxi (May 1940), 75.

box tended to warp and pull apart, and often its insulation was poor. As early as 1923 some manufacturers, including Frigidaire, were making their own cabinets, and before the end of the decade complete, integrated refrigerators with machinery and cabinet sold as a unit were the rule.[31]

The introduction of Freon-12 early in the thirties removed an important obstacle to complete acceptance of domestic mechanical refrigeration, for earlier manufacturers had been compelled to depend for refrigerants upon sulphur dioxide and methyl chloride, both of which were toxic, while one was inflammable. Freon, developed by the Dayton chemist Thomas Midgely, Jr. and his associates at the request of General Motors, makers of Frigidaire, soon achieved almost universal adoption in the household field.[32]

Important as these technical advances were, it should always be remembered that they would have meant little had it not been for the advent of cheap electric power.

During the nineteen-thirties the domestic machine was standardized. The sealed unit became the norm, and with its adoption rotary compressors were substituted for the old reciprocating type.[33] Expansion valves were replaced by capillary tubes, inexpensive small-bore tubes which effectively regulated the flow of refrigerant to the evaporator. The food compartment was chilled directly by evaporator coils rather than by the brine tank of the early models. The refrigerators, all self-contained, were built of steel and were well insulated. Finally, the mechanism was placed at the bottom of the unit. A great number of refrigerators with the mechanism mounted on top had been marketed, especially by General Electric. This loca-

---

[31] *Ice and Refrigeration*, LXIV (1923), 352; LXV (1923), 2; R. E. Ottenheimer, "The Household Refrigerator," *Ice and Refrigeration*, LXIV (1923), 495; Muffly, *loc.cit.*, 39-40; Beckman, *loc.cit.*, 11.

[32] "Chemical Industry," *loc.cit.*, 83-84; "Freon to the Front," *Time*, XLII (Aug. 16, 1943), 81-82; Beamensderfer, *loc.cit.*, 1029-1030; Muffly, *loc.cit.*, 41.

[33] The rotary compressor should not be confused with the centrifugal. In the former compression was achieved by rotating a semi-radial member within a cylinder. This system was adapted to very small refrigeration requirements. American Society of Refrigerating Engineers, *op.cit.*, 119.

tion, which permitted heat to escape up and away from the cabinet, was sound from an engineering point of view, but it failed of consumer acceptance because it necessitated a lower position of the food compartment than housewives desired.[34]

Thus a very satisfactory food-preserver was developed, but certain shortcomings were apparent by the end of the thirties. For one thing, freezing facilities were inadequate. No more than a package or two of frozen foods could be stored. The same was true with commercial ice cream, and if an attempt were made to keep any quantity of ice cubes, they would melt and fuse together. The housewife could freeze no more than a few pieces of meat. Other defects were even more serious. Foods had to be covered if dehydration and taste transfer were to be avoided, while odor-laden frost tended to collect on ice cubes and frozen desserts. Especially irritating was the necessity for periodic defrosting of the evaporator coils.[35]

Efforts were made to remedy these defects. A lamp was developed whose radiation destroyed most taste- and odor-bearing molecules and inhibited the growth of bacteria and mold. The check imposed on food spoilage permitted operating the refrigerator at humidities so high that there was no appreciable drying-out of food.[36] More important was the two-temperature refrigerator, which was introduced about 1939 and began to come into its own in the post-war period. As its name implied, it consisted of two separate compartments, one designed to maintain the very low temperatures required for freezing and the other to provide the higher levels needed for normal food storage. These two compartments were not cooled by the familiar evaporator suspended in the space itself, but by coils placed within the walls.[37] Numerous benefits were

[34] Muffly, *loc.cit.*, 40, 42; V. R. H. Greene, "Survey of the Last Twenty-Five Years in the Field of Refrigeration," *Ice and Refrigeration*, CI (1941), 21.

[35] R. F. Roider and H. W. Timmerman, "The Two-Temperature Refrigerator—Design and Construction," *Refrigerating Engineering*, LVI (1948), 134.

[36] A. W. Ewell, "The Post-War Domestic Refrigerator," *Ice and Refrigeration*, CVII (Sept. 1944), 25; Muffly, *loc.cit.*, 42; *Ice and Refrigeration*, XCVI (1939), 421.

[37] *Ice and Refrigeration*, XCVI (1939), 494; XCVIII (1940), 135,

achieved by this design. More room was provided for the storage of commercial frozen foods and for limited home freezing, and ice cubes could be made and stored in larger quantities. Since cubes and frozen desserts were kept in a separate compartment, they could not become contaminated with foreign tastes. But the arrangement of the coils within the walls was the most important key to the improved performance of these refrigerators. When the traditional evaporator was used as a means of heat transfer, its limited area had to be carried at a temperature far below that desired in the above-freezing compartment. When, however, the entire wall surface was employed, the space could be cooled without lowering the wall temperature materially below that required for preserving the food. With this reduced temperature differential there was less condensation and consequently, less dehydration. No defrosting was necessary. Although a certain amount of moisture did condense on the walls, it did so not as frost, but as water which trickled down and escaped through a trap to a pan where heat from the compressor caused it to evaporate rapidly. Because taste-bearing particles were carried away with this moisture, taste transfer was minimized. Foods stored in such a refrigerator for only two or three days did not need to be covered.[38]

The ammonia-absorption system was not ignored in domestic refrigeration. The machine most important in the United States was a Swedish invention improved by Servel, Inc., the firm which produced it here.[39] Although this ingenious unit was promoted vigorously by gas companies, its sales were small in comparison with the combined total of its compression-system competitors.

394; Muffly, *loc.cit.*, 42-43; "That Refrigeration Boom," *Fortune*, xxviii (Dec. 1943), 163. For a discussion of the systems used to maintain different temperatures in each compartment see Tressler and Evers, *op.cit.*, 122.

[38] Roider and Timmerman, *loc.cit.*, 134; Ewell, *loc.cit.*, 24.

[39] Woolrich, *Handbook of Refrigerating Engineering*, 289-290; E. E. Slosson, "Refrigeration by a Flame," *The Scientific Monthly*, xxiii (1926), 565, 567; Hull, *op.cit.*, 299-330; *Ice and Refrigeration*, lxix (1925), 263; R. S. Taylor, "Progress in Household Refrigeration," *American Gas Association Monthly*, xxvii (1945), 25-30.

The development of a machine capable of meeting the relatively small refrigeration requirements of many commercial and industrial establishments paralleled the evolution of the domestic mechanism. At the time of the first World War small commercial machinery was being built, to be sure, but its use was limited, partly because it was not automatic in operation. By 1935 a variety of light, high-speed, efficient, automatic units were available for all purposes. Automatic controls, the Freon refrigerants, and cheap electric power were factors that combined to make this advance possible, but a special impetus to progress was present in the form of the economic depression of the thirties and the new interest in air conditioning. The depression helped by shifting demand from capital to consumer goods and by forcing manufacturers to study more closely the needs of their market. One of the most pressing of these was equipment suitable for small air-conditioning installations.[40]

Pipe-line refrigeration was still a factor in supplying commercial users in the market districts of large cities. Boston, New York, Brooklyn, Philadelphia, St. Louis—all had such a utility. Just before the second World War the outmoded system in the Los Angeles Grand Central Public Market was remodeled.[41]

A close relation existed between the evolution of refrigerating machinery and the development of air conditioning. Improvements in refrigeration were largely the result of the incentive provided by air conditioning, while progress in the latter art was almost entirely dependent on advances in the former.[42]

By the early thirties air conditioning had passed through what might be termed its first phase. Although the centrifugal compressor with a special refrigerant for its use had been

[40] "Commercial Refrigeration: The Small Machine of 1935—Its Evolution and Present Application," *Refrigerating Engineering*, XXIX (1935), 175-176.

[41] Starr, *loc.cit.*, 356; *Ice and Refrigeration*, C (1941), 409.

[42] Carrier, *loc.cit.*, 332, 334; R. E. Cherne, "Developments in Refrigeration as Applied to Air Conditioning," *Ice and Refrigeration*, CI (1941), 31-32.

brought out in 1922, conventional compression units employing either ammonia or carbon dioxide were still important. The centrifugal compressor was responsible for an expansion in the number of large installations, not only for industrial purposes, but to some extent for comfort cooling. Yet the great potential market consisting of applications where the air-conditioning requirements were small remained untouched. Adequate small reciprocating machines were not yet available, and the centrifugal compressor was not adapted to systems that called for greatly reduced refrigerating capacities.[43]

The second phase of air conditioning was brought about by the appearance of thoroughly satisfactory small machines —automatic, economical, safe, and dependable. Their introduction was the signal for an extension of air conditioning to the smaller industrial, commercial, and domestic applications.[44]

The story of progress since 1917 in the techniques of refrigeration could not be told without reference to the appearance of solid carbon dioxide, or dry ice, as a commercial product. This refrigerant, capable of producing intense cold, had been known and used for laboratory purposes for a century, but not until 1925 was it marketed commercially. It caught on rapidly; between 1925 and 1930 annual consumption rose from 170 to 40,000 tons. Although the ice-cream industry was the largest market for dry ice, the new product soon was used widely in the transportation of frozen foods.[45] Before long carbon dioxide in its solid form found employ-

[43] Carrier, *loc.cit.*, 333-334; L. L. Lewis and R. VD. Dunne, "Why Dry Blast Equipment is Different Now," *Blast Furnace and Steel Plant*, xxix (1941), 1122-1124; Cherne, *loc.cit.*, 29-30; "Weathermakers: Carrier Corp.," *Fortune*, xvii (April 1938), 118; *Ice and Refrigeration*, lxiii (1922), 43-45; L. G. Huggins, "Trends in Air Conditioning," *Heating, Piping and Air Conditioning*, xiii (1941), 52.

[44] Carrier, *loc.cit.*, 333-334; Huggins, *loc.cit.*, 52; H. Sloan and W. B. Vilter, "Development of Refrigerating Equipment," *Ice and Refrigeration*, ci (1941), 25-26.

[45] C. L. Jones, "The Uses of Solid Carbon Dioxide," *Refrigerating Engineering*, xxv (1933), 331-332; C. L. Jones, "The 'Dry-Ice' Refrigerator Car," *Ice and Refrigeration*, lxxx (1931), 113; W. E. Becker, "Recent Developments in the Solid Carbon Dioxide Field," *Ice and Refrigeration*, lxxxii (1932), 61.

ment in industry, particularly in the metal-working trades, where it was helpful in making shrink fits.[46]

Quick freezing was perhaps the most spectacular development in refrigeration. Actually, the idea of freezing preservation was not new; it had been used on a significant commercial scale in the storage of meat, eggs, poultry, fish, and small fruits.[47] The focal point in the evolution of the new technique, both in the United States and Europe, was the fish industry. This was not surprising, since it was essential that its highly perishable product be marketed in a palatable condition. Existing methods of freezing were superior to canning or drying, but they still left much to be desired. That experimenters would turn to the perfection of the best available methods of preservation was to be expected.[48]

One thread in the story was the experimentation conducted in England and on the continent in the freezing of fish and other food products by direct or indirect immersion in cold brine. Though one English patent for such a process went back to 1842, most failed of practical application.[49] This was not true of the system for freezing fish by direct contact with a brine spray that a Norwegian merchant, Nikolai Dahl, developed and between 1912 and 1915 patented in England, Denmark, and the United States. Dahl-system installations actually were made at Los Angeles, San Diego, and San Francisco. About the same time J. A. Ottesen, a Dane, received patents in most of the important fishing nations for a direct-immersion process based on his discovery that little salt from the brine would penetrate the fish as long as the brine was

[46] W. Martin, "The Growing Market for Dry Ice," *Ice and Refrigeration*, XCIII (1937), 365-366.

[47] See Chapter X.

[48] I. C. Miller, "Quick Freezing Thaws Frozen Channels of Distribution," *Food Industries*, X (1938), 199; Dominick and Dominick, New York City investment bankers, quoted in *Ice and Refrigeration*, LXXIX (1930), 125.

[49] C. Birdseye and G. A. Fitzgerald, "History and Present Importance of Quick-Freezing," *Industrial and Engineering Chemistry*, XXIV (1932), 676; H. F. Taylor, *Refrigeration of Fish* (U. S. Department of Commerce, Bureau of Fisheries, *Document No. 1016*) (Washington, 1927), 506, 574.

not saturated and was cooled to the point at which the water in the brine began to crystallize and separate.[50] Of greater importance than the invention of specific methods was the research of three Germans, Plank, Ehrenbaum, and Reuter. The results of their work, published in Berlin in 1916, showed that fish frozen rapidly in brine were of superior quality to those frozen by ordinary means. After this demonstration progress in quick freezing was rapid.[51]

A second major thread in quick-freezing history was the work done in the United States. As early as the sixties Piper and Davis patented systems for freezing fish by indirect contact with a mixture of ice and salt.[52] In 1918 the United States Bureau of Fisheries imported an Ottesen brine freezer which it set up in its Washington laboratory. The Ottesen design did not become important in the United States in its own right, but its service as a demonstration plant provided an impetus for the discovery of more efficient methods.[53] One of these was worked out by Harden F. Taylor, Chief Technologist of the Bureau of Fisheries, who devised a system in which fish were frozen by a brine shower while they passed through a tunnel.[54] P. W. Petersen of Chicago contributed a method which involved placing the fish in molds or cans and freezing them by indirect contact with brine, while R. E. Kolbe of Erie, Pennsylvania, introduced methods similar in principle

[50] Taylor, *op.cit.*, 578, 580-581; *Ice and Refrigeration*, XLVI (1914), 236. The theory was that if the water and brine were separating, the water on and in the fish would not absorb salt, but instead would freeze without mixing with brine.

[51] H. F. Taylor, "Theory and Practice of Rapid Freezing in the Fish Industry," *Ice and Refrigeration*, LXXIX (1930), 113-115; G. Poole and M. T. Zarotschenzeff, "Four Years Progress in Quick-Freezing," *Ice and Refrigeration*, XCI (1936), 216; Tressler and Evers, *op.cit.*, 75.

[52] *Supra*, 62-63; Birdseye and Fitzgerald, *loc.cit.*, 676.

[53] J. M. Lemon, *Developments in Refrigeration of Fish in the United States* (U. S. Department of Commerce, Bureau of Fisheries, *Investigational Report No. 16*) (Washington, 1932), 3; L. Radcliffe, "Recent Developments in Fish Distribution," *Ice and Refrigeration*, LXXVI (1929), 501.

[54] Taylor, *Refrigeration of Fish*, 593-595.

though somewhat different in technique.[55] All of these systems were designed to freeze fish whole, but in 1924 Clarence Birdseye of Gloucester, Massachusetts, brought forward the idea of freezing in dressed, ready-to-cook form. Birdseye, who stressed the importance of heat transfer by means of direct contact between the product to be frozen and a refrigerated metal surface, devised two ways of accomplishing this. The first was his double-belt freezer, in which the packaged fish were frozen between a lower and upper metal belt as it moved through a freezing tunnel. The lower was chilled by brine sprayed against its under side, while the top belt was cooled by brine allowed to flow over its upper surface. Birdseye's second method was a multi-plate froster, in which packages were frozen by contact between two hollow metal plates cooled by the direct expansion of liquid refrigerant. Both A. H. Cooke of New York and Kolbe followed Birdseye in pioneering the freezing of dressed fish.[56]

Ultimately many different methods were used for quick freezing. Lively arguments raged as to the merits of each, but the requirements of food freezing were so varied that opportunity existed for several methods.[57] Freezers could be classified according to the cooling medium they employed. Most widely used were the several variants of the air-blast type, which featured intensely cold air blown at high velocity over the product. Some froze by means of a brine spray, while others employed liquid—such as brine or a sugar syrup—to chill the food either by direct or indirect contact. Machines which

[55] *ibid.*, 599-606; Birdseye and Fitzgerald, *loc.cit.*, 677; Lemon, *op.cit.*, 5-8; E. L. Carpenter and M. Tucker, *Farm and Community Refrigeration* (University of Tennessee Engineering Experiment Station, *Bulletin No. 12*) (Knoxville, 1936), 38.

[56] C. Birdseye, "Progress of Quick-Freezing in the United States," *Ice and Refrigeration*, LXXXIX (1935), 129; Lemon, *op.cit.*, 3-5; *Ice and Refrigeration*, LXXXI (1931), 246, 340; Carpenter and Tucker, *op.cit.*, 37-39.

[57] H. C. Diehl, "Frozen Food Production," *The Canner*, LXXXVI (Feb. 26, 1938), 24; "Quick-Frozen Foods," *Fortune*, XIX (June 1939), 128; W. J. Finnegan, "Factors To Be Considered in the Evaluation of Food Freezing Methods," *Ice and Refrigeration*, XCVI (1939), 432.

froze by the contact of metal with a packaged food were popular, while some systems found favor which used a combination of several cooling media.[58]

Although there was considerable disagreement on many phases of the new technique, particularly over the rapidity with which foods had to be frozen,[59] it was generally conceded that quick freezing resulted in a product vastly superior to one processed by the slow old methods. Perhaps the most important reasons for this were first, that the food was chilled quickly to a temperature below that at which bacterial, mold, and yeast growth took place; second, that the food was cooled rapidly to a temperature so low that enzymatic action was of no practical significance; and third, that there was less damage to the cellular structure of the food.[60]

Too much emphasis may have been placed on the importance of rapid freezing. All products did not require extreme speed. One group of investigators from the federal Department of Agriculture failed to find any marked advantages in

[58] R. C. Walther, *Methods of Quick Freezing* (American Society of Refrigerating Engineers, *Refrigerating Engineering Application Data—Section 22*) (Published as Section 2 of *Refrigerating Engineering* for February 1941), 2; C. W. Hulse, "Quick Freezing Plants and Their Operation," *Ice and Refrigeration*, cxii (May 1947), 44-47; W. J. Finnegan, "Food Freezing," *Ice and Refrigeration*, xcviii (1940), 457.

[59] J. E. Nicholas and N. B. Guerrant, "Problems Concerning Frozen Foods," *Refrigerating Engineering*, l (1945), 417; U. S. Department of Agriculture, *Technology on the Farm: A Special Report by an Interbureau Committee and the Bureau of Agricultural Economics of the United States Department of Agriculture* (Washington, 1940), 35.

[60] Tressler and Evers, *op.cit.*, 73; D. K. Tressler, "Chemical Problems of the Quick-Freezing Industry," *Industrial and Engineering Chemistry*, xxiv (1932), 682. The explanation most often given the public for the reduced damage to cellular structure was that in quick freezing the ice crystals were smaller than when slow methods were employed and did not make punctures in cell walls that caused leakage when the product was thawed. Some investigators believed that the principal cause of damage from slow freezing was the irreversible changes that occurred in the colloidal structure of perishable foods. For a summary and evaluation of the various theories of slow-freezing damage see W. R. Woolrich, "The Romance and Engineering of Food Preservation," *Science*, xcix n. s. (Feb. 11, 1944), 113.

quick freezing peaches, strawberries, and apple cider,[61] and it was reported that little advantage lay in quick freezing fruits packed in sugar.[62] Besides, it was learned that the colloidal structure of some foods was such that slow freezing did not have particularly adverse effects.[63] In general, quickness of freezing was not so important for preserving fruits as it was for meat, fish, and poultry.[64] The belief took hold that certain factors in addition to speed were important in turning out frozen food of high quality. Among these were selection of the most suitable raw material and proper care at every stage in handling, processing, packaging, storing, transporting, and finally cooking by the consumer.[65]

[61] J. M. Lutz, J. S. Caldwell, and H. H. Moon, "Frozen Pack: Studies on Fruits Frozen in Small Containers," *Ice and Refrigeration*, LXXXIII (1932), 113.

[62] D. K. Tressler, "Simple Methods for the Preparation and Freezing of Fruits and Vegetables Intended for Storage in Lockers," *Ice and Refrigeration*, XCIV (1938), 302.

[63] Woolrich, "The Romance and Engineering of Food Preservation," *loc.cit.*, 113-114.

[64] American Public Health Association, Committee on Foods, "Public Health Aspects of Frozen Foods with Particular Reference to the Products Frozen in Cold Storage Lockers and Farm Freezers," American Public Health Association, *Yearbook, 1939-1940*, 79-80; C. Birdseye, "Effect of Quick Freezing on Distribution of Fruits and Vegetables," *Ice and Refrigeration*, LXXX (1931), 131-132.

[65] G. A. Fitzgerald, "Trends in the Refrigeration of Foods," *Ice and Refrigeration*, CVI (1944), 103; H. C. Diehl, "Freezing to Preserve Vegetables and Fruits Still in Pioneer Stage," U. S. Department of Agriculture, *Yearbook . . . , 1932* (Washington, 1932), 526.

CHAPTER XIII

MECHANICAL VS. ICE REFRIGERATION,
1917-1950

THE ice plant was the foundation upon which both domestic and commercial refrigeration rested at the time of the first World War. During the next quarter-century it lost its pre-eminent position, but not because of any failure to modernize the process of ice making.

The most noteworthy technical development was the supremacy achieved by the raw-water can system, which by 1929 was employed in the production of almost eighty-six per cent of the national output. A decade later over ninety-one per cent was made by the new process. Distilled-water plants still held on in a few places, but the plate system, never important, practically disappeared.[1] As it took over the nation's ice making, the raw-water method itself underwent refinement. Improved air-agitation devices were adopted which minimized the opaque core that formed in the center of the block, while better means of purifying the water before freezing contributed to the same end. Heat transfer was made more efficient by new techniques of cooling and circulating the brine, and the entire operation was speeded by equipping plants with automatic can-fillers and multiple-harvesting devices.[2]

Electricity rapidly replaced steam as the primary source

[1] U. S. Department of Commerce, Bureau of the Census, *Fifteenth Census of the United States: 1930. Manufactures: 1929* (Washington, 1933), II, 160; *Sixteenth Census of the United States: 1940. Housing* (Washington, 1943), II, Pt. 1, 257.

[2] R. N. Cole, "Advancements in Raw Water Ice," *Ice and Refrigeration*, LXIV (1923), 384-386; W. C. Phillips, "Latest Development in Ice Manufacturing Equipment," *Ice and Refrigeration*, LXXVI (1929), 261-262; W. H. Motz, *Principles of Refrigeration . . .* (Chicago, 1926), 307-311; J. A. Moyer and R. U. Fittz, *Refrigeration* (New York, 1928), 293-294; S. E. Lauer, "Recent Developments In Ice Manufacture," *Ice and Refrigeration*, LXXVI (1929), 201; O. Luhr, "Economical Ice Harvesting," *Ice and Refrigeration*, LXII (1922), 16.

of ice-plant power.[3] The principal reason for the shift was the general adoption of the raw-water system, which did away with the necessity for having available large quantities of exhaust steam to condense into distilled water.[4] Once distilled water was no longer required, electric power was cheaper than steam. Electric motors cost less initially, they occupied less space, they needed less labor, and they cost less to operate, particularly when public-utility companies offered ice plants preferential rates in efforts to build up their loads. Electric drive even provided more latitude in the choice of locations, for a factory powered by electricity did not require a railroad siding for coal delivery.[5]

An important innovation in manufacturing methods was the automatic ice machines which the makers of refrigerating machinery began to place on the market early in the nineteen-thirties. These units, a departure from the traditional can method, were capable of the rapid production of ice not in the usual three-hundred-pound cakes but in small flakes or cylinders. They occupied only a small part of the area required by can-ice plants of equal capacity, required little labor, and were said to turn out ice in small sizes more economically than it could be made by freezing and crushing large blocks.[6]

Not technical shortcomings, but defects in the refrigera-

[3] H. Sloan and W. B. Vilter, "Development of Refrigerating Equipment," *Ice and Refrigeration*, CI (1941), 25.

[4] W. R. Woolrich, *Handbook of Refrigerating Engineering* (New York, 1938), 175-176.

[5] C. A. Stanley, "Electric Power for the Ice Plant," *Ice and Refrigeration*, XLII (1912), 219; T. Hibbard, "Synchronous Motor Drives in Ice and Refrigerating Plants," The Fourth International Congress of Refrigeration . . . , *Proceedings*, I (London [1925?]), 577; Woolrich, *op.cit.*, 176; Motz, *op.cit.*, 577.

[6] Sloan and Vilter, *loc.cit.*, 26; J. R. Watt, "The High-Speed Production of Ice," *Ice and Refrigeration*, CX (May 1946), 17; *Ice and Refrigeration*, LXXXIX (1935), 179-180; XCVII (1939), 339-340; Advertisement, *Ice and Refrigeration*, CXVIII (Feb. 1950), 10. As early as 1869 Daniel L. Holden obtained a patent for a machine which made ice by freezing it on the outer surface of a revolving cylinder. The ice was removed in flakes by a scraper and pressed into cakes. J. F. Nickerson, "The Development of Refrigeration in the United States," *Ice and Refrigeration*, XLIX (1915), 171.

tion service given the consumer were important in explaining the difficulties in which the ice industry found itself in the twenties and thirties. These defects were of primary significance, for they helped prepare the way for the enthusiastic reception which the public accorded the domestic machine.

One group of faults stemmed from the inherent characteristics of the industry and of ice as a refrigerant. For one thing, ice was expensive—at least in the minds of consumers who always tended to think that sixty cents a hundredweight was a lot to pay for frozen water. High charges prevailed not because ice companies made unconscionable profits, but because distribution costs were heavy for a product of such bulk, weight, and relatively low value and because enough money had to be made during the summer to pay fixed costs during the long off-season.[7] Although ice had great merits as a refrigerant, it was not perfect. Since it melted and had to be replaced, it made the consumer dependent upon the visits of the deliveryman. Moreover, it did not provide—at least conveniently—the freezing temperatures which were increasingly in demand among domestic users. Not a great deal could be done to correct these inherent difficulties.

Another class of defects was the result of inertia in the industry. The quality of ice was not always what it should have been; sometimes dirt, splinters, or even spiders were frozen in.[8] Refrigerators were poor, even worse, perhaps, than they had been at the turn of the century.[9] The cheap boxes that dominated the market had inferior insulation and inadequate provision for air circulation. Ice chambers were often too small in relation to the food compartment and sometimes so irregular in shape that an ice cake of standard size would not fit. The result of such deficiencies, of course, was heavy ice consumption but inadequate refrigeration. The temperatures provided were much too high and subject, be-

[7] P. A. Weatherred, "Where Persistence Is Needed," *Ice and Refrigeration*, xcvii (1939), 25; Editorial, *Ice and Refrigeration*, lix (1920), 133.

[8] *Ice and Refrigeration*, lxxiii (1927), 33.

[9] R. T. Frazier, "The Household Ice Refrigerator," *Refrigerating Engineering*, xviii (1929), 61.

sides, to wide fluctuations in direct proportion to the size of the ice charge.[10] Refrigerators were not only inefficient, but inconvenient as well, for most had no provision for the meltwater except a drip pan which had to be emptied every day.[11] The iceman, who had become a national joke, was responsible for much of the dissatisfaction occasioned by the use of his product. He was apt to be a rough, uncouth individual whose route across the kitchen floor was marked by dirty footprints and puddles of water and who too often would fail to make delivery when need was greatest. The iceman was adept at giving short weight. It was a simple matter to cut a three-hundred-pound block into seven "fifty-pound" pieces and to pocket the price of the extra fifty.[12] Some companies were partly responsible for such practices because of their laxity in dealing with dishonest employees.[13] Poor refrigerators and poor service were defects that could easily have been corrected. Progressive forces in the industry tried to do so throughout the twenties, but they were unable to make much headway against the prevailing inertia until the competition of the refrigerating machine made action imperative.[14]

[10] C. F. Belshaw, "Domestic Ice Refrigeration—Current Views on an Old Issue: Trends in Box Design," *Refrigerating Engineering*, XXVI (1933), 28; *Ice and Refrigeration*, LXIV (1923), 451; "Icemen's Shivers," *Business Week*, Nov. 29, 1941, 51.

[11] C. F. Belshaw, "Drain Water Disposal for Ice Refrigerators," *Ice and Refrigeration*, XCII (1937), 255.

[12] An old-line iceman once demonstrated to the writer how this could be done.

[13] H. P. Hill, "Development of Ice Industry," *Ice and Refrigeration*, LXX (1926), 168; H. D. Norvell, "Controlling the Ice Industry in Any Locality Legitimately and Lawfully," *Ice and Refrigeration*, LXIII (1922), 369; *Ice and Refrigeration*, LVI (1919), 385-386. Dishonesty was a long-standing charge against the iceman; even in India his sharp practices were noted as early as 1828. F. Parlby (Parks), *Wanderings of a Pilgrim* (London, 1850), I, 81. Norvell, *loc.cit.*, 369, quotes the following verse inspired by the unreliable deliveryman:

> "Gaily the gallant ice man
> Goes sporting in the sun;
> He selleth fifteen hundred weight
> And chargeth for a ton!"

[14] Hill, *loc.cit.*, 168; *Ice and Refrigeration*, LXXIII (1927), 33; LXXXI (1931), 159.

Other defects had their roots in the disorganization that characterized the industry. Duplication in delivery service by competing companies often pushed upward costs that were already high.[15] A more serious evil was overproduction, a condition which prevailed in most of the large cities of the country and in many of the smaller. Every year of good ice sales prompted the operators of existing plants to expand and encouraged outsiders to finance new units.[16] The situation was aggravated by the efforts of the manufacturers of refrigerating machinery to create a market; on occasion they even financed competition to existing plants in order to sell their product.[17] Overproduction was bad for both the producer and the consumer. It might bring price wars, or, as a division of volume raised unit costs, it might result in high rates and poor service.[18]

Vigorous efforts to deal with these conditions were made by both industry and government before the full impact of the domestic refrigerating machine was felt. Throughout the twenties trade associations sought to correct the evils that arose from an excess of competition. In the same decade a definite trend was apparent toward consolidation on a city and even a regional basis. One of the best examples of this was the City Ice and Fuel Company of Cleveland, which extended its operations to nearly one hundred Midwestern cities and towns.[19] Often consolidation took the form of city delivery companies, created by several ice plants to provide a common, noncompetitive service. Such centralized service had many advantages. It not only cut delivery costs, but those of production as well. It enabled management to pool production and either to run a plant at full speed or to shut it down,

[15] *Ice and Refrigeration*, LXV (1923), 40.

[16] *ibid.*, LXV (1923), 40; LXIX (1925), 223; Editorials, *Ice and Refrigeration*, LXVIII (1925), 520; LXVI (1924), 111.

[17] H. P. Hill, "Commission Control of the Ice Industry," *Ice and Refrigeration*, LXXVI (1929), 81.

[18] New State Ice Co. v. Liebmann, 285 U. S. 282, 292-293.

[19] *Ice and Refrigeration*, LVIII (1920), 109; LXIII (1922), 45; LXIX (1925), 194; LXXIII (1927), 33; Hill, "Development of Ice Industry," *loc.cit.*, 168; Hill, "Commission Control of the Ice Industry," *loc.cit.*, 81; "Water Still Freezes," *Fortune*, VII (May 1933), 90.

thus eliminating uneconomical operation at partial capacity.[20] A factor in furthering consolidation in the industry was the activity of public-utility corporations in acquiring ice-plant properties. Their interest, however, was not primarily in marketing ice, but in building up their sales of electric power.[21]

Various governmental units tried their hands at stabilizing conditions. After the war municipalities showed some interest in establishing publicly owned plants, but little was accomplished, partly because of the activity of the National Association of Ice Industries in opposing state legislation granting the necessary authority.[22] Early in the twenties bills designed to regulate the industry were introduced in a number of state legislatures.[23] The drive for state regulation bore important fruit in Oklahoma in 1925, when a law was passed which declared the manufacture, sale, and distribution of ice a public utility and subjected it to licensing by a state commission. A license was not to be issued if the commission found that the facilities already existing in the area concerned were sufficient to meet the public need. Adoption of this method of control elsewhere was checked in 1932, when the Supreme Court of the United States, despite a ringing dissent by Mr. Justice Brandeis, found that the Oklahoma attempt to limit the number of those engaging in the business was repugnant to the due-process clause of the Fourteenth Amendment.[24] In 1933 the federal government entered the picture. Under the supervision of the National Recovery Administration a code of fair competition was drawn up which

[20] F. B. Ostermueller, "Centralized Ice Delivery," *Ice and Refrigeration*, LXIX (1925), 356-357.
[21] W. L. Hutton, "Modern Raw Water Plant of Public Service Corporation," *Ice and Refrigeration*, LXX (1926), 183.
[22] *Ice and Refrigeration*, LV (1918), 104; LVIII (1920), 104. The Missouri Supreme Court blocked the plans of Kansas City for a municipal ice plant. "Municipal Ice Plant Melts Away," *The American City*, XXI (1919), 71. The Louisiana Constitution of 1921 permitted municipalities to own and operate ice plants. J. Dymond, "Public Utilities, Legislation and the Ice Industry and Municipal Ownership," *Ice and Refrigeration*, LXIV (1923), 35-36.
[23] *Ice and Refrigeration*, LX (1921), 224; L. C. Smith, "Legislation and the Ice Industry," *Ice and Refrigeration*, LXIII (1922), 351.
[24] 285 U. S. 262.

sought to check price cutting and which required that the code administrator be shown that the establishment of additional production facilities in any given territory was required by public necessity and convenience.[25] The leaders of the industry welcomed the NRA as an attempt to achieve what they had been working for all along, the elimination of the evils of too much competition. When the NRA was invalidated, they tried to maintain what gains they could through the agency of their trade associations.[26]

The domestic refrigerating machine won widespread acceptance in the years that the ice industry wrestled with the forces of inertia and overproduction. Only five thousand units were sold in 1921, but in 1930, the year that the mechanical established sales supremacy over the ice refrigerator, 850,000 were marketed. Save for reverses in 1932 and 1938 even adverse economic conditions did not prevent a steady annual increase in volume until a peak in the neighborhood of four million was reached in 1941.[27] In 1944 almost seventy per cent of American homes that had refrigerators were equipped with mechanicals.[28]

Many factors combined to make possible such rapid acceptance. Underlying them all was a deep-seated dissatisfaction with ice refrigeration[29] that was combined with a corresponding gratification over the convenience and efficiency of even the early household machines. Of great importance

[25] "Code of Fair Competition for the Ice Industry," *Ice and Refrigeration*, LXXXV (1933), 192-194.

[26] Editorial, *Ice and Refrigeration*, LXXXIV (1933), 414-415; M. Taylor, "The New Ice Industry Program," *Ice and Refrigeration*, LXXXIX (1935), 15.

[27] American Society of Refrigerating Engineers, *Refrigerating Data Book*, 5th edn. (New York, 1943), 341; A. W. Cruse, "The Electrical Goods Industries," U. S. National Resources Committee, Science Committee, Subcommittee on Technology, *Technological Trends and National Policy, Including the Social Implications of New Inventions* (Washington, 1937), 317; "That Refrigeration Boom," *Fortune*, XXVIII (Dec. 1943), 163.

[28] U. S. Department of Commerce, Bureau of the Census, *Characteristics of Occupied Dwelling Units, for the United States: October, 1944 (Series H-45, No. 2)* (Washington, 1945), 8.

[29] Hill, "Development of Ice Industry," *loc.cit.*, 168.

were the aggressive merchandising methods used to promote the new product.[30] Sales forces hammered home the arguments that the domestic machine was more convenient and dependable than ice, that it provided a more constant temperature, that it was more sanitary, and that it cost less to operate.[31] Not content with appealing to technical efficiency, they endeavored to make ownership of a mechanical refrigerator a badge of social prestige and a symbol of modern living.[32] Two familiar techniques of appliance retailers were brought into play—liberal credit policies to make purchasing "painless" and "loss-leaders" to attract the attention of prospective buyers. Loss-leaders in the refrigerator industry were basic models stripped of all gadgetry and selling at a strikingly low price, perhaps less than cost. They were designed to lure customers to a showroom. There a salesman could shift attention to more elaborate models whose prices were pegged at levels which yielded satisfactory profits.[33] But of more fundamental importance in putting over the idea of the household mechanical was the sharp downward trend of retail prices. The average refrigerator in 1920 cost $600, but by 1930 it had declined to $275 and by 1940 to $154.[34] The principal reason for this reduction was the lowering of unit-manufacturing costs that took place as mass production was achieved, but a partial explanation lay in the competition provided by Sears, Roebuck and Company. By the middle thirties Sears was marketing its Coldspot refrigerator for about forty dollars less than comparable models of the standard companies. This was possible, Sears claimed, partly because it had eliminated the wholesaler and was shipping

[30] The editor of *Electric Refrigeration News* believed that the role of the merchandiser was more important than that of the engineer. G. F. Taubeneck, "Refrigeration Has 'Come of Age'," *Rand McNally Bankers Monthly*, LIV (1937), 263-266.

[31] Kelvinator Corporation, *Kelvinator in Place of Ice* (Detroit, 1920). For a summary of the arguments of mechanical-refrigeration interests and an iceman's refutation of them see C. S. Johnson, "Household Refrigeration," *Ice and Refrigeration*, LXXI (1926), 157-158.

[32] Taubeneck, *loc.cit.*, 263-266.

[33] "The Nudes Have It," *Fortune*, XXI (May 1940), 73.

[34] American Society of Refrigerating Engineers, *op.cit.*, 341.

directly from the factory to its retail stores. In 1940 the Kelvinator Corporation, with the main purpose of meeting Sears competition, reduced the prices of all its models and put out a stripped loss-leader with a price as low as that of the Coldspot. The Kelvinator reductions were planned so carefully that salesmen had little difficulty in selling the more elaborate and expensive refrigerators. But other companies, forced to adopt similar techniques on short notice, did not fare so well. Their hastily devised loss-leaders looked so much like their luxury models that buyers rushed to the stripped units and ignored those on which the companies had planned to make their profits.[35] Less obvious than price reductions, but nevertheless important in explaining the acceptance of the household machine was the attitude of electric-utility companies. At first the utilities did not seem to recognize its potentialities, but about 1925 they began to promote it vigorously as a means of building up their sales of electric current.[36]

In the commercial field the refrigerating machine did not win acceptance so quickly as in the domestic, partly no doubt because the inconvenience of ice deliveries, so distressing to the housewife, did not mean much to the corner grocer or butcher.[37] In the thirties, however, the small machine was applied to a great variety of uses in establishments that retailed food and drink,[38] though there still remained many assignments such as the cooling of display cases for which ice competed on even or better-than-even terms.

The ice industry was slow to recognize the threat to its welfare presented by domestic and commercial machines. There was a flurry of excitement between 1924 and 1926, but a feeling of complacency set in again which was not

[35] For a detailed account of this episode see "The Nudes Have It," *loc.cit.*, 73-75, 102, 104, 106, 111.

[36] J. W. Beckman, "Pioneer Explains Barriers Facing Early Manufacturers," *Electric Refrigeration News*, vi (July 6, 1932), 11; E. W. Lloyd, "Electric Household Refrigeration," *Ice and Refrigeration*, lxx (1926), 655.

[37] "Water Still Freezes," *loc.cit.*, 90.

[38] S. C. Moncher, *Commercial Refrigeration and Comfort Cooling* (Chicago, 1940), 29-38.

entirely shaken until ice sales slumped sharply in 1932.[39] As the seriousness of machine competition was sensed, some reacted impractically by suggesting that a boycott be imposed upon the makers of mechanical refrigerators, but responsible leadership saw that the best hope for salvation lay in correcting the shortcomings so long associated with ice refrigeration.[40]

Numerous efforts to improve service were undertaken. Practically every improvement was suggested early in the twenties, but it took the menace of the machine to bring about anything like general adoption.[41] A change simple in principle yet of major significance was the practice of scoring. As the ice left the storage room to be loaded on a delivery truck, it was run through a battery of circular saws which cut grooves an inch or so in depth that marked out the fifty and twenty-five-pound pieces to be sold to consumers. The importance of this practice lay in the fact that a large block could be divided only into the pieces marked out by the scoring machine. Now a deliveryman could not sell three hundred and fifty pounds from a three-hundred-pound cake. This improvement not only built confidence in the industry, but it lowered delivery costs, for it shortened measurably the time a driver needed to split the ice into the sizes desired.[42] Actual delivery techniques were improved by the insistence of many companies that their servicemen be uniformed neatly

[39] Belshaw, "Domestic Ice Refrigeration," *loc.cit.*, 28; W. M. Oler, "Status of the Household Refrigerating Machine," *Ice and Refrigeration*, LXIX (1925), 334.

[40] V. R. H. Greene, "Proposed Boycott of Mechanical Refrigerator Manufacturers," *Ice and Refrigeration*, LXXVI (1929), 75-76.

[41] *Ice and Refrigeration*, LXXIII (1927), 33; LXXXI (1931), 159. Ice-cake sizes were standardized late in the twenties, but standardization, though helpful, would probably have taken place even if there had been no machine competition. U. S. Department of Commerce, Bureau of Standards, *Ice Cake Sizes (Simplified Practice Recommendation R96-28)* (Washington, 1929), 7-8.

[42] *Ice and Refrigeration*, LIX (1920), 212; Norvell, *loc.cit.*, 369. Deliverymen seem to have been aware that the scoring machine limited their opportunities for extracurricular profits. Those who worked for one Portland, Oregon, ice company struck in protest against its introduction. *Ice and Refrigeration*, LXVI (1924), 572.

and that they carry the ice in heavy canvas bags that kept water from dripping on the kitchen floor.[43] Prices were lowered by the establishment of neighborhood icehouses where sales were made on a cash-and-carry basis.[44]

One of the most vigorous features of the industry's reaction to competition from the refrigerating machine was the emphasis that it came to place on the sale by ice companies of good ice refrigerators. The decision to enter the sales field, though taken belatedly by many concerns and in some cases reluctantly, was thoroughly sound. The refrigerator was the key to the prosperity of the industry; every person who bought one was an assured market for ice, perhaps about two and a half tons a year. Moreover, only the iceman had the vested interest in the sale of the refrigerant to make sure that the equipment for using it was of the highest efficiency.[45]

Underlying the emphasis that the industry came to place on refrigerator sales were basic technical changes. Domestic units were redesigned, inside and out. The most important alteration was the placement at the top of the cabinet of an ice chamber usually large enough to hold one hundred pounds. The ice, which rested on a finned metal grid, was exposed to the air of the food chamber only at the bottom. Three features of this arrangement assured the maintenance of a steady, low temperature: the location of the ice compartment at the top meant good air circulation; the metal fins increased the surface available for the transfer of heat; and finally, the refrigerating efficiency was undiminished as long as enough ice remained—only a few pounds were required—to cover the grid upon which it rested. Good insulation and steel construction increased effectiveness, while streamlined design matched the eye appeal of the mechanical rival. Since these units required icing only at intervals ranging from four to seven days, the visits of the iceman were less frequent, and delivery costs were lowered. The old drip pan, so prone to

[43] *Ice and Refrigeration*, LXXIII (1927), 105-107; XC (1936), 346; "Ice Slicks Up," *Business Week*, Oct. 24, 1936, 35-36.

[44] Several cities had such "jitney" stations as early as 1917. *Ice and Refrigeration*, LIII (1917), 261.

[45] Editorial, *Ice and Refrigeration*, LXXII (1927), 93.

overflow, was eliminated by the simple expedient of channeling the meltwater through copper tubing into the plumbing system. Equipment was provided for cutting cubes from the ice charge, and accessories employing crushed ice and salt were furnished for the preparation of frozen desserts. An attractive feature of the improved models was their low initial cost, roughly one-half that of the mechanicals.[46] Commercial boxes were redesigned along similar lines, although in some installations head-room requirements limited the use of the top-icing principle, while under certain conditions forced air circulation was needed.[47]

The sale of refrigerators, of course, depended upon vigorous promotional work by individual companies, but the spark was provided by the National Association of Ice Industries, which stimulated activity by local concerns and assisted them in their sales work by organizing publicity on a national scale. The advertising sponsored by the association took many tacks in its efforts to promote the welfare of the industry, but increasingly its emphasis was on the merits of preserving foods with ice in a modern refrigerator.[48] This work, however, was not that of a unified industry. Even within the directorate of the national organization there were those who opposed promoting better refrigerators on the ground that since they would use less ice, they would reduce sales.[49]

The main arguments used in selling ice refrigerators were that the moist air provided checked the dehydration of foods and that the gaseous impurities given off were deposited

[46] C. F. Holske, "Domestic and Commercial Ice Refrigerators," *Refrigerating Engineering*, xxxii (1936), 150; Belshaw, "Domestic Ice Refrigeration," *loc.cit.*, 29; Belshaw, "Drain Water Disposal for Ice Refrigerators," *loc.cit.*, 255-257; *Ice and Refrigeration*, lxxxvi (1934), 319; xci (1936), 275, 374-375; xciii (1937), 434; "Ice Slicks Up," *loc.cit.*, 35-36; "Icemen's Shivers," *loc.cit.*, 51.

[47] Holske, *loc.cit.*, 152.

[48] For an early statement of the aims of the national publicity program see National Association of Ice Industries, National Publicity Committee, "Report" [at the Tenth Annual Convention], *Ice and Refrigeration*, lxxiii (1927), 302-303.

[49] *Ice and Refrigeration*, lxxxviii (1935), 187.

upon the ice and carried away with the meltwater. In an effort to capitalize on public interest in air conditioning, the industry advertised its product as "the only automatic air-conditioned refrigeration."[50] Sometimes claims were carried too far. In 1937 the Federal Trade Commission ordered the Coolerator Company, one of the leading manufacturers, to cease and desist from certain claims that unfairly disparaged mechanical units.[51]

There is no doubt that the development and promotion of efficient refrigerators helped the industry to check the inroads made upon its sales by the machine. In 1944 only about thirty per cent of American homes with refrigerators used ice,[52] but without the vigorous efforts taken, the industry's share might have been much less. Perhaps the best evidence of the effectiveness of sales promotion was testimony at the Federal Trade Commission hearings in 1936 by a representative of mechanical interests to the effect that within the past two years the competition had been felt.[53] But these observations should not obscure the fact that the best years for the sale of ice refrigerators were the late twenties, years in which the modern units were not a real factor. Paradoxically, these banner sales probably were due to the advertising of the mechanicals, which made the public refrigeration-conscious.[54]

To maintain its share of business in the commercial field, the industry pushed the sale of processed ice. Cubes sawed from large cakes were sold to drink-dispensing establishments, a market opened by the repeal of Prohibition. Crushed ice, produced either by breaking up large blocks or with the new automatic ice machines and graded according to its relative coarseness, was sold to restaurants, bars, soda fountains,

[50] See advertisement reproduced in *ibid.*, LXXXVI (1934), 321.
[51] *Ice and Refrigeration*, XCIII (1937), 93-94.
[52] U. S. Department of Commerce, Bureau of the Census, *Characteristics of Occupied Dwelling Units*, 8.
[53] *Ice and Refrigeration*, XCI (1936), 96.
[54] Review of the refrigeration industry in "The Index," a publication of the New York Trust Co., quoted in *ibid.*, LXXXV (1933), 264; L. K. Wright, *The Next Great Industry* (New York, 1939), 36.

groceries, and meat markets.[55] Emphasis was placed on the nutritive and economic value of displaying vegetables in beds of finely crushed "snow."[56] The market for processed ice was threatened after 1945 by sales of automatic machines to consumers, particularly those who needed ice for use in beverages.[57] Although the seriousness of the new competition was not clear, icemen were worried.[58]

Despite the general acceptance of mechanical refrigeration, the ice industry did not collapse before the onslaught of its rival. Its survival in a reasonable state of health was due principally to the fact that as a whole it sold less than a third of its total production to the domestic trade, where the inroads of the machine were most damaging.[59] The well-being of the industry was reflected in its annual tonnage statistics. In 1920 thirty-three million tons were sold. The next decade witnessed a steady rise, and in 1931 the all-time peak, fifty-seven million tons, was reached. A serious slump occurred in the late thirties, but during the second World War production surged upward, impelled by the increased importance of food preservation and the ban on the manufacture of mechanical refrigerators. In 1946 sales passed fifty-four million tons, but in 1948 they dropped to forty-seven million, in 1949 to forty-one million, and in 1950 to thirty-four million.[60] The significance of this recession was not entirely cer-

[55] *Ice and Refrigeration*, CI (1941), 37-38; "Iceman's Comeback," *Business Week*, Nov. 11, 1939, 41-42.

[56] "The Natural Way," *Fortune*, XXXVIII (Sept. 1948), 128.

[57] For a description of a typical machine see the advertisement of Mills Industries, Inc., in *Refrigerating Engineering*, LVIII (May 1950), 416-417.

[58] *Ice and Refrigeration*, CXVII (July 1949), 25-26.

[59] The remaining sales were to commercial users, to the operators of refrigerator cars and trucks, and to those who needed it for industrial processes. "The Natural Way," *loc.cit.*, 131. The increase in population and greater refrigeration-consciousness were other factors which helped to explain the survival of the industry. "Boom in Ice," *Business Week*, Sept. 17, 1949, 66.

[60] "The Natural Way," *loc.cit.*, 128; M. Taylor, "Annual Report of the Executive Secretary, National Association of Ice Industries," *Ice and Refrigeration*, CV (1943), 279-280; *Ice and Refrigeration*, CIII (1942), 315; CXVII (Aug. 1949), 31; CXIX (Aug. 1950), 28; CXXI (Aug. 1951), 22.

tain, for the situation was confused in 1948 by poorer economic conditions and in 1949 and 1950 by adverse weather. Naturally, the industry was concerned, but it hoped to offset its losses by increasing its sales in such fields as truck refrigeration and vegetable-display icing. The National Association was engaged actively in sponsoring research projects designed to develop new markets.[61]

Domestic refrigeration, stimulated by the advent of the household machine and by the struggle of ice to survive, became an essential part of the American standard of living long before the second World War. Perhaps no more than half of the nation's homes were equipped with some kind of refrigerator in 1930,[62] but in 1944 the number had risen to more than eighty-five per cent. Ninety-four per cent of urban dwellings had refrigerators, but only about eighty-two per cent of farm homes were so furnished.[63]

On the farm refrigeration was of special utility. It was important for the usual domestic purposes, to be sure, but in addition it was essential if fresh meat were to be available throughout the year and the commercial products of the dairy, the hen house, the orchard, and the truck patch were to be marketed in best condition. Yet in the twenties farm refrigeration was conspicuous by its absence. Ice companies generally neglected to service rural areas, for the cost of delivering to scattered customers was prohibitive.[64] A farmer

[61] M. J. Garvey, "Seasoned Progress—President's Report to Ice Industry at Annual Convention," *Ice and Refrigeration*, CXVII (Dec. 1949), 29-31; *Ice and Refrigeration*, CXV (Aug. 1948), 45-46; CXIX (Aug. 1950), 27; CXXI (Aug. 1951), 21-22. The annual research budget of the NAII was $40,000. "Boom in Ice," *loc.cit.*, 66.

[62] Estimates vary. *Ice and Refrigeration*, LXX (1926), 223; LXXI (1926), 302; LXXV (1928), 348; L. C. Smith, "Because It Is Right," *Ice and Refrigeration*, LXX (1926), 268; The President's Conference on Home Building and Home Ownership, *House Design, Construction and Equipment; Reports of Committees on Design, Construction, and Fundamental Equipment*, J. M. Gries and J. Ford, eds. (Washington, 1932), 279; "A Renaissance in Refrigeration," *The Review of Reviews*, LXXXV (May 1932), 44.

[63] U. S. Department of Commerce, Bureau of the Census, *Characteristics of Occupied Dwelling Units*, 8.

[64] *Ice and Refrigeration*, LXIV (1923), 92; LXXVIII (1930), 510; Editorial, *Ice and Refrigeration*, XCVII (1939), 193.

might harvest his own ice, but this alternative was limited to northern farmers who had available a suitable body of water. The small commercial machine had not yet been perfected, and although the household machine was satisfactory, many farms had no electric power.[65]

During the thirties a great many efforts were made to stimulate the use of refrigeration on the nation's farms. The United States Department of Agriculture was particularly active. The Bureau of Home Economics made performance studies on various types of refrigerators and prepared instructions on their proper management.[66] The Cooperative Research and Service Division sought to encourage the use of freezer lockers,[67] while the Rural Electrification Administration made a fundamental contribution by bringing power to agricultural areas and demonstrating to farmers how refrigeration could serve them.[68] State experiment stations joined in the work, some of them devoting special attention to the development of equipment adapted specifically to farm requirements.[69] No agency campaigned on more fronts for rural refrigeration than the Tennessee Valley Authority. Not content with making electricity available at low cost, it encouraged the development and sale of cheap household machines[70] and urged farmers to organize community projects

[65] Some manufacturers sought to overcome this obstacle by introducing for farm use machines powered by liquid fuel. G. F. Taubeneck, "Development of Electric Refrigeration Industry," *Electric Refrigeration News*, xix (Sept. 9, 1936), 16.

[66] U. S. Department of Agriculture, Bureau of Home Economics, *Report of the Chief of the Bureau . . . , 1936* (Washington, 1936), 7-8.

[67] *Infra*, 293.

[68] U. S. Department of Agriculture, Rural Electrification Administration, *Electricity for the Farm through REA* (Washington, 1940), 16-17, 19.

[69] H. J. Dana and R. N. Miller, *Building the Farm Freezing Plant* (State College of Washington, Engineering Experiment Station, *Engineering Bulletin No. 64*) (Pullman, 1940); M. C. Easter and M. L. Nichols, *Dairy Refrigeration on Rural Electric Lines* (Alabama Polytechnic Institute, Agricultural Experiment Station, *Bulletin 241*) (Auburn, 1934).

[70] Alabama icemen tried legal action to prevent the TVA from promoting electric refrigerators. *Ice and Refrigeration*, LXXXVII (1934), 12.

such as walk-in refrigerators and locker plants.[71] The refrigerating-machinery industry was somewhat remiss in providing equipment adapted to farm conditions and problems.[72] But no doubt indicative of a trend was International Harvester's decision to enter the postwar market with an expanded line of redesigned and restyled farm units.[73] The day did not seem far distant when the use of refrigeration on the farm would be practically as extensive as in the city.

[71] *Infra*, 316-317.

[72] W. R. Woolrich, R. B. Taylor, and M. Tucker, "Farm and Community Refrigeration in Rural Readjustment," *Refrigerating Engineering*, xxx (1935), 331, 333-334.

[73] *Ice and Refrigeration*, cviii (Jan. 1945), 62.

CHAPTER XIV

THE IMPROVEMENT OF REFRIGERATED
TRANSPORTATION, 1917-1950

WHEN the United States entered the war in 1917, the path which the development of more efficient ventilator refrigerator cars should take had been laid out by investigations of the Department of Agriculture in cooperation with private agencies. Competition in the construction of cars of high efficiency had developed among the roads, and some excellent equipment rode the rails.[1] Yet when the Railroad Administration took over, it found that most refrigerator cars were distinctly inferior in construction and performance. Many were no more than tight box cars, equipped with ice bunkers but practically uninsulated.[2]

The Administration, anxious to encourage uniformity in the interests of efficiency, appointed a committee to draw up plans for a standard refrigerator car. The effective work was done by a subcommittee of six experts so chosen that the railroads, the private-car companies, and the Department of Agriculture had equal representation.[3] In six short weeks in the spring of 1918 specifications were formulated. That the committee accomplished its objective so rapidly was due to the groundwork that had been laid since 1908 by government investigators, whose recommendations the committee followed closely. More effective insulation for floor, roof, and walls was prescribed, as were bunkers in the form of woven-wire baskets, insulated bulkheads, and racks that raised the load four inches above the floor. These specifications were accepted by the mechanical committee of Director-General McAdoo and designated as the United States Standard Refrigerator Car. The Railroad Administration required that the larger of the existing cars be made to conform to the new

[1] *Supra*, 121-122; W. A. Sherman, *Merchandising Fruits and Vegetables* (Chicago, 1928), 301.

[2] Sherman, *op.cit.*, 302.

[3] Federal Trade Commission, *Food Investigation. Report on Private Car Lines* (Washington, 1920), 32.

standard when undergoing general repairs or being rebuilt, but during the war no new construction could be undertaken.[4]

A car-building program was essential once peace had been restored. Not only was the nation's refrigerator fleet obsolete; it was in poor physical condition, for wartime service had been hard, and only the most necessary repairs had been made.[5] In the construction undertaken in the decade and a half that followed the Armistice the Standard Refrigerator Car served as a model.[6] Basket bunkers, insulated bulkheads, floor racks, and sound insulation became the rule, not the exception.[7]

A number of radical changes—among them fans to improve air circulation, solid carbon dioxide as a refrigerant, and mechanical cooling units—were suggested late in the thirties. The fan recommendation received favorable consideration,[8] but the others did not. Doubts existed among the owners and operators of refrigerator cars as to how such special equipment would fit into the regular service that had been organized around the traditional unit. Moreover, there was uncertainty as to what the shippers actually wanted. Dry ice presented serious technical difficulties and cost too much. Refrigerating machinery was generally considered to be impractical. Its cost was too high, not only because of the initial investment required, but also because the average refrigerator car was used as such perhaps only thirty per cent of the

[4] M. E. Pennington, "Standard Refrigerator Car Development," *Ice and Refrigeration*, LVI (1919), 28; M. E. Pennington, "The Development of a Standard Refrigerator Car," *A.S.R.E. Journal*, VI (1919-1920), 1, 10-11; Federal Trade Commission, *op.cit.*, 32.

[5] I. C. Franklin, "The Effect of the War Upon the Cold Storage Industry," *A.S.R.E. Journal*, VI (1919-1920), 249.

[6] J. M. Kelley, "Perishable Transport of the Future," *Railway Age*, CXVIII (Feb. 10, 1945), 304.

[7] W. V. Hukill and D. F. Fisher, "Present Practice with Refrigerator Cars," *Refrigerating Engineering*, XXX (1935), 76-77. For an excellent discussion of the situation in 1922 see W. H. Winterrowd, "Design and Construction of Refrigerator Cars," *Ice and Refrigeration*, LXIII (1922), 144-148, 185-188.

[8] In 1945 about seven thousand refrigerator cars were equipped with fans. C. C. Rowland, "Transportation of Frozen Foods—Refrigerator Car Service," *Ice and Refrigeration*, CX (Jan. 1946), 49.

time. This infrequent use gave the advantage to ice, which was cheap and was carried only when necessary. Besides, ice was more flexible. When great amounts of refrigeration were required, as in precooling, nothing needed to be done but to put additional ice in the bunkers. To attain the same flexibility mechanically, a machine with a large, uneconomical capacity would have to be installed. Other obstacles were the physical beating imposed on machinery by the severe jolts of railroad car-handling methods and the heavy loss that would be suffered should units fail.[9] It was suggested that the railroads opposed the adoption of mechanical refrigerator equipment because of a desire to protect their interests in profitable icing-station monopolies.[10] This may or may not have been true, but no such explanation was necessary to find ample reason for reluctance to abandon existing cars.

Fundamental changes did not take place, but the traditional car was improved in a building program begun in 1935 within the framework of new specifications set forth by the Association of American Railroads and embodying recommendations of the Department of Agriculture.[11] One improved version of the standard type was the steel-sheathed unit adopted by the General American Transportation Corporation for its subsidiary, the Union Refrigerator Transit Company.[12]

At the end of the second World War, as after the first, a great opportunity presented itself to modernize the ventilator refrigerator car. Much of the equipment was ancient, and practically all of it was badly worn from the rigors of

[9] Kelley, *loc.cit.*, 304; L. I. Denton, "Refrigerator Car Precooling," *Ice and Refrigeration*, XCIX (1940), 105-106; A. M. Casberg, "Refrigeration Applications In the Fruit and Vegetable Industry," *Ice and Refrigeration*, CI (1941), 84; *Chicago Sunday Tribune*, June 22, 1947; P. H. Montgomery, "Railway Transportation of Perishable Products," *Refrigerating Engineering*, L (1945), 109; *Ice and Refrigeration*, CX (April 1946), 64.

[10] D. A. Munro, "Civil War Splits America's Railroad Empire," *New Republic*, CXVI (Jan. 20, 1947), 23.

[11] Kelley, *loc.cit.*, 304.

[12] "Steel Refrigerator Adopted as Standard," *Railway Age*, CI (July 18, 1936), 122-123.

wartime service.[13] In 1944 the United Fresh Fruit and Vegetable Association moved to take advantage of this opportunity by appointing a refrigerator-car committee composed of representatives of shipper and receiver interests from all over the country. The accomplishments of the committee were threefold. First, it gave the fruit and vegetable industry a unified voice in its call for improvement. Second, it drew up recommendations for changes which it considered currently practical and believed should be included in all new cars. These recommendations required no basic alteration in methods of refrigeration, but they did call for changes that would make service more satisfactory. The bunkers were to be collapsible so that a full load could be carried when ice was unnecessary. Insulation was to be heavier, with the roof and floor having the greatest protection. Fans were to be provided and so arranged that they would reverse normal air circulation, forcing the air up through the bunkers and out over the top of the load. To make possible higher speeds and less damage to produce, easy-riding trucks, improved draft gears, steel wheels, and lighter construction were prescribed. The third achievement of the committee was to stimulate the beginning of a test program under the auspices of the Association of American Railroads and the United States Department of Agriculture which promised to yield new recommendations for improved efficiency.[14]

The major trend after the war was construction of cars along the lines suggested by the United Fresh Fruit and Vegetable Association. Both the railroads and the private-car lines accepted enthusiastically the new ideas, perhaps because they saw the necessity of improving their service if they were to check the inroads of motor-truck competition.[15]

[13] P. B. Reed, "Refrigerated Transport," *Refrigerating Engineering*, LII (1946), 121.

[14] J. N. Kelley, "Postwar Redesign of the Refrigerator Car," *Refrigerating Engineering*, LIII (1947), 112, 114-115; *Ice and Refrigeration*, CVII (Oct. 1944), 24-25; CIX (Aug. 1945), 20; CXII (May 1947), 38; *Chicago Sunday Tribune*, June 22, 1947.

[15] Kelley, "Postwar Redesign of the Refrigerator Car," *loc.cit.*, 112-113.

Of the 110,000 general-service refrigerator cars in operation in January 1950 about 35,000 units, both new and rebuilt, conformed to the suggestions of the fruit and vegetable industry.[16] Ice remained supreme as the cooling agent and promised to maintain its position so long as it was cheaper and easier to use.[17] But a great deal of interest in mechanical refrigerator cars still prevailed, and a variety of methods were proposed. Of these the most promising was probably the split-absorption system, so-called because only half of the usual ammonia-absorption cycle took place on the car. The ammonia in the absorber tank was not continuously recovered as in normal usage; instead the ammonia-laden water was withdrawn at a station, where the refrigerant in solution was separated by heating for addition to the generator of another car. The system had the advantage of containing no moving parts except a valve. Furthermore, it required fewer stops for servicing on a cross-country run than did ice. It was more expensive than the ice-and-salt method, but its promoters believed that this disadvantage would be eliminated if operations were conducted on a large scale.[18]

Brine-tank refrigerator cars continued to be used for meat, poultry, eggs, and certain other products despite some opinion that the basket-bunker type charged with ice and salt provided sufficient cooling.[19] In the twenties some cars were equipped with pipes suspended from the ceiling through which brine was circulated,[20] but more important was the brine-

[16] E. A. Gorman, Jr., "Produce Protection in Rail Transit," *Refrigerating Engineering*, LVIII (July 1950), 670.

[17] D. F. Fisher, "Current Investigations on the Transit Refrigeration of Perishables," *Ice and Refrigeration*, CX (Feb. 1946), 22; Gorman, *loc.cit.*, 671.

[18] "Rolling Warehouse," *Chemical Industries*, LXII (1948), 387-388. For a report on tests by the United States Department of Agriculture see H. D. Johnson and D. H. Rose, "Mechanical Refrigerator Car Announced—Tests Conducted by U. S. Department of Agriculture," *Ice and Refrigeration*, CXIII (July 1947), 21.

[19] Winterrowd, *loc.cit.*, 145-146.

[20] *ibid.*, 187; R. M. Kramer, *Transportation in Relation to the Export Trade in Agricultural Products* (U. S. Department of Commerce, Bureau of Foreign and Domestic Commerce, *Trade Information Bulletin No. 216*) (Washington, 1924), 23.

retention system introduced in the thirties. The usual battery of four tanks was kept, but drainage from them was so arranged that it could be controlled by a single valve. With the drain cock closed, the tanks were charged with crushed ice and rock salt. After the car had received its load of meat, salt brine was added to within four inches of the tops of the tanks. The brine served two purposes: it added to the amount of refrigeration available, but much more important, by filling the spaces between the ice and establishing solid contact with the tank walls, it improved their heat-transferring qualities. Fewer re-icings were required; better refrigeration was assured at reduced cost.[21]

A word should be said about efforts to adapt refrigerated rail transport to the requirements of the frozen-food industry. Such adaptation was essential, not only because close control of temperature was necessary if frozen products were to reach the consumer with their original high quality unimpaired, but because of motor-truck competition, which was particularly strong in this field. After 1930 a great amount of experimentation took place. Overhead brine tanks were developed. These provided uniform low temperature at relatively low cost, but they were expensive to build and required considerable time to be iced properly. Especially heavy insulation was devised, and end-bunker cars were equipped with fans.[22] It appeared that such modifications of the traditional ice-and-salt unit would carry the brunt of the load until satisfactory mechanical equipment was adopted.[23]

The motor truck entered the refrigerated-transportation picture in the nineteen-twenties. The first attempts were primitive, sometimes involving no more than an ice-filled vinegar barrel and a pad or blanket hung on the side of the body for

[21] H. A. Glenn, "New Brine Refrigeration System for Refrigerator Cars," *Ice and Refrigeration*, xci (1936), 6-8.

[22] A. L. Reneau, "Frozen Food Transportation and Distribution," *Ice and Refrigeration*, cxvii (Oct. 1949), 17; A. E. Stevens, "Optimum Conditions for Frozen Foods in Refrigerated Storage," *Ice and Refrigeration*, ci (1941), 209; Rowland, *loc.cit.*, 49.

[23] The Chicago, Burlington, and Quincy Railroad undertook experiments with dry ice with a view to its possible use in the frozen-food traffic. *Ice and Refrigeration*, cxii (June 1947), 23.

insulation.[24] But improved techniques soon made their appearance. First came ice, sometimes mixed with salt, placed either in iron drums or in bunkers. As in railroad cars the principal advantage of ice was its low cost, particularly when a vehicle was not used constantly for refrigerated shipments. This was counterbalanced to some extent by its weight and the fact that meltwater was apt to penetrate and ruin the insulation. Solid carbon dioxide was tried in the later twenties. Despite its expense and the difficulty of controlling it, dry ice was used widely, especially on smaller trucks. Mechanical units appeared about 1930. With the perfection of the small commercial machine they became increasingly important, much more so than in rail transport, for trucks could be operated with so high a refrigerated-load factor that the cost was not prohibitive. Moreover, the driver was always on hand at no extra expense to check the performance of the machine. Another development was hold-over plate refrigeration, which involved flat metal cells filled with a solution of low-freezing point which were mounted on the ceiling and walls of the truck body. The solution was frozen at night by pumping refrigerant from a stationary plant through coils built into the cells. The plates remained cold throughout the next day. Though this system was not expensive to operate, its initial cost was high, its weight excessive, and its heat absorption uneven.[25]

Marine refrigeration was a difficult field, for the machinery had to be capable of maintaining the desired degree of cold in the cargo space despite considerable fluctuations in the temperature of seawater and atmosphere.[26] After the first

[24] B. D. Davidson, "Refrigerated Motor Truck Service for Frozen Foods," *Ice and Refrigeration*, CX (Jan. 1946), 50.

[25] H. W. Krotzer, "Truck Refrigeration," *Refrigerating Engineering*, L (1945), 324; D. Albert, "Truck-Trailer Refrigeration," *Refrigerating Engineering*, LV (1948), 32-33; "Refrigerated Trucks Essential in Sale of Frozen Foods," *Refrigerating Engineering*, XL (1940), 286; D. K. Tressler and C. P. Evers, *The Freezing Preservation of Foods*, 2nd edn. (New York, 1947), 689-691; Letter, D. Albert, San Mateo, Calif., to writer, May 21, 1952.

[26] L. L. Westling, "New Food Ships to Aid Development of Foreign Markets," *Food Industries*, XVII (Aug. 1945), 102.

World War, during which the importance of refrigerated vessels was dramatized, better equipment and techniques were introduced. The most common refrigerant was carbon dioxide, but some ammonia was still used, and the Freon refrigerants, like carbon dioxide quite safe, were enlisted. Most new installations were powered by electricity, but Diesel engines sometimes were relied upon. Cargo space was cooled by brine coils located along the bulkheads behind baffles or placed in batteries through which air was forced.[27] As in the early days of marine refrigeration the British held the lead, for their very existence depended on perishables brought in from overseas.[28] Most of the shipboard machinery was of British make, but the number of installations by American manufacturers increased.[29] Perhaps the most important work done by Americans in this field was that accomplished by the engineers of the United Fruit Company, who devised satisfactory vessels for bringing bananas north from the plantations of the Caribbean.[30]

[27] "A Survey of Marine Refrigerating Machinery," *Refrigerating Engineering*, XXXI (1936), 168; "A Silent Service without which Britain Would Starve: Modern Methods in the Refrigerated Food Ships of To-Day," *The Illustrated London News*, CCXII (May 15, 1948), 540-541; S. F. Dorey and D. Gemmell, "Developments in Marine Refrigeration," *Actes du VIIe Congrès International du Froid . . . :* , IV (Utrecht, 1937), 450-473; Westling, *loc.cit.*, 908; A. C. Rohn and J. H. Clarke, "Refrigeration Equipment for Merchant Ships," *Refrigerating Engineering*, XLIV (1942), 10-11; S. W. Brown, "Modern Developments in Marine Refrigeration," *Refrigerating Engineering*, LV (1948), 36-41, 86, 88, 90; A. R. T. Woods, "Transport of Refrigerated Cargoes Under Modern Marine Practice," *Engineering*, CXXXIX (1935), 636-640.

[28] For an indication of British consciousness of the importance of marine refrigeration see "A Silent Service," *loc.cit.*, 540-541.

[29] "A Survey of Marine Refrigerating Machinery," *loc.cit.*, 168-169.

[30] C. F. Greeves-Carpenter, "Refrigeration Has Built the Banana Industry," *Refrigerating Engineering*, XXVII (1934), 67-68.

CHAPTER XV

COLD STORAGE: IMPROVEMENT' AND
ACCEPTANCE, 1917-1950

THE improvements that took place in the cold-storage field after 1917 were of great significance because they tended to enlarge the role played by refrigerated preservation in the distribution of our food resources.[1] Yet it is difficult to generalize about the advances in warehouses and equipment. Although the universal objective was the preservation of foods under optimum conditions, the methods used to attain this goal varied greatly.[2]

Refrigerated warehouses generally were fireproof structures built largely of concrete. Many different kinds of insulation were employed, but the board type, composed of cork, glass, or rock fibers, became most common. In choosing insulation, great attention was directed toward its resistance to infiltration by moisture, for if it became water-soaked, it did not offer effective resistance to heat transfer. The best warehouses featured a "curtain wall," an outer shell entirely independent of the rest of the building. To its inner face some material such as cork board was applied so that the entire storehouse was enclosed by an unbroken sheath of insulation.[3]

The cooling equipment probably most generally used was two-stage ammonia compression. For distributing refrigeration to the actual storage rooms, the brine-circulation method held sway in the twenties and thirties. Later, brine installa-

[1] W. Broxton, "Cold Storage Reporting," *Proceedings of the Fiftieth Annual Meeting of the American Warehousemen's Association* (Chicago, 1941), 283; J. D. Black, *Introduction to Production Economics* (New York, 1926), 818.

[2] E. Brandt, "Refrigerated Warehouses—1949 Plants and Methods," *Ice and Refrigeration*, CXVI (Jan. 1949), 32.

[3] F. A. Horne, "Cold Storage Warehousing in the United States," *Ice and Refrigeration*, LXVII (1924), 148; G. A. Horne, "New Cold Storage Warehouse of Merchants Refrigerating Co.," *Ice and Refrigeration*, LXIX (1925), 378; Brandt, *loc.cit.*, 32-33; J. A. Moyer and R. U. Fittz, *Refrigeration* (New York, 1928), 343-344; D. K. Tressler and C. P. Evers, *The Freezing Preservation of Foods*, 2nd edn. (New York, 1947), 49-51.

tions continued to be used because of their inherent stability, safety, and ease of control, but more and more direct-expansion systems were introduced. A major reason for this change was the greater demand that existed for freezer space.[4] The most important innovation after 1935 no doubt was the introduction of unit coolers, which consisted of coils and fans enclosed in a casing in the room to be cooled. The fans forced the air across the coils and through louvers which directed its delivery. Such units not only saved space, but made possible higher relative humidity, better air distribution, and more rapid chilling.[5]

That low temperature alone was enough for proper food preservation was an erroneous principle which too often formed the basis of early operations. But gradually it was understood that ideal refrigerated storage involved accurately controlled temperatures and close attention to the factors of relative humidity, air motion, and air purity.

Greater concern with the exact temperature needed for the storage of each commodity was the rule. A strong tendency was evident toward the use of lower levels all along the line. The range at which eggs were carried dropped from between thirty-two and thirty-four degrees Fahrenheit to between twenty-nine and thirty-one. Ten degrees above zero used to be considered satisfactory for freezer space, but now five to ten below or even lower was preferred.[6]

Maintenance of the proper relative humidity was a problem of prime importance. Too much moisture created an environment conducive to the rapid growth of molds and bacteria. Too little moisture resulted in the desiccation of the products stored.[7] Both conditions were troublesome, but in most cases

[4] Brandt, *loc.cit.*, 33; T. W. Heitz, *The Cold Storage of Eggs and Poultry* (U. S. Department of Agriculture, *Circular No. 73*) (Washington, 1929), 9.

[5] H. C. Hoover, "Air Movement in Refrigeration," *Refrigerating Engineering*, XLIII (1942), 285-286.

[6] M. E. Pennington, "Refrigerated Warehousing of Tomorrow," *Ice and Refrigeration*, CVI (1944), 97.

[7] M. W. Browne, "Humidity Control in Cold Storage Warehouses," *Ice and Refrigeration*, LXIII (1922), 142; J. A. Hawkins, "Cold Storage Operation," *Ice and Refrigeration*, XC (1936), 82.

in actual practice the principal danger was excessive loss of water content. In 1920 cold-storage operators were just beginning to see the importance of the humidity factor, but by 1930 its necessity was recognized generally.[8] At first, methods of increasing the relative humidity were often crude. Floors were sprinkled, and wet blankets suspended in the storage; sometimes low-pressure steam was blown in.[9] Soon a great number of methods were used. One was reduction of the temperature differential between the air of the storage and the refrigerating surfaces, while another was introduction of moisture by means of mist nozzles.[10]

Forced air circulation became a feature of most modern cold-storage plants. Fans, of course, were used to set the air in motion. A variety of arrangements were employed to direct it to all parts of the refrigerated space. The principal advantage of forced air circulation was the better and more rapid distribution of available refrigeration that was achieved. It also prevented the accumulation of gases, the product of the respiration of stored fruits, which if not removed served as a blanket of insulation that interfered with proper cooling.[11]

Air purity was essential. Were it not maintained, conditions would be favorable for the transfer and propagation of bacteria and mold spores. An important aid was good circulation, for it assured the deposition of impurity-laden moisture on the cold surfaces of the refrigerating coils. Ventilation, or the

[8] Browne, *loc.cit.*, 100; M. W. Browne, "Humidity and Air Circulation in Cold Storage," *Refrigerating Engineering*, xxi (1931), 27.

[9] M. O. Soroka, "Air Conditioning in Industry—1—Handling of Foods," *Heating and Ventilating*, xxxiii (Nov. 1936), 57; J. R. Magness, "Handling of Apples and Pears in Storage," *Proceedings of the Thirty-second Annual Meeting of the American Warehousemen's Association* . . . (Pittsburgh, 1923), 388-389; "Report of the Relative Humidity Experimentation Committee," *Proceedings of the Thirty-sixth Annual Meeting of the American Warehousemen's Association* . . . (Chicago, 1927), 177.

[10] Brandt, *loc.cit.*, 34; P. B. Christensen, "Humidity in Cold Storage," *American Egg and Poultry Review*, v (1944), 426, 430; M. Kalischer, "The Role of Humidity in Air Conditioning and Refrigeration," *Refrigerating Engineering*, xxxvii (1939), 177.

[11] Moyer and Fittz, *op.cit.*, 351-357; W. H. Motz, *Principles of Refrigeration*, 280-281; Hawkins, *loc.cit.*, 82.

gradual introduction of outside air, also was helpful.[12] Removal of moisture assisted, but did not assure absolute purification, for some gases had little or no affinity for water. To eliminate these, ozone, formed by the passage of an electrical discharge through dry air, was employed. Since ozone contained an extra atom of oxygen which readily separated and combined with any carbonaceous material present, its effect was to oxidize organic impurities.[13] A late development was the removal of the ethylene given off by ripening fruits by passing air through canisters of coconut-shell activated carbon.[14]

The amount of refrigerated warehouse space in the United States increased rapidly in the decade and a half after the Armistice. Between 1923 and 1931 almost 120,000,000 cubic feet were added to the capacity of public warehouses alone. This was an increase of about fifty-eight per cent.[15] The war itself was in part responsible. Capacity had been enlarged during the emergency, and with the peace the government released the space it had occupied. These disturbances were aggravated by Prohibition, which prevented breweries that had been converted to cold storage during the war from being returned to their normal function. The abundance of capital seeking investment during the twenties did not help the situation, since it tempted promoters to sponsor large terminal warehouses. A final factor was competition among

[12] E. L. Carpenter and M. Tucker, *Farm and Community Refrigeration* (University of Tennessee Engineering Experiment Station, *Bulletin No. 12*) (Knoxville, 1936), 12; Browne, "Humidity Control in Cold Storage Warehouses," *loc.cit.*, 294-295; Motz, *op.cit.*, 258; Moyer and Fittz, *op.cit.*, 357-358.

[13] P. R. Moses, "The Heating, Ventilating and Air-Conditioning of Factories," *The Engineering Magazine*, xxxix (1910), 880; A. W. Ewell, "The Use of Ozone in Cold Storage Plants," *Ice and Refrigeration*, xci (1936), 295-296; P. N. Gliddon, "Ozone and Cold Storage," *Ice and Refrigeration*, lxvii (1924), 7-8; Hawkins, *loc.cit.*, 82; Browne, "Humidity Control in Cold Storage Warehouses," *loc.cit.*, 294.

[14] R. M. Smock, "Controlled Atmospheres Improve Refrigerated Storage," *Food Industries*, xxi (April 1949), 107; A. Van Doren, "Air Purification in Refrigerated Storage Rooms," *Ice and Refrigeration*, cxvi (May 1949), 61.

[15] U. S. Department of Agriculture, *Agricultural Statistics, 1944* (Washington, 1944), 571; Hawkins, *loc.cit.*, 82.

the railroads, whose efforts to attract perishable-freight shipments resulted in the construction of huge railroad-owned refrigerated warehouses at the major terminals.[16] Overcapacity disorganized the cold-storage industry, for it forced managers to strive desperately to attain higher levels of occupancy and thus made rate cutting inevitable.[17]

During the second World War an unprecedented demand for space developed. An indication of its extent was the fact that on August 1, 1944, cold-storage holdings were roughly sixty per cent above the average for the five years 1936 through 1940. Existing storage capacity was taxed severely, particularly in the invasion year of 1944. Warehouses frequently were overcrowded, and at times food was stored where facilities were not adequate. Yet spoilage of purchases by the War Food Administration amounted to less than one-tenth of one per cent. First among the factors which explained the space shortage was the war-induced expansion of the production of perishables. The handling of these products was complicated by the curtailment of canning preservation necessitated by short tin supplies, by the increase in egg drying, and by wartime limitations on the construction of new space.[18] It must be pointed out that the cold-storage industry was not enthusiastic about undertaking new construction. In 1940 the National Association of Refrigerated Warehouses appointed a defense committee to work with the appropriate govern-

[16] F. A. Horne, "Present Outlook for the Cold Storage Industry," *Ice and Refrigeration*, LXIII (1922), 90; F. A. Horne, "Summary of Conditions and Trends in the Cold Storage Industry Today," *Refrigerating Engineering*, XIX (1930), 105; F. Chow, *Railroad Warehousing Service* [Shanghai, 1931?], 63-64, 66-67.

[17] For summaries of conditions in the industry in the thirties see *Ice and Refrigeration*, LXXXIV (1933), 33; LXXXVI (1934), 49; LXXXVIII (1935), 46; XCI (1936), 403. The code of fair competition for the refrigerated warehousing industry approved in August 1934 contained provisions aimed at control of new capacity and the elimination of price cutting. "Code of Fair Competition for Refrigerated Warehousing Industry," *Ice and Refrigeration*, LXXXVII (1934), 125-126.

[18] *Ice and Refrigeration*, CIII (1942), 223-224; CVII (Aug. 1944), 25; CVII (Oct. 1944), 33; M. Jones, *How Food Saved American Lives* (Washington [1947]), 56, 156.

ment agencies in an effort to see that space was not inflated as it had been in 1917-1918.[19]

The shortages of the war were met in part by new construction, some of which was financed privately and some by the Defense Plant Corporation.[20] The most ingenious project of the War Food Administration was its conversion of a limestone cave at Atchison, Kansas, into a storage with a capacity of between seven and eight million cubic feet.[21] Other measures designed to ease the emergency were conversion of cooler to freezer space, prohibition by the WFA of the storage of certain commodities for whose preservation refrigeration was not essential, and more intensive utilization of existing room.[22] It seemed unlikely that the additional space created would give rise to a new problem of overcapacity.[23]

Refrigerated-warehouse space in the United States was concentrated in three general locations: producing areas, intermediate areas between the points of production and consumption, and terminal areas near the large centers of population.[24]

A trend toward more storage in producing areas was apparent, particularly in the case of apples.[25] As early as 1914, though apples still were stored mainly in metropolitan warehouses, the new tendency was evident.[26] As late as 1925 almost all New England apples kept under refrigeration were

[19] E. R. Curry, "The Refrigerated Warehousing Industry—From Early Days to 1949," *Ice and Refrigeration*, cxvi (Jan. 1949), 27-28.

[20] *ibid.*, 27-28; *Ice and Refrigeration*, cvii (Aug. 1944), 25, 27.

[21] *Ice and Refrigeration*, cvii (Oct. 1944), 35. For sharp criticism of this project see "Investigation of the War Food Administration," *House Report*, 79 Cong., 1 sess., no. 816, p. 10-11.

[22] J. R. Shoemaker, "A Year of Industry Performance—What It Has Been and What It Needs To Be," *Ice and Refrigeration*, civ (1943), 163; *Ice and Refrigeration*, cv (1943), 122; cvii (Aug. 1944), 27; cviii (Jan. 1945), 31.

[23] Curry, *loc.cit.*, 28.

[24] Horne, "Cold Storage Warehousing in the United States," *loc. cit.*, 149.

[25] E. A. Duddy, *The Cold-Storage Industry in the United States* (Chicago, 1929), 73; J. C. Folger and S. M. Thomson, *The Commercial Apple Industry of North America* (New York, 1921), 334-335. Hawkins, *loc.cit.*, 81.

[26] M. Cooper, *Practical Cold Storage*, 2nd edn. (Chicago, 1914), 342.

warehoused in the larger cities, but ten years later almost forty-four per cent were placed in cold storage at country points. Two technological factors which made this decentralization possible were the perfection of the small commercial machine and the extension of electric facilities to rural areas. An economic factor of importance was the easy credit made available by the federal government.[27] Storage near the point of production had definite advantages. It reduced the danger of deterioration between harvest and storage. It was apt to be cheaper, in part because land values and construction costs were lower in rural areas. But most important to the grower, it enabled him to control profitably the flow of his harvest to market.[28]

Warehousing at intermediate points, which was encouraged by railroad storage-in-transit tariffs, gave the grower flexibility in the choice of a market, while storing in terminal areas enabled him to take advantage of sudden price fluctuations and to make prompt deliveries.[29] The dominance of terminal storage in the nation's food-distribution system was shown by the concentration of space in urban districts. In 1945 forty-eight per cent of the public-warehouse capacity was located in ten city areas, while sixty-five per cent was to be found in twenty-seven cities.[30]

The edibles kept in cold storage in greatest volume con-

[27] J. Howard, "Cold Storage—a Burning Issue," *American Fruit Grower*, LVI (July 1936), 5-6. Some apples were kept at country points without the benefit of refrigeration. W. V. Hukill and E. Smith, "The Cold Storage of Apples," U. S. Department of Agriculture, *The Yearbook of Agriculture, 1943-1947* (Washington, 1947), 868.

[28] Duddy, *op.cit.*, 49-50; Folger and Thomson, *op.cit.*, 334-335; Hawkins, *loc.cit.*, 81; A. W. McKay et al., "Marketing Fruits and Vegetables," *Agriculture Yearbook, 1925* (Washington, 1926), 674; "The Growth of Modern Cold Storage on the Fruit Farm," *American Fruit Grower*, LXI (June 1941), 13.

[29] Horne, "Cold Storage Warehousing in the United States," *loc. cit.*, 149; Hawkins, *loc.cit.*, 81; Duddy, *op.cit.*, 48-49.

[30] U. S. Department of Agriculture, Production and Marketing Administration, Marketing Facilities Branch, *A Survey of Cold Storage Space as of October, 1945*, by V. M. Queener and A. S. Walker (Washington, 1946), 10, 23. For a map showing total gross storage space by states see *ibid.*, 22.

tinued to be meats, fish, fruits, eggs and poultry, and dairy products. Among the newcomers were dehydrated foods and canned goods. The former had to be kept at cool temperatures until one to three months from the time they were to be consumed. The latter did not have to be refrigerated to keep, but it was discovered that cold storage checked losses in both vitamin content and flavor.[31] Many nonedibles were still preserved in the nation's refrigerated warehouses.

Public hostility to the cold-storage industry flared up after the first World War. The basis of the renewed attack was partly no more than lingering prejudice against refrigeration as a means of food preservation. Most of this prejudice was unreasonable, but to a limited extent it was justified, for in the early days methods and techniques were poor. The public still regarded storage food as suspect; it had not learned to judge a product by its inherent quality. But the major basis for the renewed assault on the industry was a widespread belief that the speculation in perishable food products made possible by the cold-storage warehouse was raising food prices.[32]

Efforts to pass a federal cold-storage law were the outcome of public antagonism. Both in 1919 and 1920 President Wilson included an appeal for legislation in his annual message to Congress.[33] Several bills were introduced with objectives quite similar in character. They aimed at supervision over sanitary conditions. They sought to prevent deception by requiring that cold-storage foods be labelled as such and marked as to the length of the storage period. They called for a limitation, usually twelve months, on foods kept under refrigeration, and

[31] C. Birdseye, "The Effect of Perishable Food Preservation on the Electric Utility," *Edison Electric Institute Bulletin*, XII (1944), 143.

[32] F. A. Horne, "Economic Aspects of the Development of Cold Storage," *Ice and Refrigeration*, LXI (1921), 6-7; J. O. Bell and I. C. Franklin, *Reports of Storage Holdings of Certain Food Products* (U. S. Department of Agriculture, *Bulletin No. 709*) (Washington, 1918), 3; U. S. Congress, House, Committee on Agriculture, *Cold-Storage Legislation. Hearings . . . August 11-26, 1919*, 66 Cong., 2 sess. (Washington, 1919), passim; "Cold Storage Bill," *House Report*, 66 Cong., 1 sess., no. 337 (Sept. 25, 1919), 1-2.

[33] *Congressional Record*, 66 Cong., 2 sess., 30; 66 Cong., 3 sess., 26.

they required full reports to federal authorities on total holdings.[34]

Warehousemen were not opposed to the principle of federal regulation. Pleased by the results of wartime supervision by the Food Administration, the Cold Storage Subdivision of the American Warehousemen's Association went on record at its 1918 convention to the effect that a continuance of federal regulation would be wise. It was felt that it would stabilize the industry and relieve the inconvenience it suffered as a result of the divergencies of state laws.[35] Industry spokesmen criticized certain aspects of the House and Senate bills that seemed likely to become law in 1920 and 1921, but they did not try to block the main provisions.[36]

No federal law was adopted, though it is not clear why, for each house of the Sixty-sixth Congress passed a bill. Considerable difficulty was experienced in reconciling the two,[37] but on March 3, 1921, a conference report was submitted which would seem to have resolved the principal differences. The Senate accepted it, but the House failed to act before adjournment.[38]

[34] Horne, "Economic Aspects of the Development of Cold Storage," *loc.cit.*, 9-10; "Cold Storage Bill," *loc.cit.*, 2-3; U. S. Congress, Senate, Committee on Agriculture and Forestry, *Cold Storage: Hearing ... on H. R. 9521 ...* , 66 Cong., 2 sess. (Washington, 1920), 37-40.

[35] It was resolved that "in our judgment it would be wise at this time to continue a measure of Federal regulation under some governmental department known to be constructive in work, in order to promote conservation, facilitate distribution, stimulate production, and eliminate wasteful, discriminatory, and unfair practices in the handling of the Nation's food supply, and in order to relieve the industry from the unnecessary annoyances and losses incidental to divergent State laws." "Cold Storage Bill," *loc.cit.*, 3-4.

[36] U. S. Congress, Senate, Committee on Agriculture and Forestry, *op.cit.*, 7-35, 40-57; F. A. Horne, "The Present Status of Federal Cold Storage Legislation," *Ice and Refrigeration*, LIX (1920), 188-189.

[37] Western representatives feared that the products of their section would be designated as "cold storage" as a result of their preservation by refrigeration during the long trip to market. *Congressional Record*, 66 Cong., 3 sess., 3692-3697.

[38] *ibid.*, 4400-4403, 4493-4496; "Combined Report of the Committee on Legislation of the American Warehousemen's Association and Joint Committee Representing Cold Storage Warehousemen and Affiliated Industries," *Proceedings of the Thirty-first Annual Meeting*

Cold Storage: Acceptance, 1917-1950

Though no federal law directed specifically against refrigerated warehousing came into being, the national government nevertheless exercised a considerable amount of supervision and control. One of the devices used was the cold-storage reporting of the Department of Agriculture. Statements on holdings by warehousemen, begun on a voluntary basis in 1914 and expanded in 1916 and 1917, were made mandatory during the war by the Food Administration. When the emergency passed, voluntary reporting was resumed. Gradually its scope was broadened. Many additional commodities were covered, and space and occupancy surveys were initiated. The publicity thus given to cold-storage operations not only protected the public from possible economic abuses, but also provided warehousemen with needed business information.[39] Another federal control was the United States Warehouse Act of 1916, designed to strengthen the integrity and acceptability of warehouse receipts. To accomplish this, the Secretary of Agriculture was authorized to license public warehouses storing certain agricultural commodities provided that prescribed storage conditions prevailed. Cold-pack fruit was the most important cold-storage product placed under the scope of the act.[40] Control over sanitary conditions was accomplished through the work of the Food and Drug Administration and the meat-inspection service of the Department of Agriculture. This control, of course, extended only to foods that entered interstate commerce.[41]

---

of the American Warehousemen's Association . . . (Pittsburgh, 1922), 287-289.

[39] Broxton, *loc.cit.*, 280-284. For a typical report see U. S. Department of Agriculture, Production and Marketing Administration, Marketing Facilities Branch, *Cold Storage Report*, xxxi (Dec. 1, 1946).

[40] J. H. Frederick, *Public Warehousing: Its Organization, Economic Services and Legal Aspects* (New York, 1940), 228, 231n. See *Regulations for Warehousemen Storing Cold-Pack Fruit; Approved July 2, 1940 Amended September 20, 1940* (U. S. Department of Agriculture, Agricultural Marketing Service, *Service and Regulatory Announcements No. 159*) (Washington, 1940).

[41] H. F. Taylor, "Legislation Affecting Refrigerated Foods," *Ice and Refrigeration*, lxxx (1931), 456-457. See Federal Security Agency, *Annual Report . . . for the Fiscal Year 1946. Section 1, Food and Drug Administration* (Washington, 1946), 18-19, 21-22.

Several states passed cold-storage laws in the twenties, but in 1940 only twenty had legislation directed specifically toward refrigerated warehousing. The others relied on pure-food laws for the protection of the consumer. The statutes did not depart significantly from the pattern set before the first World War. Time limits usually were set at twelve months, with ample provision made for extension if the foods were found to be in good condition.[42] In some states lack of enforcement made these laws for practical purposes dead letters; in no state were they particularly troublesome to warehousemen.[43] But the industry believed, quite understandably, that state laws should be uniform. It objected to the customary requirement that poultry, eggs, and butter be sold as cold-storage products. It held that eggs should be marketed by grade alone and that poultry, meat, butter, and fish should be designated merely as "frozen" or "frosted."[44]

A marked decline in public prejudice against cold storage took place after the early twenties. Though hostility did not disappear entirely, it was no longer a serious threat.[45] Among the forces responsible for this change were the public-relations efforts of the industry[46] and the contributions of refrigeration during the two world wars.[47] The drop of food prices in the

[42] For summaries of state laws see "Store Frozen Foods for Proper Distribution," *Quick Frozen Foods*, I (April 1939), 24-25; Frederick, *op.cit.*, 137, 140-143.

[43] R. H. Switzler, "Refrigerated Warehousing over the Years," *Proceedings of the Fiftieth Annual Meeting of the American Warehousemen's Association* (Chicago, 1941), 67.

[44] "Report of Committee on State Legislation," *Proceedings of the Fiftieth Annual Meeting of the American Warehousemen's Association* (Chicago, 1941), 378.

[45] F. A. Horne, "Cold Storage and Our Food Supply," *Refrigerating Engineering*, XI (1924-1925), 429; American Public Health Association, Committee on Foods, "Refrigeration of Foods and Its Relation to Present-Day Public Health," American Public Health Association, *Yearbook, 1935-1936*, 48; V. O. Appel, "Freezing and Refrigerated Storage of Poultry and Poultry Products in the United States," *Ice and Refrigeration*, CXV (Sept. 1948), 40.

[46] For an account of this see remarks by F. A. Horne in discussion following J. Raymond, "Some Notes on Propaganda on Behalf of Refrigeration," The Fourth International Congress of Refrigeration . . . , *Proceedings*, II (London [1925?]), 1614.

[47] Horne, "Cold Storage and Our Food Supply," *loc.cit.*, 430; Appel, *loc.cit.*, 40.

nineteen-twenties no doubt undermined the conviction that speculators were using the warehouse to gouge the consumer. But more important were the factors that convinced the public that there was nothing wrong with refrigerated preservation as such. Improvements in the technique of storage resulted in more attractive and palatable products. Frozen foods and vegetables became symbols of high quality and modern living.[48] Perhaps most significant in explaining the decline of prejudice was the widespread use of domestic refrigeration. The housewife learned by experience in her own kitchen that most of her old fears were groundless.

[48] Appel, *loc.cit.*, 40.

CHAPTER XVI

FOOD PRODUCTION AND DISTRIBUTION,
1917-1950

IN the meat-packing industry refrigeration retained its basic importance. As soon as life was extinct in a butchered animal, its natural heat had to be removed rapidly.[1] The growing recognition of this fact among packers[2] was reflected in the adoption of improved facilities for chilling promptly the warm carcasses. The most common system, a holdover from an earlier day, featured coils banked in overhead bunkers and natural air circulation. Expensive, inefficient, and unsanitary, it was seldom used in new installations. Sometimes, however, its functioning was improved by providing for a brine spray over the coils and by adding a fan to force circulation. More satisfactory were systems in which brine sprays mounted in overhead decks were substituted for coils. The greatest change in chill-room practice was the introduction of unit coolers, compact affairs equipped either with coils or brine sprays and with a fan. They won favor not only because of their refrigerating efficiency, but also because of their low initial cost and their small size, which by permitting lower ceiling heights, afforded savings in building costs.[3]

When thoroughly chilled, carcasses were either shipped, as was the case with most beef and fresh pork, stored in holding coolers at temperatures above thirty-two degrees, or sent to freezing rooms. The function of the holding cooler, where refrigeration was applied in much the same way as in the chill

---

[1] H. J. Koenig, "Handling, Packing and Shipping of Freezer Meats," *Ice and Refrigeration*, LXXVI (1929), 25.

[2] Letter from H. C. Gardner in American Association of Ice and Refrigeration, Educational Committee, "Report" [at the Twelfth Annual Meeting], *Ice and Refrigeration*, LXIV (1923), 472.

[3] V. F. Self, "Air Motion in Refrigerated Spaces," *Refrigerating Engineering*, XLII (1941), 293-294; J. S. Bartley, K. E. Wolcott, and J. P. McShane, *Meat Packing Plant Refrigeration. Part I. Carcass Coolers* (American Society of Refrigerating Engineers, *Refrigerating Engineering Application Data—Section 45*) (Published as Section 2 of *Refrigerating Engineering* for March 1949), 2-3; *The Annual Meat Packers Guide, 1946* (Chicago, 1946), 71.

room, was to keep the meat from spoiling while it aged or cured or awaited processing. After the development of air conditioning much attention was given to the maintenance of relative humidities which would check dehydration and the development of mold. The length of time meats had to be kept in the holding cooler was reduced. Hams were cured more quickly by a process which featured pumping the pickle through an artery so that it penetrated at once to the interior of the cut. Fresh meat was aged more rapidly by permitting aging to proceed at temperatures between forty and sixty degrees and at a relative humidity of ninety per cent. The surface deterioration that might be expected at such high levels of temperature and humidity was checked by the bactericidal action of ultraviolet light.[4]

Although freezing preservation was used for only a small amount of the meat that was retailed as fresh, more meat was frozen each year than any other food. Some of the frozen product went to the hotel and restaurant trade. Some, divided into consumer cuts and packaged before freezing, was destined for the retail market. During each of the world wars vast quantities were boned and frozen for the military forces. But the bulk of the meat frozen was held for subsequent manufacture into sausage and other meat foods or for curing and smoking.[5]

Beginning in the late twenties the packaging and quick

[4] Bartlett, Wolcott, and McShane, *op.cit.*, 7-8; H. D. Tefft, "Refrigeration Functions In a Packing Plant," *Refrigerating Engineering*, xxxvi (1938), 170; L. B. Jensen, *Meat and Meat Foods; Processing and Preservation from Meat Plant to Consumer* (New York, 1949), 19, 38-39; P. B. Christensen, "Modern Beef Chilling: Tenderizing and Chill Room Design, Control and Operation," *Refrigerating Engineering*, xxxix (1940), 296-297.

[5] Jensen, *op.cit.*, 22; D. K. Tressler and C. P. Evers, *The Freezing Preservation of Foods*, 2nd edn. (New York, 1947), 491, 505; V. H. Munnecke, "Operations: Beef, Lamb, and By-Products," *The Packing Industry* (Chicago, 1924), 152; L. D. H. Weld, "The Packing Industry: Its History and General Economics," *The Packing Industry* (Chicago, 1924), 83; G. A. Fitzgerald, "A Look at the Future of Quick Frozen Foods," *Quick Frozen Foods and The Locker Plant*, vi (Dec. 1943), 23; C. W. Towne and E. N. Wentworth, *Pigs, From Cave to Corn Belt* (Norman, 1950), 267-268. For the methods used in freezing meat see Tressler and Evers, *op.cit.*, 503-507.

freezing of meats for sale in their frozen state to the consumer aroused considerable interest.[6] This method of distribution promised several advantages. To the buyer it offered convenience and the opportunity to select purchases from lines of standardized quality. To the packer it made possible savings on shipping costs, greater opportunities for the utilization of waste, and the profits that attended the marketing of a branded product.[7] Despite these advantages, the trade was slow to develop; dreams of large-scale volume failed to materialize. A combination of obstacles seemed responsible. For one thing, the packing plants would have had to be reorganized to take over the retail functions of cutting, trimming, weighing, and packaging, and new equipment would have had to be installed. For another, the entire distribution system extending from refrigerator cars and trucks through dealer display cases to home-storage cabinets would have had to be adapted to the maintenance of freezing temperatures. Packers feared an adverse consumer reaction to higher prices and had no wish to become involved with the butchers' unions, which were hostile to the whole idea.[8]

The earlier trend toward geographic centralization, for which the refrigerator car was so largely responsible, was re-

[6] L. P. Beebe et al., "Developments in the Production and Distribution of Frozen Foods," New York Food Marketing Research Council, *Proceedings of Sixteenth Regular Meeting, December 9, 1930,* 2; Koenig, *loc.cit.*, 25; *Ice and Refrigeration,* LXXIX (1930), 277-278.

[7] "Frozen Meat Future Dim," *Business Week,* March 29, 1947, 62, 65; L. Corey, *Meat and Man; A Study of Monopoly, Unionism, and Food Policy* (New York, 1950), 157; C. O. Brannen, A. C. Hoffmann, and F. L. Thomsen, "Technological Developments in Agricultural Marketing," *Journal of Farm Economics,* XXIX (1947), 312.

[8] "Frozen Meat Future Dim," *loc.cit.*, 62, 65-66; L. B. Mann, *Refrigerated Food Lockers, A New Cooperative Service* (U. S. Department of Agriculture, Farm Credit Administration, Cooperative Division, *Circular No. C-107*) ([Washington] 1938), 26; "Geographic, Seasonal Analysis of 1949 Frozen Food Volume," *Quick Frozen Foods and The Locker Plant,* XII (March 1950), 67; F. W. Specht, "A Packer Looks at Pre-Packaged Meats," *The National Provisioner,* CXV (Nov. 9, 1946), 19; B. Savesky, "The Frozen Food Industry—Modern Merchandizing Turns Dreams into Reality," *Ice and Refrigeration,* CXI (Aug. 1946), 62; "Frozen Foods: Interim Report," *Fortune,* XXXIV (Aug. 1946), 182; Corey, *op.cit.*, 176-177.

versed. The new departure became apparent during the first World War and the early twenties with the establishment of small, independent concerns. In the thirties the big packers themselves began to decentralize either by building plants of their own or by purchasing those of their new competition.[9] Many factors contributed to this shift in the character of the industry. Foremost among them was the motor truck, which gave packers located outside the big centers excellent connections both with livestock-producing areas and with markets. New processes and equipment made smaller plants at least as efficient as the larger.[10] Changes in freight-rate structures resulted in heavier shipments of livestock to Eastern slaughtering plants. Finally, decentralization was encouraged by the growth of large population centers in the Midwest, the South, and on the Pacific coast.

Concentration of control in the industry continued to be a serious problem. The monopolistic position of the Big Five packers, reported the Federal Trade Commission in 1919, depended mainly upon their control of stockyards and of refrigerator cars and cold-storage facilities.[11] The Commission recommended that refrigerator cars either be owned and operated as a government monopoly or by the railroads under federal licensing, and that the government acquire such cold-storage space as was necessary to provide for competitive marketing in the principal centers of distribution and consumption.[12] These recommendations were not accepted by Congress, but when the Department of Justice appeared ready to bring suit under the anti-trust laws against the Big Five, the packers began negotiations with the Attorney-General

[9] A. C. Hoffman, *Large-Scale Organization in the Food Industries* (Temporary National Economic Committee, *Investigation of Economic Power, Monograph No. 35*) (Washington, 1940), 16-17. For the distribution of the meat-packing industry among the states in 1939 see American Meat Institute, *Reference Book of the Meat Industry* (Chicago [1941]), 31.

[10] Hoffman, *op.cit.*, 17; Corey, *op.cit.*, 209-211.

[11] Federal Trade Commission, *Report on the Meat-Packing Industry* (Washington, 1918-1920), I, 31.

[12] *ibid.*, I, 77; Federal Trade Commission, *Report on Private Car Lines* (Washington, 1920), 14-15.

that eventuated in the Packer Consent Decree. This agreement of 1920, to which the big packers subscribed on condition that it did not constitute an admission of guilt nor a judgment against them, required the packers among other things to give up the wholesale distribution of certain "unrelated lines" which consisted mainly of groceries. It did not, however, require them to give up their refrigerator cars and cold-storage warehouses. Though the decree enabled the government to subject the packers to greater restraints than otherwise would have been possible, it was criticized on the ground that it left the Big Five in control of their distribution network.[13] After 1920 the proportion of the slaughtering business done by the Big Four (Nelson Morris and Company merged with Armour in 1923) declined slightly, but in 1935 it still stood at over sixty per cent. It was significant, nevertheless, that the packing giants, though operating in an age of corporate consolidation, did not increase their power. That they did not was due partly to legal restrictions, especially the sale of stockyard interests required by the Consent Decree, and partly to economic factors.[14] Of these last the declining importance of packer ownership of refrigerator cars should be singled out. Non-packer car lines acquired beef cars for which an independent could contract with the assurance that he would receive adequate, dependable transportation.[15] So satisfactory was this service that one of the Big Four, Swift and Company, completely abandoned car ownership.[16] Furthermore, refrigerated trucks in many instances furnished the independent with an attractive alternative to rail transport.

What of the power of the large packers in the distribution of non-meat food products? From the distribution of unrelated lines they were barred by the Consent Decree. But shortly after they attacked its validity. When the Supreme Court up-

[13] G. O. Virtue, "The Meat-Packing Investigation," *The Quarterly Journal of Economics*, xxxiv (1919-1920), 677-678; "The Packer Consent Decree," *Yale Law Journal*, xlii (1932-1933), 83-85, 91-92; Corey, *op.cit.*, 87-89.

[14] Hoffman, *op.cit.*, 16-19; Corey, *op.cit.*, 89-90.

[15] Charles B. Heinemann, National Independent Meat Packers Association, interview with writer, Aug. 22, 1947.

[16] Corey, *op.cit.*, 188.

held it in 1928, the Big Four renewed their assault on the ground that conditions had so changed that the decree, unjust to the industry and unnecessary for the protection of the public, should be modified. In 1930 the Supreme Court of the District of Columbia removed the restrictions on the distribution of unrelated lines, but in 1932 the Supreme Court of the United States reversed the decision and gave the packers a year to rid themselves of the stocks they had on hand.[17] In the distribution of dairy products and of eggs and poultry, which had not been affected by the Consent Decree, the packers maintained a strong position. In 1934 ten of them sold nineteen per cent of the butter produced in the United States and forty-six per cent of the cheese.[18]

A vast increase in fruit and vegetable production took place after 1918.[19] Intimately connected with this phenomenon was the progress made in refrigerated transportation and storage.

Transit refrigeration retained its basic importance, for the greater part of the fresh fruits and vegetables consumed in the big cities still was shipped in from areas of specialized production. Such produce commonly was better than that grown locally. It came from places well adapted to its cultivation; it usually was graded more carefully than local supplies; and most important, the prompt and adequate refrigeration that was the rule often made it fresher than home-grown products, which too frequently did not receive proper care.[20]

Refrigerator cars, which handled the bulk of the long-dis-

[17] "The Packer Consent Decree," *loc.cit.*, 85-87, 91; Swift and Company *v.* United States, 276 U. S. 311; United States *v.* Swift and Company, 286 U. S. 106.

[18] Hoffman, *op.cit.*, 35.

[19] Between 1919 and 1948 the production of eighteen selected fruits increased seventy-five per cent, while between 1929 and 1948 the production of truck crops increased fifty-two per cent. U. S. Department of Agriculture, *Agricultural Statistics, 1949* (Washington, 1949), 170, 236.

[20] D. F. Fisher, "Commercial Handling and Shipping of Fruits and Vegetables," *Ice and Refrigeration*, CXVII (Sept. 1949), 20. For a map showing the regions from which New York City received its fruit and vegetable supplies see W. C. Crow, *Wholesale Markets for Fruits and Vegetables in 40 Cities* (U. S. Department of Agriculture, *Circular No. 463*) (Washington, 1938), 93.

tance trade, were managed as before by specialized organizations. Equipped with elaborate icing stations,[21] they shepherded long trains of perishables across the continent with such speed that ninth-morning delivery in New York was guaranteed on shipments from southern California. A measure of their efficiency was the fact that by the mid-thirties payments on claims of improper refrigeration averaged less than twenty-five cents a car.[22]

The availability of adequate car supplies to shippers of fruits and vegetables owed much to the United States Railroad Administration, which instituted a pooling system during the war emergency. After the authority of the Administration lapsed, the Car Service Division of the Association of American Railroads took over the work. The efficiency of its service was demonstrated in 1925 and 1926 when serious shortages were avoided in the movement of unusually large crops of grapes from California and of peaches from Georgia.[23] During the second World War a shortage was threatened, largely because the rationing of certain canned and frozen foods increased the market for fresh fruits and vegetables.[24] To make sure that the limited equipment available was used with maximum efficiency, the Interstate Commerce Commission, at the request of the Office of Defense Transportation, named as its agent the manager of the Refrigerator Car Section of the Car Service Division of the Association of American Railroads. Assisted by an advisory committee which included a representative of the ODT, this agent was given authority to control the movement of the cars. More difficult were the shortages of bunker ice. Labor scarcity and heavy

[21] The icing facilities of the Pacific Fruit Express at Laramie, Wyoming, were capable of servicing 248 cars at one time. *Ice and Refrigeration*, CXIII (Aug. 1947), 24-26.

[22] "Perishable Handling—A Mighty Transportation Spectacle," *Railway Age*, XCIX (Dec. 28, 1935), 854-855.

[23] W. A. Sherman, *Merchandising Fruits and Vegetables* (Chicago, 1928), 174, 308-309.

[24] *Ice and Refrigeration*, CIV (1943), 215. Other factors were the discontinuance of coastwise shipping of perishables, the curtailment of long-distance truck movements, and the scarcity of materials for additional car construction. *ibid.*, CIII (1942), 231.

military and civilian requirements were in part responsible, but perhaps even more important were the demands of shippers of fruits and vegetables. Lured by high prices, they insisted on more refrigeration than was customary in an effort to deliver as much as possible in salable condition. The ICC dealt with this situation by restricting the service that perishable shipments might receive. Some spoilage took place as a result of these orders, but probably no alternative was open.[25]

Improved icing techniques were an important factor in refrigerator-car efficiency. Precooling, introduced earlier,[26] came into its own in the twenties and thirties.[27] Both the warehouse and the car system were employed. The warehouse method, used mainly for citrus fruits, did a better job and permitted holding for favorable marketing conditions, but required greater capital investment and double handling costs. Cooling was accomplished either by ice or by mechanical units. The car method was most rapid and effective when done at stationary plants, where refrigeration was supplied from a central installation. But more popular was the simple expedient of placing portable fans in iced and loaded cars. When the load was cooled through the agency of forced air circulation, the fans were removed. The major field of this method was deciduous fruits.[28] In the forties a novel system of precooling vegetables was developed. Products such as asparagus, celery, cauliflower, and sprouts were run on a conveyor through a "hydro-cooler" in which they were sprayed with recirculated

[25] *ibid.*, civ (1943), 215-216, 236; cviii (June 1945), 33; Interstate Commerce Commission, *57th Annual Report . . . , 1943* (Washington, 1943), 52-53; "Ice for Cars Cut," *Business Week*, July 3, 1943, 38, 40.

[26] *Supra*, 153-155.

[27] A. M. Casberg, "Refrigeration Applications In the Fruit and Vegetable Industry," *Ice and Refrigeration*, ci (1941), 83; L. I. Denton, "Refrigerator Car Precooling," *Ice and Refrigeration*, xcix (1940), 103; F. W. Allen, "Refrigerated Transportation of Deciduous Fruits from California," *Actes du VIIe Congrès International du Froid*, iv (Utrecht, 1937), 526.

[28] J. C. Rear, "Precooling Practices in California," *Ice and Refrigeration*, cxiii (Nov. 1947), 32; W. J. Allison, "Present Trends in Car Icing," *Ice and Refrigeration*, xci (1936), 228; Denton, *loc.cit.*, 103; Allen, *loc.cit.*, 517; "Perishable Handling," *loc.cit.*, 856; *Ice and Refrigeration*, xcvii (1939), 51.

water cooled by crushed ice. So efficient was this method that the precooling of asparagus, which required ten hours with the car method, was achieved in ten to twelve minutes.[29] A system based on the creation of a vacuum by the action of high-pressure steam was tried commercially by California lettuce-growers. The evaporation of water from the surface and tissues of lettuce placed in the vacuum chamber resulted in rapid chilling.[30]

The most fundamental change in technique was the introduction of body icing in the vegetable traffic. The idea of placing crushed ice in direct contact with the load was not new —it had been used in the transport of fish—but for many years it was not applied to vegetables, which were hauled in cars refrigerated by the conventional bunker method. Early in the twenties body icing was adapted to shipping spinach from Texas, and a 1927 report by the United States Department of Agriculture resulted in equipping most refrigerator cars with the necessary waterproofed floors and meltwater outlets. The key event, however, in the history of body icing was the development about 1930 of a crusher-slinger, a machine which could crush ice as fine as snow and sling it through a flexible hose to all parts of a car. Body icing was used extensively to refrigerate shipments of melons and many vegetables, including lettuce, carrots, broccoli, and spinach. Its advantages were high cooling effectiveness and great economy of ice.[31]

The United States Department of Agriculture was untiring in its efforts to improve the techniques of refrigerated rail transportation. Its investigation of Bartlett pears grown on the Pacific coast showed the relation between ripeness at time of picking and carrying qualities and stressed the value of prompt refrigeration.[32] Studies by the Bureau of Plant In-

[29] Rear, *loc.cit.*, 33.

[30] *The New York Times*, March 5, 1950, 9E.

[31] E. D. Mallison and W. T. Pentzer, *Body Icing in Transit Refrigeration of Vegetables* (U. S. Department of Agriculture, *Technical Bulletin No. 627*) (Washington, 1938), 1-2; Rear, *loc.cit.*, 34; Casberg, *loc.cit.*, 82-83; "Perishable Handling," *loc.cit.*, 856.

[32] J. R. Magness, *The Handling, Shipping, and Cold Storage of Bartlett Pears in the Pacific Coast States* (U. S. Department of Agriculture, *Bulletin No. 1072*) (Washington, 1922), 16.

dustry in the transportation of peaches resulted in the development of a ventilated package that speeded the cooling process.[33] So constant was the Department's interest in methods that body-icing work, begun in 1924, was still proceeding in 1946.[34] Shippers benefited greatly from the efforts of the government experts. Perhaps the best example was that of the orange-growers of southern California, who in 1928 persuaded the Department to see if handling and transportation methods might be adopted that would enable them to deliver their fruit to eastern cities in good condition but at lower costs. The twenty-nine tests made between 1928 and 1933 showed that if cars of oranges were properly cooled before starting, they usually needed but one re-icing en route; the old system of costly re-icing every twenty-four hours was unnecessary. This meant a saving to citrus men of thirty to forty dollars a car and of millions each year.[35]

Regional specialization, far advanced by 1917, remained a striking characteristic of the fruit and vegetable industry. The established produce areas were supplemented by new districts such as the Salinas Valley of California and the valley of the Lower Rio Grande.[36]

The privately owned refrigerator car no longer was the major source of friction it had been early in the century, for icing charges had been brought under effective supervision of the Interstate Commerce Commission, and the packers withdrew from private-car operations in the fruit and vegetable field.[37]

[33] U. S. Department of Agriculture, *Report of the Secretary . . . , 1937* (Washington, 1937), 84.

[34] Mallison and Pentzer, *loc.cit.*, 2-3, 27, 35, 41; D. F. Fisher, "Current Investigations on the Transit Refrigeration of Perishables," *Ice and Refrigeration*, cx (Feb. 1946), 22-23.

[35] C. W. Mann and W. C. Cooper, *Refrigeration of Oranges in Transit from California* (U. S. Department of Agriculture, *Technical Bulletin No. 505*) (Washington, 1936), 1-3, 84-85; D. F. Fisher and C. W. Mann, "Transit-Refrigeration Charges on Fruit Reduced by Recent Discoveries," U. S. Department of Agriculture, *Yearbook . . . , 1935* (Washington, 1935), 318-319.

[36] Sherman, *op.cit.*, 184; E. J. Foscue, "Agricultural History of the Lower Rio Grande Valley Region," *Agricultural History*, viii (1934), 136-137.

[37] Armour and Company sold the Fruit Growers Express to a group of railways.

Serious problems had risen before 1917 from the fact that the fruit- and vegetable-grower who participated in the long-distance trade made possible by the refrigrator car had no direct personal contact with his market. In the absence of generally recognized standards of quality confusion and misunderstanding were inevitable. These troubles largely disappeared in the twenties and thirties. Cooperative marketing associations were organized to assist the individual grower in the selling function.[38] The market-news work begun by the Department of Agriculture in 1913 was expanded into a complete and efficient service that left the producer no longer at so great a disadvantage in bargaining with the buyer.[39] Of major importance was the Department's work in recommending standards and in making available at cost a federal inspection service.[40]

City market facilities with few exceptions remained inadequate. Separate installations were too numerous, while suitable terminals for motor-truck receipts were nonexistent. In the vast majority of American cities direct rail connections were not available. Produce arriving by rail had to be unloaded, hauled by truck to the wholesale and jobbing market, and then unloaded once more. Such procedures not only were wasteful, but created paralyzing traffic snarls.[41]

The truck became an important factor in the transportation of fruits and vegetables. Its employment varied with the different producing regions. Highway shipments from the Midwest were much greater than from the Pacific coast,

[38] Sherman, *op.cit.*, 104-106. In 1937 California alone had six hundred such associations. J. R. Smith and M. O. Phillips, *North America, Its People and the Resources, Development, and Prospects of the Continent as the Home of Man* (New York, 1942), 689n.

[39] M. S. Eisenhower and A. P. Chew, *The United States Department of Agriculture* (U. S. Department of Agriculture, *Miscellaneous Publication No. 88*, rev. edn.) (Washington, 1934), 44; Sherman, *op.cit.*, 257-292.

[40] R. L. Spangler, *Standardization and Inspection of Fresh Fruits and Vegetables* (U. S. Department of Agriculture, *Miscellaneous Publication No. 604*) (Washington, 1946), 4-8; Sherman, *op.cit.*, 185-256.

[41] W. C. Crow, W. T. Calhoun, and J. W. Park, *The Wholesale Fruit and Vegetable Markets of New York City* (Washington, 1940), 48-53; Crow, *op.cit.*, 2-24.

where distance and mountains gave the advantage to the rail-roads.[42] Short truck hauls were seldom refrigerated, but many loads destined for longer runs were cooled either by machines or by ice.

The refrigerator ship, aside from the importation of bananas, had two areas of utility: the coastwise and the export trade. In the first its use was limited to traffic that required no extensive rail haul at either end of the voyage. Since fruit shipments from Florida and California to cities along the Atlantic seaboard fell within this limitation, it was between these points that the refrigerator ship became a feature of the domestic trade in perishables.[43] In the export trade refrigeration played a larger role after the Armistice.[44] One reason for this was the availability of the Panama Canal for direct shipments from the Pacific coast to Europe.[45] Between 1928 and 1938 alone refrigerated exports from the west coast increased one hundred per cent. But so complete was the lack of refrigerated capacity on American merchant ships that most of our exports were carried in foreign bottoms.[46] More effective refrigeration on shipboard as well as in railroad cars was an objective of investigations by the Department of Agriculture.[47]

[42] *Moody's Manual of Investments, American and Foreign: Railroad Securities, 1943* (New York, 1943), a5.
[43] A. W. McKay et al., "Marketing Fruits and Vegetables," U. S. Department of Agriculture, *Agriculture Yearbook, 1925* (Washington, 1926), 691; R. G. Hill and L. A. Hawkins, *Transportation of Citrus Fruits from Porto Rico* (U. S. Department of Agriculture, *Bulletin No. 1290*) (Washington, 1924), 1-2; "Survey of Marine Refrigeration," *Refrigerating Engineering*, XXI (1931), 354-355; E. L. Overholser, "History of Fruit Storage and Refrigeration in the United States," *Better Fruit*, XXIX (Aug. 1934), 9; "South's Pre-Cooling and Refrigerating Projects," *Manufacturers Record*, CII (1933), 48.
[44] For statistics of the export trade in oranges and apples, the fruit crops shipped abroad in greatest quantities, see U. S. Department of Agriculture, *Agricultural Statistics, 1938* (Washington, 1938), 147, 388. Not all of these shipments, of course, moved under refrigeration.
[45] McKay et al., *loc.cit.*, 691.
[46] A. C. Rohn and J. H. Clarke, "Refrigeration Equipment for Merchant Ships," *Refrigerating Engineering*, XLIV (1942), 10.
[47] L. A. Hawkins, "Investigations on Subjects Relating to Refrigeration," *Ice and Refrigeration*, LXVII (1924), 153-154; Hill and Hawkins, *op.cit.*, 2, 19; P. L. Harding and C. L. Powell, *Transportation of Apples from the Shenandoah Cumberland Section to Overseas Mar-*

The possibility that air transport might alter substantially existing methods of shipping fruits and vegetables was brought to the fore by the wartime demonstration of its capabilities. Air transport, it seemed clear, would make possible definite economies. Lighter and cheaper packing containers would be practical. Some waste would be eliminated, for the damaging jolts of rail travel would not be suffered, and refrigeration costs could be reduced. But the total expense of transportation would still be high, a disadvantage that would become increasingly serious as improvements were made in refrigerator cars and trucks. The future of air transport in the fruit and vegetable industry was not yet clear. To a degree it was likely to furnish competition to traditional methods, for it was demonstrated that some products could be sold in certain markets at prices sufficiently high to cover costs, but the airplane was not likely to move a substantial volume of the existing traffic. It was possible its main service might lie in hauling highly perishable tropical and subtropical commodities that were barred from important markets by the inadequacy of other transportation, but this probably depended on whether or not such commodities were adapted to freezing preservation. Perhaps the most important benefit of the airplane would be the improvements its competition might be expected to encourage surface carriers to adopt.[48]

The cold-storage warehouse, like the refrigerator car, retained its importance in the distribution of fresh fruits and vegetables. Not only were warehouses more efficient in themselves, but significant progress was made in managing them so that produce was stored under optimum conditions. Care was given to maintenance of the exact temperature and rela-

---

kets (U. S. Department of Agriculture, *Technical Bulletin No. 523*) (Washington, 1936), 1-2, 25.

[48] U. S. Department of Agriculture, Interbureau Committee on Postwar Programs, Working Group on Conversion of Marketing Facilities and Methods, *Air Transport of Agricultural Perishables* (U. S. Department of Agriculture, *Miscellaneous Publication No. 585*) (Washington, 1946), passim; U. S. Department of Agriculture, Bureau of Agricultural Economics, and Edward S. Evans Transportation Research, *Florida's Production of Agricultural Perishables in Relation to the Development of Air Freight* (Washington, 1945), 5.

tive humidity required by each commodity and to adequate packaging.[49] Perhaps the greatest triumph of proper packaging was the development of oiled wraps to control apple scald.[50] In Great Britain much attention was devoted to storing fruit in concentrations of carbon-dioxide gas, but in the United States this practice was not adopted generally.[51]

Constant research was a necessity in the refrigerated preservation of fruits and vegetables, for each product had different requirements, and new problems resulting from changing conditions continually arose.[52] Agencies of both federal and state governments were active.[53] One of the most elaborate projects of the federal Department of Agriculture was the United States Horticultural Station at Beltsville, Maryland, where cold-storage facilities and the laboratories necessary for related horticultural investigations were available.[54] For many years the cold-storage industry did not actively promote research, but in 1943 the National Association of Refrigerated Warehouses established and endowed the Refrigeration Research Council with the mission of discovering new methods

[49] O. C. Mackay, "The Place of Cold Storage in the Marketing of Fruits and Vegetables," *Ice and Refrigeration,* LXXVIII (1930), 513.

[50] C. Brooks, J. S. Cooley, and D. F. Fisher, "Nature and Control of Apple-Scald," *Journal of Agricultural Research,* XVIII (Nov. 15, 1919), 239-240; C. Brooks, "Control of Apple Scald," *Ice and Refrigeration,* LXIX (1925), 48.

[51] M. E. Pennington, "Adaptation of Refrigerated Gas Storage to Perishables in the United States," *Proceedings of the Forty-Eighth Annual Meeting of the American Warehousemen's Association . . .* (Chicago, 1939), 300-305; F. W. Allen, "Fruit Handling and Storage Investigations," *Ice and Refrigeration,* XCVI (1939), 451-453; R. M. Smock, "Controlled Atmospheres Improve Refrigerated Storage," *Food Industries,* XXI (April 1949), 106.

[52] McKay et al., *loc.cit.,* 682-683.

[53] American Association of Ice and Refrigeration, Committee on State and National Experimental Investigations, "Report" [at the Twelfth Annual Meeting], *Ice and Refrigeration,* LXIV (1923), 476-477; "Report" [at the Fifteenth Annual Meeting], *Ice and Refrigeration,* LXIX (1925), 20-21; American Institute of Refrigeration, Committee on State and National Experimentation, "Report Presented at Annual Convention . . . Held at Washington, D.C., May 2, 1935," *Ice and Refrigeration,* LXXXVIII (1935), 368-369.

[54] D. F. Fisher, "A National Cold Storage Laboratory," *Refrigerating Engineering,* XL (1940), 143-144.

and new applications. The Council, staffed with men of wide experience, subsidized, mainly in the form of grants to colleges and universities, a full research program.[55]

Most of the cold-storage space devoted to the preservation of fruits and vegetables at temperatures above freezing was occupied by apples.[56] One of the most outstanding developments in the apple trade transpired in rural areas, where there was a large-scale conversion from common to refrigerated storage.[57]

The most revolutionary change in the employment of cold storage in the fruit and vegetable field was the development of freezing preservation. Early in the century the cold- or frozen-pack method was employed to hold small fruits in bulk for out-of-season use by manufacturers of jellies, ice creams, and pies, but volume was comparatively small before the World War.[58] After the war the frozen pack increased sensationally. Only about three thousand fifty-gallon barrels of strawberries were packed in the Pacific Northwest in 1918; a decade later the output rose to seventy thousand barrels. Other fruits commonly preserved in this fashion were red and black raspberries, blackberries, currants, gooseberries, and cherries.[59] Though the technique originated in the East, the

[55] "Report of Research Committee," *Proceedings of the Forty-Eighth Annual Meeting of the American Warehousemen's Association* (Chicago, 1939), 316-317; U. S. Department of Agriculture, Agricultural Research Administration, Office of Experiment Stations, *Experiment Station Record*, xci (1944), 112; V. O. Appel, "The Refrigeration Research Council—Its Accomplishments and Future Plans," *Ice and Refrigeration*, cxvi (Jan. 1949), 41-42; *Ice and Refrigeration*, cvi (1944), 93, 151; H. C. Diehl, "The Refrigeration Research Foundation, a Progress Report," *Ice and Refrigeration*, cx (April 1946), 51-52, 62.

[56] Overholser, *loc.cit.*, 8.

[57] W. V. Hukill and E. Smith, "The Cold Storage of Apples," U. S. Department of Agriculture, *The Yearbook of Agriculture, 1943-1947* (Washington, 1947), 868; F. W. Swan, "Air Conditioned Apples: Control of Temperature and Humidity with Blower Coils," *Refrigerating Engineering*, xl (1940), 78-80; "Converting Common Storages," *American Fruit Grower*, lvii (June 1937), 19-20.

[58] *Supra*, 164-165.

[59] H. C. Diehl et al., *The Frozen-Pack Method of Preserving Berries in the Pacific Northwest* (U. S. Department of Agriculture, *Technical Bulletin No. 148*) (Washington, 1930), 1-2.

greater part of the annual pack came from Washington and Oregon. The method was basically quite simple. Fruit that had been sorted, washed, and hulled was placed in a container, usually in a sugar solution which served to check enzymatic action, and frozen in a room where the temperature was maintained at ten to fifteen degrees above zero.[60] In later years systematic rolling of containers was practiced as a means of assuring more thorough mixing of the sugar syrup with the fruit, more rapid freezing, and consequently, a better product.[61]

The quick-freezing processes first developed in the fish industry were not well adapted to the preservation of fruit, for they involved either direct immersion in salt brine or indirect contact through the clumsy media of molds and cans. These disadvantages, however, did not apply to the systems developed by Clarence Birdseye, and in 1929 one of his double-belt frosters was installed at Hillsboro, Oregon, and devoted mainly to the freezing of cherries and strawberries.[62] Other methods were soon applied, and the quick freezing of fruit assumed the proportions of a major industry.

The freezing of vegetables, mainly peas, had been the subject of experiments as early as 1917, and by the late twenties many concerns were conducting trials. But the results of these early attempts were generally poor, for enzymatic action was not checked sufficiently by low temperatures to prevent deterioriation and the development of off-flavors.[63] A turning point came in 1930, when H. C. Diehl and C. A. Magoon of the Bureau of Plant Industry suggested to packers in the Northwest that before freezing peas, they first scald them

[60] Tressler and Evers, *op.cit.*, 383; D. K. Tressler, "Methods of Freezing Fruits and Fruit Juices," *Ice and Refrigeration*, LXXXVIII (1935), 275.

[61] R. Ireland, "The Freezing and Shipping of Cold-Pack Strawberries Sugared in Barrels," *Proceedings of the Fiftieth Annual Meeting of the American Warehousemen's Association* (Chicago, 1941), 359-361. The U. S. Department of Agriculture was active in the improvement of the frozen-pack method. Diehl et al., *op.cit.*, 10.

[62] Tressler and Evers, *op.cit.*, 397-398; Tressler, *loc.cit.*, 275.

[63] F. W. Knowles, "How Foods Are Frozen in the Northwest," *Food Industries*, XII (April 1940), 54-55; Tressler and Evers, *op.cit.*, 309.

briefly.[64] Experience soon showed that scalding, or "blanching" as the trade rather illogically termed it, was essential. Accomplished either with steam or hot water, its value lay in the fact that it inactivated such enzymes as might cause spoilage during a reasonable storage period. Basically it was a process of partial cooking.[65] Although the Diehl-Magoon suggestion came in 1930, it was not until 1937 that the freezing of vegetables began to expand rapidly.[66]

Improved methods soon were developed for both fruits and vegetables. Better blanching technique permitted storage at somewhat higher temperatures. The addition of ascorbic acid to fruit packs proved effective in checking browning. Eventually it was recognized generally that careful attention had to be paid to every phase of the production and distribution of frozen foods. The varieties most adaptable to freezing had to be selected. The temperatures during storage, transportation, and retail display had to be uniformly low.[67]

A vast amount of research lay behind these improvements. State agricultural experiment stations were active, as was the Department of Agriculture itself, which in 1931 established its Frozen Pack Laboratory at Seattle, Washington, to investigate all phases of the problem of producing high-quality frozen foods. In 1941 the Seattle station was incorporated into the Western Regional Research Laboratory at Albany, California, which, under the supervision of the Agricultural

[64] J. J. Hoey, "Food Freezing Research," *Ice and Refrigeration*, CVI (1944), 100. Diehl and Magoon were not the first to conclude that scalding was the answer to the problems of the vegetable-freezers. Tressler and Evers, *op.cit.*, 309.

[65] J. G. Woodruff et al., *Studies of Methods of Scalding (Blanching) Vegetables for Freezing* (University System of Georgia, Georgia Agricultural Experiment Station, *Bulletin 248*) (Experiment, Georgia, 1946), 3; G. A. Fitzgerald, "The Effects of Freezing on the Vitamin Content of Vegetables—A Review," *Refrigerating Engineering*, XXXVII (1939), 33; Hoey, *loc.cit.*, 100.

[66] Tressler and Evers, *op.cit.*, 4.

[67] G. Poole and M. T. Zarotschenzeff, "Four Years Progress in Quick Freezing," *Ice and Refrigeration*, XCI (1936), 217; D. K. Tressler, "Recent Advances in Frozen Food Technology," *The Fruit Products Journal and American Food Manufacturer*, XXV (1946), 228; C. R. Mundee and F. C. Porcher, *Quick-Frozen Foods* (Washington, 1938), 2.

Research Administration, collaborated with state agricultural colleges. In addition to the work of governmental agencies was the investigation that continually took place under the auspices of frozen-food-packers and container-manufacturers.[68]

The investments in scientific research paid dividends in the form of increases both in the volume and in the variety of products frozen. The commercial frozen-fruit pack in 1935 was only 77,000,000 pounds, but in 1950 it was 481,000,000. During the unusual conditions of 1946 the nation's fruit pack actually had gone up to 519,000,000. The volume of frozen-vegetable production lagged behind that of fruit for many years, but in 1947 it assumed the lead and in 1950 rose to 587,000,000 pounds.[69] The quantity of fruits and vegetables frozen was still extremely small in comparison with the amount that was canned or distributed fresh. In 1948, for example, over twelve and a half times more vegetables were canned than frozen.[70]

The variety of products frozen was bewildering,[71] but among the vegetables peas, lima beans, snap beans, cut corn, and spinach were most important, while strawberries, cherries, raspberries, apples, and peaches led among the fruits. The few products which did not yield to successful freezing included bananas, pears, tomatoes, and lettuce and related vegetables.[72]

Orange juice was by far the most spectacular addition to the list of foods frozen commercially. Interest in it always was higher than in other juices. For one reason, the pasteurization

[68] Mundee and Porcher, *op.cit.*, 8; U. S. Department of Agriculture, Production and Marketing Administration, *Instructions on Processing for Community Frozen-Food Locker Plants* (U. S. Department of Agriculture, *Miscellaneous Publication No. 588*) (Washington, 1946), 48-52; Hoey, *loc.cit.*, 99n., 100-102, 157-159.

[69] U. S. Department of Agriculture, *Agricultural Statistics, 1951* (Washington, 1951), 214, 273; Tressler and Evers, *op.cit.*, 3. The figures for fruit do not include frozen citrus juices.

[70] *Agricultural Statistics, 1949*, 291; "Frozen Foods: Interim Report," *Fortune*, xxxiv (Aug. 1946), 105.

[71] *1946-1947 Directory of Frozen Foods Processors of Fruits, Vegetables, Seafoods, Meats, Poultry, Specialties* (New York, 1946), 305-479.

[72] *Agricultural Statistics, 1949*, 231, 292; C. L. Walker, "What's in the Deep Freeze?" *Harper's Magazine*, cxcviii (June 1949), 47.

involved in the canning process destroyed some of the flavor and aroma of the fresh product. For another, the public had been so thoroughly sold on the merits of orange juice that a large market for it existed. For a decade and a half after 1930 there were many commercial experiments with freezing it, but none of the methods developed turned out a product that was successful at the retail level. The frozen juice was inconvenient to defrost and not sufficiently better than the canned, particularly in view of the fact that the latter did not have to be stored under refrigeration.[73] Successful frozen orange juice stemmed from experience during the second World War in the dehydration, under high vacuum and without the addition of heat, of penicillin, blood plasma, and orange juice. A satisfactory orange-juice powder was developed, but its manufacturing costs were too high to make it practical for the peace-time civilian market. Promoters, however, hit upon the idea of freezing the orange concentrate that was available at one stage of the powder-making process. The result was the frozen concentrated juice which was to grow so remarkably in popularity.[74] In the 1945-1946 season only 226,000 gallons of concentrate were frozen; in 1948-1949 production soared to over 12,000,000 gallons and in 1949-1950 to over 25,000,000.[75] The explanation of this no doubt lay mainly in its convenience and flavor, which might not have been exactly the same as fresh juice, but apparently so closely approximated it in the minds of many consumers that it made little difference. The economy of the frozen juice and the assurances given the public that it contained more Vitamin C than fresh

[73] R. V. Grayson, "Freezing of Orange Juice in National Juice Corp. Plant, Tampa, Fla.," *Ice and Refrigeration*, LXXX (1931), 416-417; W. J. Finnegan, "Preserving Citrus Juice by Freezing," *Ice and Refrigeration*, LXXXVIII (1935), 51; Tressler, "Methods of Freezing Fruits and Fruit Juices," *loc.cit.*, 275; M. Wharton, "New Methods of Citrus Juice Packing," *Ice and Refrigeration*, XCII (1937), 281-283; W. J. Finnegan, "Food Freezing in Cans," *Ice and Refrigeration*, XCIX (1940), 203-204; Tressler and Evers, *op.cit.*, 421.

[74] T. A. Rector, "Frozen Concentrated Orange Juice—Its Research Background," *Refrigerating Engineering*, LVIII (1950), 349-352; Walker, *loc.cit.*, 46-47.

[75] *Quick Frozen Foods and The Locker Plant*, XIII (March 1951), 134; *Agricultural Statistics, 1951*, 183.

fruit bought in the market also added to its appeal.[76] Among the repercussions of this innovation were increased consumption, rises in the prices of raw fruit, and a revolution in citrus marketing so great that in 1949-1950 about one-third of Florida's orange crop was frozen. The concentrate technique soon was extended to lemon, grapefruit, and grape juice, and experiments were being conducted with other juices, including tomato and apple.[77]

The fruit-freezing industry was concentrated largely in the West, which in 1948 produced one-half of the total pack. From the Midwest came one-fifth, from the Northeast one-sixth, and from the South one-eighth. In vegetables the West held a similar position of leadership.[78]

Food freezing offered many advantages to a producing area. By extending market opportunities, it tended to reduce unsalable surpluses and to stabilize prices. By making possible the elimination of waste at points of origin, it afforded savings in transportation and storage costs and encouraged the utilization of by-products. By creating a local processing industry, it brought to rural areas a larger share of the consumer's dollar.[79]

It was possible in the years to come that freezing might affect materially the volume of fruits and vegetables preserved by canning. Canners realized this and sought protection by improving both their raw materials and their processing meth-

[76] Walker, *loc.cit.*, 47.

[77] "Frozen Orange Juice: Postwar Bonanza," *Business Week*, Feb. 4, 1950, 62; *Quick Frozen Foods and The Locker Plant*, XIII (March 1951), 133.

[78] *Agricultural Statistics, 1949*, 231, 292. For an analysis of producing areas by state and commodity see Tressler and Evers, *op.cit.*, 6-8. H. Carlton, *The Frozen Food Industry* (Knoxville, 1941), 71-73, has maps of the main areas of frozen-food production.

[79] H. C. Diehl, "Freezing Preservation and Frozen Pack Research," *Ice and Refrigeration*, XCVIII (1940), 495; R. V. Grayson, "What the Frozen Food Industry Means to the United States," *Ice and Refrigeration*, LXXXII (1932), 61; G. Poole, "Refrigeration as a Factor in Eliminating Wastage in Food Production and Distribution," *Ice and Refrigeration*, LXXXII (1932), 76; L. W. Waters, "Development of the Quick-Freezing Process," *Chemical and Engineering News*, XX (Dec. 10, 1942), 1561; E. T. Gibson, "Increased Volume of Frozen Foods," *Ice and Refrigeration*, CIV (1943), 165.

ods. Some entered the freezing field themselves.[80] Additional safeguards were the important advantages that canned goods still retained. Not only were they low in price, but they were comparatively imperishable and did not require refrigerated storage by the retailer or the housewife. Though some were stored under refrigeration at the wholesale level, the temperatures employed were relatively high. Finally, some products had distinctive characteristics that were likely to assure them a market. Canned peaches, for example, did not taste like fresh peaches, but they had a flavor and texture that were especially attractive.

Freezing seemed to pose a threat to the climatically favored areas that specialized in supplying fresh produce out-of-season to northern markets. If the public came to prefer frozen fruits and vegetables, it was reasoned, why could not local producers freeze a considerable part of their crop to satisfy the off-season demand being met by shipments from southern and western points?[81] Whatever the real gravity of this threat might have been, fresh-produce interests took steps to forestall any danger. They tried to improve the quality of their product and experimented with prepackaging as a means of duplicating the consumer-convenience features of frozen foods. They were, as we have seen, trying to improve the refrigerator-car facilities upon which they so heavily depended, and they planned an advertising campaign to persuade the public to insist on fresh foods. Entry into frozen-food production was, of course, an alternative,[82] and it should have been reassuring that the public preferred some commodities fresh, such as strawberries, even though they could be frozen satisfactorily.

In the dairy industry refrigeration was an absolute necessity, for milk as it came from the cow provided an ideal en-

[80] Carlton, *op.cit.*, 152; Walker, *loc.cit.*, 45.

[81] U. S. Department of Agriculture, *Report of the Secretary . . . , 1940* (Washington, 1940), 64; "Do Canners Need to Worry about Frozen Foods?" *The Canner*, xc (March 23, 1940), 13; Carlton, *op. cit.*, 172.

[82] Carlton, *op.cit.*, 152; "Do Canners Need to Worry about Frozen Foods?" *loc.cit.*, 14; "Frozen Foods: Interim Report," *loc.cit.*, 182; *The New York Times*, April 30, 1950, F1.

vironment for the growth of bacteria. Unless their rate of multiplication were retarded, milk soon became unfit to drink. Nature's dictate that milk be kept cool was reinforced by regulations of local governments whose effect was to make refrigeration mandatory.[83]

The best practice called for refrigeration to begin at once on the dairy farm. Usually milk still was cooled by allowing it to flow over cold surfaces or to stand in cans immersed in cold water.[84] A new development was the stainless-steel storage tank, which held the cooled yield of the daily milkings or, if equipped with evaporator coils in its walls, performed the chilling job itself. This method of managing milk in bulk, used in conjunction with tank trucks, eliminated the high costs involved in handling cans.[85] Spring or well water, ice, and mechanical refrigeration were employed to furnish the necessary cooling effects, but after 1930 the machine came rapidly to the fore. Its adoption, desirable because of the difficulties many dairy farms experienced in obtaining an adequate ice supply, was made possible by the appearance of perfected models and the extension of rural electric lines.[86]

Milk destined for marketing in liquid form was shipped to city plants to be pasteurized, bottled, and distributed. In transportation methods the most important change was the introduction of insulated tank trucks and railroad cars able to haul

[83] W. R. Sanders, "Industrial Application of Refrigeration in the United States," The Fourth International Congress of Refrigeration, *Proceedings*, II (London [1925?]), 969-970. Milk was kept fresh on an experimental basis without refrigeration by preventing it from coming in contact with air, which was always laden with bacteria. "Fresh Milk from Cans," *Life*, XXVII (Nov. 7, 1949), 59.

[84] J. E. Nicholas, "Requirements of Farm Electric Milk Coolers," *Ice and Refrigeration*, XCV (1938), 324; J. T. Bowen, *Refrigeration in the Handling, Processing, and Storing of Milk and Milk Products* (U. S. Department of Agriculture, *Miscellaneous Publication No. 138*) (Washington, 1932), 36-39.

[85] D. C. Lightner, "Tank Truck Hauling of Milk From Farm to Plant," *The Milk Dealer*, XXXVIII (May 1949), 43, 98, 100; R. W. Blackburn, "A New Milk Transport," *Hoard's Dairyman*, XCII (Sept. 25, 1947), 766.

[86] R. P. Hotis and J. R. McCalmont, *Cooling Milk on the Farm with Small Mechanical Outfits* (U. S. Department of Agriculture, *Circular No. 336*) (Washington, 1934), 1-3.

chilled milk in bulk for considerable distances. This equipment not only made possible reductions in handling costs, but permitted such extensions of milksheds that New York City could tap Midwestern supplies and milk-short Florida could buy in Wisconsin.[87] In modern city plants elaborate refrigerating machinery was used to cool incoming shipments, to chill after pasteurization, and to hold for distribution.[88] Many plants even manufactured the ice needed to pack around the bottles during delivery. For this the new automatic ice machines proved both convenient and cheap.[89]

Freezing preservation was tried, but since the result was not particularly palatable, milk was frozen only for areas where it was scarce and expensive. In the Canal Zone, for example, the Army found it useful. After 1946 there was much experimental interest in the same process that proved so successful with orange juice. No product suitable for commercial exploitation was developed, but should this be achieved it was foreseen that the market-milk industry might well be revolutionized, for distribution costs would no doubt be cut and greater consumption encouraged.[90] Frozen cream was stored in considerable volume, though not for the retail market. Its great utility was as a means of holding surplus production for manufacturers of ice cream and cream cheese.[91]

[87] W. P. Hedden, *How Great Cities Are Fed* (Boston, 1929), 49, 86-87; R. P. Hotis, *Transporting and Handling Milk in Tanks* (U. S. Department of Agriculture, *Technical Bulletin No. 243*) (Washington, 1931), 1, 16-17, 23-24; H. E. Ross, *The Care and Handling of Milk*, rev. and enl. edn. (New York, 1939), 368; "Demountable Milk Tank to Cut Borden Costs 16 Per Cent," *Food Industries*, XII (Jan. 1940), 59-61; *Ice and Refrigeration*, LXX (1926), 145; LXXI (1926), 90-91.

[88] Bowen, *op.cit.*, 46-47; R. J. Ramsey, "Refrigeration in the Dairy Industry," *Refrigerating Engineering*, LV (1948), 358-361; B. McKenna, "Refrigeration in The Dairy Plant," *Ice and Refrigeration*, CXVI (April 1949), 37-39.

[89] C. L. Roadhouse and J. L. Henderson, *The Market-Milk Industry* (New York, 1941), 327.

[90] Tressler and Evers, *op.cit.*, 630; Walker, *loc.cit.*, 47; "Frozen Orange Juice: Postwar Bonanza," *loc.cit.*, 62; R. S. McBride, "Frozen Milk May Change Existing Marketing Methods," *Food Industries*, IX (1937), 4-6; *The New York Times*, Feb. 11, 1951, E9.

[91] Tressler and Evers, *op.cit.*, 627; J. H. Nair, "Preservation Problems in the Dairy Industry," *Industrial and Engineering Chemistry*,

In the butter and cheese industries improved refrigeration equipment and technique made possible output of superior quality. Modern creameries used mechanical units to afford them accurate control at all stages of the manufacturing process, while careful attention was given to the maintenance of below-zero temperatures and high relative humidities during storage.[92] Establishments engaged in the curing and storing of cheese likewise were concerned with temperature and humidity, for upon these depended control over the proper growth of bacteria and mold.[93]

In the manufacture of ice cream the refrigerating machine supplanted almost entirely the mixtures of ice and salt which before 1910 had been the mainstay of the industry. The quick-hardening processes of modern factories, which were accomplished in thirty to sixty minutes at temperatures that ranged as low as fifty-five degrees below zero, would have been impossible without the powerful mechanical units available. In the distribution of the finished product the old techniques gave way before the efficiency and convenience of the small machine and of solid carbon dioxide.[94]

The handling of eggs in the shell was improved all along the line by greater awareness of the importance of refrigeration and by advances in the methods of applying it. Particularly important were improved treatment on the farm, for which the more general availability of refrigerating machinery was responsible, and the motor truck, which eliminated much of the deterioration that used to occur when eggs were hauled by wagon from farm to shipping point. Equally significant was the evolution of superior storage techniques. Shrinkage was checked by new case liners, by high relative humidities,

---

xxiv (1932), 674; O. C. Mackay, "The Place of Cold Storage in the Marketing of Butter and Cream," *Ice and Refrigeration*, lxxix (1930), 18-19.

[92] Bowen, *op.cit.*, 46-47.

[93] *ibid.*, 47-49; E. L. Sylvester, "Mechanical Refrigeration—A Boon to the American Cheese Industry," *National Butter and Cheese Journal*, xxvii (Dec. 10, 1936), 41.

[94] Bowen, *op.cit.*, 50-58; R. V. Jones, "Use of Solid Carbon Dioxide in Distribution of Ice Cream," *Ice and Refrigeration*, lxxv (1928), 334-335.

by the practice of dipping eggs in mineral oil, and by strict adherence to the no-washing rule. Spoilage was reduced by closer attention to optimum temperatures and by care to store infertile eggs only, while the development of off-flavors was brought under control by better provision for air circulation and by the introduction of ozone. Thanks largely to Department of Agriculture demonstrations, storage of undergrade eggs in the shell was not attempted.[95]

The quantity of shell eggs stored each year declined in the two decades after 1930. At the peak of the season in 1930 over eleven million cases were in storage, but in 1941 only about six and one-half million. The quantity placed in refrigerated warehouses rose briefly during the war, but in 1947 and 1948 sank below pre-war levels.[96] One explanation of this was the trend toward increased production during the fall and winter months, which lowered the cost of out-of-season eggs and enabled more to afford them. Another was the expansion of the frozen-egg industry.[97] A new element in the situation was the buying program begun by the federal government in the spring of 1947. Governmental purchases kept prices so high at the season of top production that many egg men were

[95] M. A. Jull et al., "The Poultry Industry," U. S. Department of Agriculture, *Yearbook, 1924* (Washington, 1925), 436-437; T. W. Heitz, *The Cold Storage of Eggs and Poultry* (U. S. Department of Agriculture, *Circular No. 73*) (Washington, 1929), 15, 26-28, 31; C. L. Brown, "Present Day Cold Storage," *American Poultry Journal* (Eastern and Southern Edition), LXI (Nov. 1930), 4-18; M. E. Pennington, "Effect of Ventilation on Keeping Qualities of Eggs in Cold Storage," *Ice and Refrigeration*, LXIV (1923), 538; J. A. Hawkins, "Cold Storage Operation," *Ice and Refrigeration*, XC (1936), 82; M. E. Pennington, "Fifty Years of Refrigeration In the Egg and Poultry Industry," *Ice and Refrigeration*, CI (1941), 47; O. C. Mackay, "The Place of Cold Storage in the Marketing of Eggs," *Ice and Refrigeration*, LXXVIII (1930), 514; A. W. Ewell, "The Use of Ozone in Cold Storage Plants," *Ice and Refrigeration*, XCI (1936), 295-296; M. K. Jenkins, *Commercial Preservation of Eggs by Cold Storage* (U. S. Department of Agriculture, *Bulletin No. 775*) (Washington, 1919), 34-35.

[96] *Agricultural Statistics, 1938,* 379; *Agricultural Statistics, 1949,* 484.

[97] C. C. Warren, "The Cold Storage Egg," *Refrigerating Engineering*, XL (1940), 342.

afraid to store and to take a chance on getting their money back with adequate profit in the fall.[98]

The egg-freezing industry began about the turn of the century, primarily as a means of salvaging eggs not fit for marketing in the shell. Before the first World War freezing was a comparatively minor means of preservation, but about 1927 it assumed major importance. Between 1930 and 1950 the quantity of frozen eggs in storage on August 1, the peak of the season, more than doubled. Now first-quality eggs were broken for freezing, and the techniques worked out earlier by Mary E. Pennington of the Department of Agriculture became standard practice. The growth of the industry was attributed to two factors. One of these was the better quality. The other was the change in the cooking habits of American housewives which saw home-baked goods supplanted by the products of commercial bakeries. Such establishments found frozen eggs more convenient and more economical than those in the shell.[99]

The methods of refrigerating dressed poultry underwent significant changes. Initial cooling by exposing birds in mechanically refrigerated rooms long was considered best, but later the use of mixtures of crushed ice and water, for which important advantages were claimed, became popular.[100] The greater part of the poultry slaughtered in the United States continued to be frozen, though but a small proportion was delivered to the consumer before thawing. As late as 1938 about ninety-nine per cent was frozen without the removal of feet, head, and entrails. This was in accord with Department of Agriculture recommendations, but these recommendations were modified by the discovery that eviscerated or full-drawn

---

[98] *Ice and Refrigeration*, CXIII (Nov. 1947), 62; CXVII (Sept. 1949), 31.

[99] V. O. Appel, "Freezing and Refrigerated Storage of Poultry and Poultry Products in the United States," *Ice and Refrigeration*, CXV (Sept. 1948), 39-40; J. H. Radabaugh, "Economic Aspects of the Frozen-Egg Industry in the United States," *Ice and Refrigeration*, XCVII (1939), 265-266; *Agricultural Statistics, 1938*, 379; *Agricultural Statistics, 1949*, 484.

[100] R. L. Bailey, G. F. Stewart, and B. Lowe, "Ice Slush Cooling of Dressed Poultry," *Refrigerating Engineering*, LV (1948), 369-371.

poultry kept satisfactorily if adequate sanitary measures were employed, and it was likely that evisceration before storing or marketing would become increasingly common. Better packaging, the employment of a variety of the new quick-freezing techniques, and more adequate control of temperature and humidity during storage paid dividends in the form of better quality.[101] In the thirties and forties alone the storage holdings of frozen poultry almost trebled.[102]

In the fish industry advanced methods were developed for exploiting the benefits that refrigeration offered. On the vessels themselves machinery was installed to supplement the crushed ice that for decades had been relied upon to preserve the catch.[103] Abroad there was considerable interest in factory ships equipped to freeze fish at sea, but in this department the United States lagged. Although an American vessel, the *Pacific Explorer*, was fitted out elaborately after the second World War, it did not prove particularly satisfactory.[104] But highly successful American examples of the shipboard use of mechanical refrigeration were the tuna clippers of California. Before about 1925 vessels of the tuna fishery needed only ice, for they were engaged in catching the albacore, which were to be found in a narrow strip of water just off the coast of southern California. In the middle twenties the albacore haul began to decline, and it became necessary to seek other varieties, the bluefin, the yellowfin, and the skipjack, whose

[101] G. Poole, "Full-Drawn Poultry and Quick Freezing," *Ice and Refrigeration*, XCII (1937), 205-206; V. R. H. Greene, "Quick Freezing of Poultry," *Refrigerating Engineering*, XXXVI (1938), 97-98; "New Quick-Freezing," *Business Week*, May 15, 1937, 20, 22; Appel, "Freezing and Refrigerated Storage of Poultry and Poultry Products in the United States," *loc.cit.*, 40; Brown, *loc.cit.*, 18-19; Carlton, *op. cit.*, 153; Heitz, *op.cit.*, 21-25, 29; Tressler and Evers, *op.cit.*, 69, 545-546.

[102] *Agricultural Statistics, 1938*, 375; *Agricultural Statistics, 1949*, 484.

[103] J. M. Lemon, *Developments in Refrigeration of Fish in the United States* (U. S. Department of Commerce, Bureau of Fisheries, *Investigational Report No. 16*) (Washington, 1932), 17-19.

[104] J. M. Lemon and C. B. Carlson, "Freezing Fish at Sea," *Refrigerating Engineering*, LIV (1947), 528, 530, 580-581; Walker, *loc.cit.*, 45; *Ice and Refrigeration*, CXVIII (April 1950), 24.

habitat was off the west coasts of Mexico and Central America and as far south as the Galápagos Islands. On voyages to such distant tropical waters ice alone was obviously inadequate, so the holds of the clippers were equipped with refrigerating coils. In the middle thirties a satisfactory system for lengthy voyages was developed. The freshly caught fish were thrown into wells filled with brine kept at a low temperature by mechanical units. When the fish were frozen, the brine was pumped out, and the tuna were held in dry storage until the boat reached port, where the catch was unloaded, thawed, and preserved by canning. This method proved eminently satisfactory, but had the fish been destined for sale as fresh, it probably would not have been practical.[105]

In freezing, the old methods did not pass entirely into the discard. "Sharp" freezing continued to be employed for much of the fish preserved in the round, but for those dressed or filleted the new high-speed processes ruled supreme.[106] Under their impetus the volume of frozen fishery products grew phenomenally. In 1932 it stood at only 92,000,000 pounds, but by 1948 it had increased to almost 292,000,000.[107]

American fisheries in the thirties and forties achieved noteworthy gains both in volume and in value of product.[108] Among the factors responsible were the introduction of power boats and the otter trawl, which reduced catching costs, and the improvement of transportation facilities, for which the motor truck was largely responsible. But more fundamental than these were quick freezing and its companion technique, filleting. Their benefits were numerous. They permitted the

[105] Lemon and Carlson, *loc.cit.*, 529-580; H. M. Hendrickson, "Tuna Boat Refrigeration," *Refrigerating Engineering*, LVIII (1950), 456-460, 465; A. W. Ponsford, "Refrigeration in the Tuna Fishing Industry," *Ice and Refrigeration*, LXXXII (1932), 23-25.

[106] Tressler and Evers, *op.cit.*, 563-571. For a survey of quick-freezing processes see H. J. Placanica, "Quick Freezing Seafoods," *Ice and Refrigeration*, C (1941), 387-388.

[107] A. W. Anderson and E. A. Power, *Fishery Statistics of the United States, 1942* (U. S. Department of the Interior, Fish and Wildlife Service, *Statistical Digest No. 11*) (Washington, 1946), 43; *Agricultural Statistics, 1949*, 760.

[108] *Agricultural Statistics, 1949*, 756-757.

utilization of by-products, the elimination of freight payments on inedible parts, and the marketing of whiting or silver hake, which formerly were used for fertilizer. Infinitely more important, they extended markets by making possible delivery of fish over a wide area in a more palatable and convenient form than had been known before.[109]

The American diet underwent marked changes after the first World War. Some of the old staples lost favor. The per capita consumption of potatoes, for example, dropped from 192 pounds in 1916 to 115 in 1948, while that of grain products declined from 270 pounds to 171. But the consumption of dairy products, citrus fruits, tomatoes, and leafy, green, and yellow vegetables grew astonishingly.[110] Among the factors responsible were the increasingly sedentary nature of our lives, which diminished both our need and our desire for the heavy foods of our fathers, and dietary indoctrination, which was undertaken sometimes in the interest of public health and sometimes in the interest of a particular product. No adequate explanation, however, could omit refrigeration, for without its preservative action these basic changes could not have transpired.

[109] E. A. Ackerman, *New England's Fishing Industry* (Chicago, 1941), 38, 150-151, 153, 226-229; *The Boston Daily Globe*, Oct. 14, 1946; H. F. Taylor, "Science in the Distribution of Fish," *Scientific American*, cxl (1929), 252.

[110] U. S. Department of Agriculture, Bureau of Agricultural Economics, Division of Statistical and Historical Research, *Consumption of Food in the United States, 1909-48* (U. S. Department of Agriculture, *Miscellaneous Publication 691*) (Washington, 1949), 120. It should be remembered that food habits varied in accord with income. For a striking demonstration of this see *ibid.*, 143.

CHAPTER XVII

FROZEN FOODS: THE INTRODUCTION OF A
NEW TECHNOLOGY, 1929-1950

THE freezing of food in consumer packages ready to cook or serve was a long step forward in the techniques of preservation[1] that promised to bring far-reaching changes in the nation's diet and economy. The story of consumer-package frozen foods was interesting, not only as an inherent part of the history of refrigeration, but also for the demonstration it afforded of the complexities involved in the introduction of new technologies.

About 1930 there was good reason to believe that a large potential market for such frozen foods awaited exploitation. Their intrinsic advantages formed the principal basis for optimism. In the first place frozen foods approximated the appearance and flavor of fresh. In fact, they were likely to be superior in these respects to fresh products which had been subjected to improper handling. Besides, they were high in nutritive value. Investigations indicated that vitamin loss varied with product, treatment before freezing, length and temperature of storage, and cooking methods, but that in general the record of frozen foods was very good indeed.[2]

[1] Although meats, fish, poultry, and berries had long been subject to freezing preservation, deliveries to the consumer in the frozen state did not take place until the development of quick freezing in the mid-twenties. H. F. Taylor, "Theory and Practice of Rapid Freezing in the Fish Industry," *Ice and Refrigeration*, LXXIX (1930), 112.

[2] M. W. Rose, "The Effect of Quick Freezing on the Nutritive Values of Foods," *The Journal of the American Medical Association*, CXIV (April 6, 1940), 1361; V. D. Greaves and M. M. Boggs, "Trends in Freezing Preservation of Foods," *Journal of Home Economics*, XXXVII (1945), 23-24; D. K. Tressler and C. P. Evers, *The Freezing Preservation of Foods*, 2nd edn. (New York, 1947), 727-741; J. A. McIntosh, D. K. Tressler, and F. Fenton, "The Effect of Different Cooking Methods on the Vitamin C Content of Quick-Frozen Vegetables," *Journal of Home Economics*, XXXII (1940), 695; G. Poole and M. T. Zarotschenzeff, "Four Years Progress in Quick-Freezing," *Ice and Refrigeration*, XCI (1936), 216-217; R. R. Jenkins et al., "Storage of Frozen Vegetables: Vitamin C Experiments," *Refrigerating Engineering*, XXXIX (1940), 381-382; G. A. Fitzgerald, "The

Most studies dealt with vitamin retention, but what information was available on such elements as proteins and minerals revealed no material loss in nutrients.[3] Another advantage was convenience of handling and preparation. They made possible considerable savings of both time and labor in the kitchen and presented no problems of unnecessary bulk and waste disposal. Since inedible portions were removed at the point of production, freight and storage charges on waste were eliminated.[4] To the institutional trade, which included all establishments serving meals to the public, frozen foods offered special advantages. They permitted great savings in labor and space requirements, an attraction of extraordinary importance, for example, to railroad dining cars. They made it possible, by virtue of their relative imperishability, to adjust without loss to great variations in demand. Finally, they made it easy to offer a variety of attractive foods which could be passed off as fresh.[5]

But formidable obstacles stood in the way of commercial exploitation of this potential market. It was expensive to construct fixed freezing plants, especially for crops so seasonal in character as fruits and vegetables. A product of high quality had to be turned out if consumer acceptance were to be won. Proper packaging methods and materials had to be developed, and provisions had to be made for the maintenance of low temperatures in railroad cars and trucks, in warehouses, and in retail stores. Furthermore, it was necessary to educate the public, for freezing preservation did not have a very good reputation.[6]

Effects of Freezing on the Vitamin Content of Vegetables—A Review," *Refrigerating Engineering*, xxxvii (1939), 38.

[3] Tressler and Evers, *op.cit.*, 724.

[4] For discussions of the advantages of frozen foods to the consumer see H. Carlton, *The Frozen Food Industry* (Knoxville, 1941), 14, and J. S. Larson, J. A. Mixon, and E. C. Stokes, *Marketing Frozen Foods —Facilities and Methods* (Washington, 1949), 2-3.

[5] Poole and Zarotschenzeff, *loc.cit.*, 390; *Ice and Refrigeration*, xci (1936), 134; "Railroad's Use of Frozen Foods Increased 12 Fold Since 1934," *Quick Frozen Foods and The Locker Plant*, xii (Aug. 1949), 82.

[6] L. P. Beebe et al., "Developments in the Production and Distribu-

Frozen Foods, 1929-1950

Between 1930 and 1937 the essential pioneering in the introduction of frozen foods was accomplished, mainly by the General Foods Corporation, formerly the Postum Company, which purchased the quick-freezing patents of Clarence Birdseye and began marketing a line to which it affixed the brand name "Birds Eye" through a subsidiary, the Frosted Foods Sales Corporation.[7]

To achieve high quality without involving itself in heavy plant investment, the Birds Eye company organized a unique system of production. In the case of the fruit and vegetable packs, for example, its first step each year was to estimate its needs far in advance. Then production areas were selected which were adapted to the growth of the desired crops and sufficiently scattered to provide a hedge against damage by drought or flood or frost. When this had been done, the company entered into contracts with packers, usually canners. The packer agreed to do the required work at a guaranteed profit; Birds Eye agreed to furnish freezers and packaging equipment and in some cases even to provide seed. It met its obligation to supply freezers by following the harvest with its portable multi-plate frosters. Practically every phase of production was supervised by its agents. They not only watched over the entire process of preparation and freezing, but they supervised relations between the packer and the farmer, specifying that proper cultivation techniques be followed. Behind all this was systematic research in the company's laboratories.[8]

To introduce frozen foods to the public required ingenious

---

tion of Frozen Foods," New York Food Marketing Research Council, *Proceedings of Sixteenth Regular Meeting, December 9, 1930*, 2; W. P. Hedden, "Refrigeration," *Encyclopaedia of the Social Sciences*, XIII, 199; H. F. Taylor, "Legislation Affecting Refrigerated Foods," *Ice and Refrigeration*, LXXX (1931), 458.

[7] L. F. Church, "Trends in the Commercial Freezing of Fruits and Vegetables," *Ice and Refrigeration*, CX (April 1946), 22; Carlton, *op.cit.*, 4; "Let Them Eat Cake," *Fortune*, X (Oct. 1934), 124, 135.

[8] "Quick-Frozen Foods," *Fortune*, XIX (June 1939), 118, 120, 122; C. Francis, *A History of Food and Its Preservation, A Contribution to Civilization* (Princeton, 1937), 34-36; Poole and Zarotschenzeff, *loc. cit.*, 300-301; E. T. Gibson, "Quick Frozen Industry," *Ice and Refrigeration*, CI (1941), 123-124.

measures. First, tests were conducted in selected retail-market areas to prove that the public would buy them. Then in 1934 Birds Eye began a program which featured concentration on retail markets, one at a time, in the larger cities of the Northeast, where storage facilities were best and demand for food of high quality was believed largest.[9] That merchants might be encouraged to carry the new products, an inexpensive display case, developed by the American Radiator Company, was made available on a low-rental basis. The program was rounded out by granting some established wholesaler exclusive jurisdiction and by beaming an advertising campaign directly at the housewife.[10]

To distribute the frozen foods which it turned out in increasing variety,[11] the Birds Eye organization had no choice but to make the best possible use of the unsatisfactory facilities available for low-temperature transport and storage and to push as vigorously as possible the development of improved means.[12]

While Birds Eye experimented, other firms made their entry. At first their operations were limited to the institutional trade, but by 1937 several were devoting their attention to the retail market. Birds Eye remained the most important single force in the industry, but in the late thirties and early forties its proportion of total retail sales declined.[13]

Between 1937 and 1941 the distribution of frozen foods, first institutional and then retail, became national in character. But though the number of retail outlets grew rapidly, from ten in 1931 to perhaps twelve thousand in 1940, distribution

[9] Francis, *op.cit.*, 36-38; Carlton, *op.cit.*, 6, 8. Institutional customers were serviced throughout the country. "Frosted Foods Breaks the Ice," *Business Week*, Feb. 15, 1936, 12.

[10] Francis, *op.cit.*, 38-39; G. Poole, "A New Type of Low Temperature Cabinet," *Ice and Refrigeration*, xc (1936), 145; "Quick-Frozen Foods," *loc.cit.*, 118, 120; "Frosted Foods Breaks the Ice," *loc.cit.*, 12; Carlton, *op.cit.*, 8.

[11] C. Birdseye, "Progress of Quick-Freezing in the United States," *Ice and Refrigeration*, lxxxix (1935), 129-130.

[12] Carlton, *op.cit.*, 5-6.

[13] Church, *loc.cit.*, 22; Gibson, *loc.cit.*, 124; "Quick-Frozen Foods," *loc.cit.*, 126; "Frozen Foods: Interim Report," *Fortune*, xxxiv (Aug. 1946), 108.

facilities were far too thin, for that twelve thousand constituted no more than about two per cent of the nation's food stores.[14] Factors which accounted for this included the indifference of established food industries, the expensive equipment for proper refrigeration needed at all levels of distribution, and the failure of consumer demand to materialize as hoped.[15] Production of commercially packaged frozen foods increased from 205,000,000 pounds in 1937 to 568,000,000 in 1941. The meaning of these figures, however, was limited somewhat, for they did not distinguish adequately between the pack of fruit for processors and for the retail and institutional trade.[16] Before Pearl Harbor the more optimistic dreams of the industry did not come true. As early as 1939 warnings of overproduction were sounded.[17] Nevertheless, even though consumer demand developed slowly, it seemed clear that the public generally recognized frozen foods as products of superior quality.

Significant expansion took place during the war and the period that immediately followed. From 568,000,000 pounds in 1941 production rose to 1,028,000,000 pounds in 1945 and then in 1946 to 1,317,000,000.[18] The requirements of the armed services were partly responsible, as was the de-rationing of frozen fruits and vegetables seven months before canned goods. Most important of all was the simple fact that the shortages of the war and postwar period made many turn to frozen foods who otherwise might not have tried them.[19]

[14] Carlton, *op.cit.*, 14, 35; L. W. Waters, "Development of the Quick-Freezing Process," *Chemical and Engineering News*, xx (Dec. 10, 1942), 1559; J. P. Ferris and R. B. Taylor, "Immersion Quick Freezing," *Ice and Refrigeration*, xcvii (1939), 178. H. Carlton, "Frozen Foods in the Retail Market," *Food Industries*, xviii (1946), 697, estimates the number of prewar retail outlets at 30,000.

[15] Carlton, "Frozen Foods in the Retail Market," *loc.cit.*, 696-697; R. E. Ottenheimer, "The 'Bottle Neck' of Frosted Food Distribution," *Ice and Refrigeration*, xcvii (1939), 181.

[16] *1946-1947 Directory of Frozen Foods Processors of Fruits, Vegetables, Seafoods, Meats, Poultry, Specialties* (New York, 1946), 6.

[17] Ottenheimer, *loc.cit.*, 181.

[18] *1946-1947 Directory of Frozen Foods*, 6; "QFF's Annual Report On the Industry," *Quick Frozen Foods and The Locker Plant*, xii (Nov. 1949), 37.

[19] "Frozen Foods: Interim Report," *loc.cit.*, 105; *Ice and Refrigeration*, civ (1943), 52, 171; U. S. Department of the Interior, *Annual*

A wave of new processors, filled with visions of easy profits induced by the extravagant publicity frozen foods had received, flowed into the industry. By 1946 there were over six hundred separate concerns in forty-two states. Unfortunately, many of these latecomers lacked the economic and technical resources essential for success.[20]

New products found their way into the expanding market. Among them were batters and doughs, frozen raw and ready to bake. It was hoped that their labor-saving features would make them attractive to harassed housewives.[21] Another new idea was frozen precooked foods, either some special dish, such as chicken à la king, or an entire dinner. Precooked foods had been on the market since early in the thirties, but not until 1944 or 1945 did their production assume consequential proportions. Wartime demand and the entry of new processors into the industry probably accounted for the upsurge. The freezing preservation of cooked foods was a touchy proposition, largely because of the tendency of cooked fats and oils to become rancid during storage, but it offered certain advantages. For the homemaker it permitted savings in time and labor and the serving of food that was cooked properly, not overcooked, as was likely to be the result of canning preservation. For the institutional user it had special benefits. It enabled a small kitchen to supply a large demand, and it allowed great restaurant chains to provide for centralized cooking and thus to maintain a uniform standard of high quality. After 1947 the production of frozen precooked foods declined sharply. A few specialties, however, survived,[22] and entire

---

Report of the Secretary . . . , *1946* (Washington, 1946), 38. A limited expansion of production facilities was permitted by the War Production Board. "A Frozen Orphan," *Business Week*, April 3, 1943, 19.

[20] L. V. Burton, "Brutal Facts About Freezing," *Food Industries*, xviii (May 1946), 75; *1946-1947 Directory of Frozen Foods*, 480-504.

[21] G. L. Sunderlein, O. D. Collins, and M. Acheson, "Frozen Batters and Doughs," *Journal of Home Economics*, xxxii (1940), 381-382; *Ice and Refrigeration*, cx (Jan. 1946), 59-60.

[22] "Frozen Foods Sprout New Growth," *Food Industries*, xxii (Jan. 1950), 187; G. A. Fitzgerald, "Trends in the Refrigeration of Foods," *Ice and Refrigeration*, cvi (1944), 104; Greaves and Boggs, *loc.cit.*,

precooked and frozen dinners found a place especially on commercial airliners. The dinner, developed in the New York area by William L. Maxson, came frozen hard on a sectional paper plate. After a few minutes in an electric oven it was ready to serve. So admirably did it fulfill airline requirements that Pan American Airways adopted it.[23]

New methods of retailing also made their appearance. In metropolitan areas shops that specialized in frozen foods sprang up to take advantage of the opportunities offered by the short supply of food, while a number of leading department stores added the new line to their heterogeneous wares. Some venturesome firms even inaugurated businesses devoted to direct home delivery.[24]

In 1947 the boom came to an abrupt halt. Overproduction resulted in large inventories of frozen foods, some so poor in quality as to be inedible. At this juncture demand fell off, in part because of the restoration of normal conditions and in part because of the injury done consumer acceptance by the presence of inferior products on the market. In the squeeze that followed many processors who had entered the industry without adequate resources of capital and experience were forced out, as were most of the specialized frozen-food stores and home-delivery services that had mushroomed after the war.

26; Tressler and Evers, *op.cit.*, 642; D. K. Tressler, "Recent Advances in Frozen Food Technology," *The Fruit Products Journal and American Food Manufacturer*, xxv (1946), 228; J. G. Woodruff and I. S. Atkinson, *Freezing Preservation of Cooked Foods* (University System of Georgia, Georgia Agricultural Experiment Station, *Bulletin 242*) (Experiment, Georgia, 1945), 3, 15; Carlton, "Frozen Foods in the Retail Market," *loc.cit.*, 800, 802.

[23] "Precooked, Quick-Frozen Meals Offer Varied Airline Menus," *Air Transport*, iii (May 1945), 87; C. L. Walker, "What's in the Deep Freeze?" *Harper's Magazine*, cxcviii (June 1949), 44; *The New York Times*, July 9, 1950, F1, F7. An electronic oven was developed capable of thawing and heating frozen precooked meals in a matter of seconds. "Electronic Oven for Frozen Foods," *Electronics*, xx (Aug. 1947), 160, 162.

[24] *Ice and Refrigeration*, xcix (1940), 277; J. C. Bennett, "Ice Company Sells Frozen Foods," *Ice and Refrigeration*, cviii (March 1945), 23; "That Refrigeration Boom," *Fortune*, xxviii (Dec. 1943), 164, 244; Larson, Mixon, and Stokes, *op.cit.*, 123-125.

The winnowing, though painful to those eliminated, was beneficial to the industry as a whole, for those who survived were generally the ones whose product excelled in quality. The setback, moreover, was only temporary, for in the next year sales revived and climbed twelve per cent above the 1947 level. In 1949 production reached a new high, 1,516,000,000 pounds, and in 1950 passed this to attain a total of 2,055,000,000.[25]

Problems still confronted the industry. Some were technical in nature; others economic. Of the technical the most basic was the maintenance of a superior product. Even a small quantity of inferior goods could cause trouble, as was demonstrated in 1947. It seemed clear that existing techniques, properly applied, were adequate, though further improvement no doubt would take place. But suitable methods alone were not enough, for selection of only the best raw material was essential. In quality control governmental agencies took a hand. The Food and Drug Administration moved against spoiled food,[26] while the Department of Agriculture issued standards for a number of fruits and vegetables. Recommendations only, these often were ignored, but in 1948 federal authorities began hearings on standards that would be enforceable under the Food, Drug, and Cosmetic Act.[27] Related closely to the problem of quality was the public-health

[25] "Frozen Food Squeeze," *Business Week*, Feb. 22, 1947, 68, 70; "Comeback for Frozen Foods," *Business Week*, March 19, 1949, 23; "Yardstick for Distributors," *Business Week*, Dec. 3, 1949, 72; P. Lyons, "Trends in Frozen Food Distribution," *Ice and Refrigeration*, cxvi (May 1949), 66-67; "How 1949 Sales Volume Compares with 1948," *Quick Frozen Foods and The Locker Plant*, xii (Dec. 1949), 35; Walker, *loc.cit.*, 40; Larson, Mixon, and Stokes, *op.cit.*, 123, 125; E. W. Williams, "The Present Status of the Frozen Foods Industry," *Quick Frozen Foods and The Locker Plant*, xiii (May 1951), 41.

[26] Federal Security Agency, *Annual Report . . . , 1949. Food and Drug Administration* (Washington, 1950), 13-14.

[27] *1946-1947 Directory of Frozen Foods*, 4; Tressler and Evers, *op.cit.*, 814; "Frozen Food Squeeze," *loc.cit.*, 70, 72; *Ice and Refrigeration*, cxv (Sept. 1948), 66; cxvi (June 1949), 66. In the fiscal year 1950-1951 the Food and Drug Administration issued a tentative order proposing definitions and standards of identity and fill of container for frozen fruits in both bulk and consumer-sized packages. Federal Security Agency, *Annual Report . . . , 1951. Food and Drug Administration* (Washington, 1951), 15.

aspect of frozen foods, which became a matter of some concern, for freezing was not, like canning, a sterilizing process. But investigations showed there was little danger of food poisoning from products that were properly handled. Even should spoilage occur, frozen foods soured quickly and were not likely to be eaten when dangerous. There was, however, some possibility that the germs of infectious diseases might survive freezing and storage. In such cases no spoilage would be present to warn the consumer. The only protection against this was careful sanitation in processing plants. Other food industries had the same problem, but special attention was necessary in the frozen field, for a new industry could not afford to be blamed for an epidemic.[28]

Other technical problems that confronted the industry—packaging, transportation, and storage—were connected intimately with the maintenance of quality.

Packaging was important. Not only was it necessary to protect the food from dehydration and contamination during storage; it also had to be attractive enough to help sell the product. To meet these requirements, the industry adopted cardboard containers covered with specially developed moisture-vapor-proof wrappers which were sealed to make them airtight. Cans offered the same advantages and dominated the frozen-concentrate field, but their general use was checked by their higher cost. For the retail trade the small package of one pound or under became standard, but there was growing doubt as to whether this size was sufficiently large, particularly for those who had plenty of zero storage space in their homes.[29]

[28] G. A. Fitzgerald, "Are Frozen Foods a Public Health Problem?" *American Journal of Public Health*, xxxvii (1947), 700-701; F. W. Tanner, "Microbiological Examination of Fresh and Frozen Fruits and Vegetables," *American Journal of Public Health and The Nation's Health*, xxiv (1934), 485-492; American Public Health Association, Committee on Foods, "Refrigeration of Foods," American Public Health Association, *Yearbook, 1935-1936*, 49; American Public Health Association, Committee on Foods, "Public Health Aspects of Frozen Foods," American Public Health Association, *Yearbook, 1939-1940*, 80-81; Greaves and Boggs, *loc.cit.*, 23; H. C. Diehl, "Frozen Food Production," *The Canner*, lxxxvi (Feb. 26, 1938), 16.

[29] C. R. Mundee and F. C. Porcher, *Quick-Frozen Foods* (Washing-

Facilities for low-temperature transportation were essential. Deterioration in quality occurred if temperatures were allowed to rise unduly at any stage in distribution. Various expedients were tried to supply adequate refrigerated rail transport. The standard refrigerator car was unsatisfactory, but when equipped with additional insulation, air-circulating fans, or both, it served quite well. Clearly more suitable was the heavily insulated car with an overhead brine tank, for it was cheaper to operate, permitted heavier lading, and was more uniformly cooled. The industry depended on ice-and-salt mixtures, but experiments with dry ice and with refrigerating machines were undertaken, and in 1950 for the first time a limited number of mechanically refrigerated cars were placed on an operational basis. By the end of the forties there was more good equipment available than ever before, but additional facilities were still a major need.[30] Truck shipments increased in importance after the war and even became a factor in transcontinental movements. Lower rates, better refrigeration, more maneuverability, and greater adaptability to less-than-car-lot shipments often were advantages that motor transport enjoyed over rail.[31]

Storage at constant low temperatures was imperative, not only in the warehouse, but in the retail store and in the kitchen of the consumer. If proper storage were neglected, unfortunate

ton, 1938), 5; C. Birdseye, "Effect of Quick Freezing on Distribution of Fruits and Vegetables," *Ice and Refrigeration*, LXXX (1931), 132; Larson, Mixon, and Stokes, *op.cit.*, 175; "Quick-Frozen Foods," *loc. cit.*, 128. For a report on the 1949 status of packaging see "QFF's Annual Report On the Industry," *loc.cit.*, 40.

[30] *ibid.*, 42, 45, 170-171; C. C. Rowland, "Transportation of Frozen Foods—Refrigerator Car Service," *Ice and Refrigeration*, CX (Jan. 1946), 50; A. E. Stevens, "Optimum Conditions for Frozen Foods in Refrigerated Storage," *Ice and Refrigeration*, CI (1941), 209; G. M. Ellig, "Transport Facilities Best in Industry's History But Problems of Small Shippers Still Persist," *Quick Frozen Foods and The Locker Plant*, XII (Sept. 1949), 49; "Transportation Review," *Quick Frozen Foods and The Locker Plant*, XIII (Oct. 1950), 51. For a survey of the transportation situation see C. R. Havighorst and H. C. Diehl, "Frozen Foods Report No. 2: Transportation," *Food Industries*, XIX (Jan. 1947), 61-67.

[31] "QFF's Annual Report On the Industry," *loc.cit.*, 39-40; "Transportation Review," *loc.cit.*, 51.

changes in texture, appearance, and flavor, as well as losses in vitamin content, occurred. Most frozen-food experts advocated storage at zero degrees Fahrenheit or below.[32]

In cold-storage warehouses better construction and machinery were introduced to yield the necessary temperatures, while new methods, such as improved piling procedures, were devised to assure the maximum benefit from the equipment.[33] Shortages of freezer space proved an obstacle to growth of the frozen-food industry. The expansion of zero storage during the second World War helped, but the location of the new facilities was not always ideal for retail distribution. The situation in small towns, where for years freezer space was practically nonexistent, was no doubt helped by the spread of refrigerated locker plants.[34]

In retail stores cabinets with open tops were most popular. Such units had higher initial and operating costs than closed-top models, but this disadvantage was counter-balanced by the display of merchandise and easy self-service they made possible.[35] Despite the advantages of carrying a line of frozen foods, the shortage of retail storage cabinets remained serious. As late as 1949 even many large outlets did not have adequate equipment, and in 1950 retail space was still the number-one problem of the industry.[36]

[32] American Public Health Association, Committee on Foods, "Public Health Aspects of Frozen Foods," *loc.cit.*, 79-80; M. A. Joslyn, "Factors Influencing the Keeping Quality of Frozen Foods," *Ice and Refrigeration*, xcix (1940), 63; Tressler and Evers, *op.cit.*, 247.

[33] Stevens, *loc.cit.*, 208-209.

[34] Editorial, *Ice and Refrigeration*, xcvii (1939), 290; J. H. Frederick, *Public Warehousing* (New York, 1940), 117; H. Carlton, "Frozen Foods in the Retail Market," *Food Industries*, xviii (May 1946), 216. For a survey of the warehousing situation see C. R. Havighorst and H. C. Diehl, "Frozen Food Report No. 2: Warehousing," *Food Industries*, xix (Jan. 1947), 72-80.

[35] Larson, Mixon, and Stokes, *op.cit.*, 130-131; Carlton, *The Frozen Food Industry*, 47-48.

[36] J. I. Moone, "How the Distribution Picture Looks to Me," *Quick Frozen Foods and The Locker Plant*, xii (Dec. 1949), 80; "Frozen Foods: Interim Report," *loc.cit.*, 178-180. To deal with this situation, most frozen-food distributors and many packers took direct steps to stimulate the installation of retail cabinets. "What Does 1951 Hold?" *Quick Frozen Foods and The Locker Plant*, xiii (Dec. 1950), 38-39.

Home storage was a bottleneck. To achieve the maximum convenience from frozen foods, a housewife should have been able to buy in considerable quantities. But this was impossible if she did not have some place to store her purchases. The evaporator of the pre-war refrigerator would hold a few cartons, though not even these at optimum temperatures. The frozen-food compartments of modern two-temperature units were very satisfactory, but the number in use was small. One of the best things that could have happened for the industry would have been the widespread purchase of home freezers, for possession of one could have made a consumer a quantity buyer. Freezers, however, represented too heavy an investment for any immediate general availability, and retailers were reluctant to give owners the discounts for large orders that seemed necessary to promote sales volume.[37]

An economic problem, the high price of frozen foods, was one of the most crucial concerns of the industry. Potentially, great savings were possible by the distribution of food in frozen form, yet in practice high costs necessitated high prices which prevented mass acceptance.[38] The industry was able to advertise that certain frozen foods, mainly vegetables, were cheaper when equivalent weight was considered than their counterpart on the fresh market.[39] However this may have been, it seemed clear that American housewives, both those who bought frozen foods and those who did not, thought them too expensive.[40] There were several ways in which costs could be reduced. Careful attention could be given to locating and equipping processing plants so they could be used for a number of products. More adequate transportation facilities and better warehouse design and handling methods offered help. But the most fruitful field for cost reduction lay in improved distribution procedures. To lower the portion of the

[37] Larson, Mixon, and Stokes, *op.cit.*, 160-161, 166; "QFF's Annual Report On the Industry," *loc.cit.*, 39.

[38] Larson, Mixon, and Stokes, *op.cit.*, 4-8; Moone, *loc.cit.*, 80.

[39] "Birds Eye Stresses Savings," *Quick Frozen Foods and The Locker Plant*, xi (Feb. 1949), 70.

[40] "What the American Housewife Thinks of Frozen Foods," *Quick Frozen Foods and The Locker Plant*, xi (April 1949).

consumer's dollar that went to the wholesaler, food freezers experimented with more efficient methods of supplying their distributors and in some cases undertook to eliminate the wholesaler and to take over his function themselves.[41] If once a mass market could be built up, of course, the economies of large-scale production would tend to lower unit costs.

The volume that might be achieved by the frozen-food industry was clearly in the realm of speculation. Although there seemed little firm basis for prophecy, some informed quarters predicted that the dollar volume of sales might reach the neighborhood of ten billion by the mid-fifties.[42] But the horoscopes cast for the future by the more ardent enthusiasts had to be viewed with suspicion. It was said, for example, that frozen foods might soon constitute sixty per cent of all foodstuffs sold in the United States. This was not within the realm of possibility, for if all meats, poultry, fruits, and vegetables were sold in frozen form, the money received in payment would be only about fifty per cent of total retail dollar volume.[43] The extent to which the reasonable hopes of the industry might be realized depended in large part upon the success of efforts to reduce the cost to the consumer and upon the expansion of zero-storage facilities, particularly in the home. An imponderable factor was the possible development of new and more satisfactory methods of food preservation. One such possibility on the horizon was dehydration. Experiments were being made in "freeze" or "sublimation" drying, which involved evaporation of ice from a frozen product in a vacuum. This avoided the application of heat and promised a food of superior

[41] Larson, Mixon, and Stokes, *op.cit.*, 170-172; "Frozen Foods Sprout New Growth," *loc.cit.*, 186; "Yardstick for Distributors," *Business Week*, Dec. 3, 1949, 72, 75. For a survey of distribution problems see C. R. Havighorst and H. C. Diehl, "Frozen Food Report No. 2: Marketing," *Food Industries*, XIX (Feb. 1947), 73-80.

[42] E. W. Williams, "Don't Under-Estimate the Future," *Quick Frozen Foods and The Locker Plant*, XI (Oct. 1948), 45; Walker, *loc.cit.*, 40; Tressler and Evers, *op.cit.*, 22.

[43] C. O. Brannen, A. C. Hoffmann, and F. L. Thomsen, "Technological Developments in Agricultural Marketing," *Journal of Farm Economics*, XXIX (1947), 311.

quality, but the process, with its high-vacuum requirements, was inherently expensive.[44]

By 1950 it was apparent that should frozen foods achieve mass acceptance, they would force great changes in the economics of food supply. Markets for fresh and canned goods would not be destroyed, but they would decline, at least relatively, and production would have to be adapted to the new situation. The channels of food distribution would undergo readjustment. Wholesalers would tend to distribute a general line of frozen foods rather than to specialize in fresh fruits and vegetables or in meats or fish. This might reduce the number of distributors and eliminate much of the duplication in sales forces and deliveries which added to the cost of food.[45] Some felt that frozen foods, with their inevitable emphasis on brand names, would increase the trend toward monopolistic competition in agriculture by giving the edge to large marketing organizations, which were better able than small to exploit the advantages of brand-consciousness.[46] Spokesmen of the industry itself voiced the opinion that national distribution would be concentrated among a comparatively few brands, but that a situation would develop somewhat like that prevailing in the brewing industry, where, though national brands were strong, local products were able to give them vigorous competition in their own areas.[47] The most important effect, however, of frozen-food production on a truly volume basis would be a national diet at once more varied, more palatable, and more nutritious.

[44] E. W. Flosdorf, "Freeze Drying—Its Application to Penicillin, Blood Plasma, and Orange Juice," *Refrigerating Engineering*, LIV (1947), 226, 268; *Ice and Refrigeration*, CIV (1943), 55; CV (1943), 241; C. Birdseye, "Future Developments in Food Preservation," *Refrigerating Engineering*, LIV (1947), 220-221.

[45] Carlton, *The Frozen Food Industry*, 12.

[46] Brannen, Hoffmann, and Thomsen, *loc.cit.*, 317-318; "Frozen Foods: Interim Report," *loc.cit.*, 182.

[47] "QFF's Annual Report On the Industry," *loc.cit.*, 38.

CHAPTER XVIII

RURAL REFRIGERATION: LOCKER PLANTS
AND HOME FREEZERS, 1917-1950

THE American farmer received benefits of inestimable value from the refrigerating machine. It facilitated the marketing of his produce; it added variety and nourishment to his dinner table. Among its most significant applications to his uses were the locker plant and the home freezer, both of which made available the advantages of freezing preservation. The benefits of these innovations were by no means restricted to the farmer, but they appeared to be most useful to him, and he was their most ardent devotee.

The origin of the locker plant, an establishment that rented small amounts of freezing-storage space to its patrons, was obscure, but its appearance was inevitable. An owner of cold-storage space would be approached by men, often no doubt personal friends, who had small amounts of food they wished to store. To accommodate them and to preserve the identity of their goods, he would work out a system of lockers or baskets.[1] The practice started on the Pacific coast early in the twentieth century, and one of the first concerns to follow it was the Chico Ice and Cold Storage Company of Chico, California. At the outset this firm merely rented upstairs space to merchants who desired it for their eggs and apples. About 1908 it persuaded farmers to store their meat at the plant in boxes. Since this proved inconvenient and encouraged pilferage, boxes that could be locked were installed in 1913. By 1917 business had grown so large that a special room equipped with drawer-type wooden lockers arranged in tiers was provided. Now farmers used the lockers not only for meats, but for vegetables, fruits, eggs, and butter as well. It is not clear that the Chico plant provided freezing preservation, but certainly it was a forerunner of the modern cold-

[1] W. Carver, "History and Development of the Locker Plant Industry," H. D. Brown, L. E. Kunkle, and A. R. Winter, eds., *Frozen Foods; Processing and Handling* [Columbus, Ohio, 1946], 10-12.

storage locker.[2] In the next few years a number of plants appeared, mostly in the Pacific Northwest and frequently as side lines of creameries and ice factories.[3]

The spread of the locker idea was slow until 1935. In the twenties the few plants were confined mainly to the Pacific Northwest, to California, and to Kansas and Nebraska. By the end of 1935, though no more than 250 had been built, their foothold in the Midwest was broadened, and they had begun to penetrate the eastern half of the country. By the first of July, 1941 their number had soared to 3,623; five years later it stood at 8,025; and by mid-summer, 1949 it reached 11,245. In 1946 locker plants were to be found throughout the country, but their distribution was uneven. By far the greatest concentrations were in the Corn Belt and in the Pacific Northwest; comparatively few were located in New England and the South. The distribution followed roughly the availability of products to freeze. The concentration in the Corn Belt reflected the heavy beef-cattle and hog production of the region, while in the Pacific Northwest it mirrored the great quantities of fruits and vegetables grown.[4] That the locker plant was predominantly a rural phenomenon was shown by the facts that in 1946 almost one-quarter were lo-

[2] L. B. Mann, *Refrigerated Food Lockers, A New Cooperative Service* (U. S. Department of Agriculture, Farm Credit Administration, Cooperative Division, *Circular No. C-107*) ([Washington] 1938), 1-2; E. E. Jackson, "Some Highlights in the History of Refrigerated Locker Plants," *The Locker Patron*, I (Aug. 1939), 7; C. Crow, "The Farmers' Own Cold Storage," *The Country Gentleman*, LXXXII (Oct. 27, 1917), 1627.

[3] L. B. Mann, "A Widening Field for co-op service," U. S. Department of Agriculture, Farm Credit Administration, *News for Farmer Cooperatives*, V (July 1938), 15; *Ice and Refrigeration*, CI (1941), 50-51.

[4] L. B. Mann and P. C. Wilkins, *Frozen Food Locker Plants: Location, Capacity, Rates, and Use, January 1, 1946* (U. S. Department of Agriculture, Farm Credit Administration, Cooperative Research and Service Division, *Miscellaneous Report No. 105*) (Washington, 1947), iv; S. T. Warrington, *Frozen Food Locker Plants in the United States, January 1, 1941* (U. S. Department of Agriculture, Farm Credit Administration, Cooperative Research and Service Division, *Miscellaneous Report No. 41*) (Washington, 1941), 1-3, 6; L. B. Mann, "Locker Plants Having Readjustment Pains," *News for Farmer Cooperatives*, XVI (Jan. 1950), 17. See Figure 10.

cated in towns under five hundred and that eighty-four per cent were to be found in communities under ten thousand.[5]

An increase in the size of plants accompanied their extension throughout the nation. In 1940 the average capacity of each was 348 lockers, but by 1946 it had risen to 500. Expanding demand was partly responsible, as were increasing costs, which made larger and more efficient establishments desirable. Plants tended to be larger in the northeastern and western states, for in these areas they often were affiliated with ice factories and cold-storage warehouses, where space was available to facilitate expansion.[6]

About 3,300,000 families including 13,000,000 individuals were served by locker plants in 1946. Seventy-three per cent of these families were farmers. Although the war resulted in the construction of a considerable number of units in cities with a population over ten thousand, the locker plant remained a rural institution. It should be observed, however, that city dwellers in the western states were more prone to use lockers than elsewhere.[7] Perhaps the explanation of this lay in the greater extent of home gardening or in the proximity of producing areas to urban centers.

Fundamental to an understanding of the rapid expansion after 1935 was knowledge of the advantages of locker-plant storage. For one thing, its wise use assured a better diet. Meats, poultry, fruits, and vegetables generally were more palatable when preserved by freezing than by curing or canning. Greater variety and a balanced died could be maintained throughout the year. Even in winter a farm family could enjoy fresh foods.[8] Much of the vitamin content was retained, and in the case of pork the organism that produced trichinosis was killed.[9] Furthermore, the locker plant made possible cer-

---

[5] This estimate of distribution according to size of town was based on the reports of 2,882 plants. Mann and Wilkins, *op.cit.*, 19.

[6] *ibid.*, 1-3.   [7] *ibid.*, i-ii, 3, 40.

[8] In the late thirties another means of supplementing rural winter diet appeared. Some farmers began buying fresh foods at supermarkets located conveniently on the outskirts of cities.

[9] J. W. Emig, "Community Cold Storage Lockers in the Northwest," *Ice and Refrigeration*, xcii (1937), 103; American Public Health Association, Committee on Foods, "Public Health Aspects of Frozen Foods

tain economies. For the farmer it did away with selling livestock at wholesale and buying meat at retail, an expensive practice which formerly was necessary if he were to have fresh cuts on his table at all seasons. It reduced the loss from spoilage that was always possible when canning or curing was employed. For the townsman it offered the savings of quantity buying. Finally, the locker was a convenience, for if the plant offered slaughtering and processing services, the labor and bother of performing such functions on the farm would be eliminated. Even if fruits and vegetables were processed at home, there was an advantage, for preparation for freezing was less work than canning.[10] There were, however, some

---

with Particular Reference to the Products Frozen in Cold Storage Lockers and Farm Freezers," *American Public Health Association Yearbook, 1939-1940*, 82; S. E. Gould and L. J. Kaasa, "Low Temperature Treatment of Pork," *Refrigerating Engineering*, LVII (1949), 183.

[10] A. L. Blatti, "Cold Storage Locker Plants," *Ice and Refrigeration*, XCII (1937), 371; R. P. Calt and H. K. Smith, "Food Banks of the Future," *The Atlantic Monthly*, CLXVII (1941), 362-363; Mann, *Refrigerated Food Lockers*, 24-25; M. A. Schaars, "How Cold Storage Locker Plants Benefit the Public," *National Butter and Cheese Journal*, XXX (Jan. 1939), 37-38; G. W. Kable, "Some Development Trends in the Farm Freezing and Storing of Food," *Agricultural Engineering*, XXII (1941), 145.

FIGURE 10

The distribution of locker plants

The spread of lockers was slow before 1935. In the twenties they were confined mainly to the Pacific Northwest, to California, and to Kansas and Nebraska. By the end of 1935 their foothold in the Midwest had been broadened, and they had begun to appear east of the Mississippi. By 1946 locker plants were to be found throughout the country. The greatest concentrations were in the Corn Belt and the Pacific Northwest; comparatively few were located in New England and the South.

From S. T. Warrington, *Frozen Food Locker Plants in the United States, January 1, 1941* (U. S. Department of Agriculture, Farm Credit Administration, Cooperative Research and Service Division, *Miscellaneous Report No. 41*) (Washington, 1941), 3, 6; and L. B. Mann and P. C. Wilkins, *Frozen Food Locker Plants: Location, Capacity, Rates, and Use, January 1, 1946* (U. S. Department of Agriculture, Farm Credit Administration, Cooperative Research and Service Division, *Miscellaneous Report No. 105*) (Washington, 1947), iv.

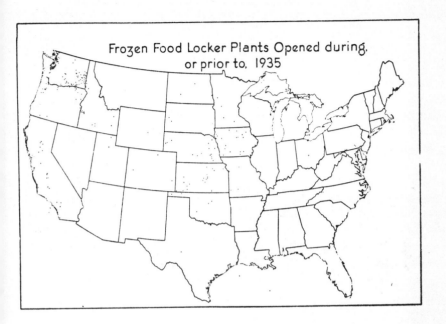

Frozen Food Locker Plants Opened during, or prior to, 1935

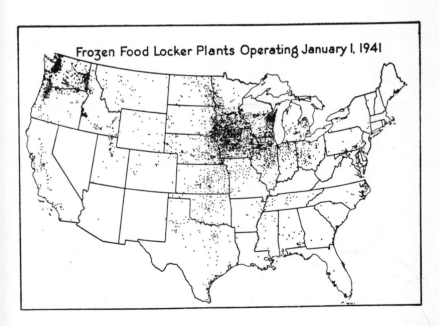

Frozen Food Locker Plants Operating January 1, 1941

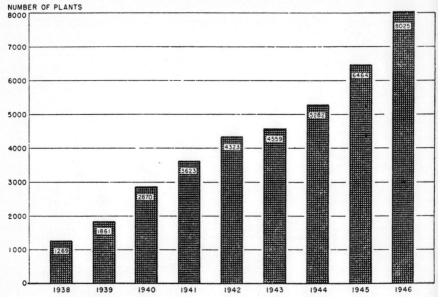

Frozen-Food Locker Plants in the United States July 1, 1938 to July 1, 1946. (Source: Extension Service)

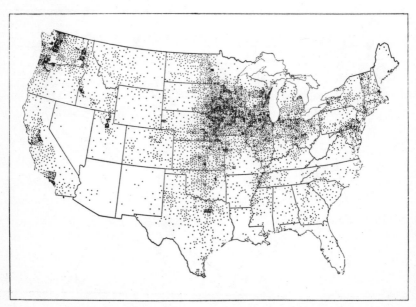

Frozen-Food Locker Plants Operating January 1, 1946.

drawbacks. A farmer might find that it was inconvenient to use a locker perhaps fifteen miles away and that a cash outlay was required greater than needed when he canned or cured his foods at home. A city family might find a locker less convenient than a nearby store; it might be reluctant to spend large sums for quantity purchases; it might discover that a quarter or a side of beef contained relatively few of the choice cuts to which it had become accustomed; and it might hesitate to purchase meats not slaughtered under federal inspection.[11]

The unsettled economic conditions of the thirties and forties contributed to the growth of the industry. The depression years made the refrigerated locker, a device for promoting self-sufficiency, especially attractive to farmers. Even the war years provided a favorable environment. Such varied factors as high consumer income, food shortages, rationing, and victory gardens combined to create an extraordinary demand, while the government, viewing lockers as means of conserving food supplies, transportation facilities, and labor, issued priorities for new plants.[12] A doubling of the number between 1941 and 1946 was the result.[13]

Additional encouragement was afforded by the activities of the Department of Agriculture. The Farm Credit Administration not only financed plants, but also provided cooperative locker associations with the information on organization, construction, technical procedures, and management they needed to make their projects successful.[14] The Extension Service

[11] Mann, *Refrigerated Food Lockers*, 25.

[12] W. R. Woolrich, "Locker Storage and Related Freezing Facilities for Community Storage Plants," *Ice and Refrigeration*, c (1941), 199-200; L. B. Mann and P. C. Wilkins, *Cooperative Frozen Food Locker Associations* (U. S. Department of Agriculture, Farm Credit Administration, Cooperative Research and Service Division, *Miscellaneous Report 116* (Washington, 1948), ii; L. B. Mann, "Locker Plants Grow Up," *News for Farmer Cooperatives*, XII (Jan. 1946), 13.

[13] Mann and Wilkins, *Frozen Food Locker Plants*, iv, 40.

[14] U. S. Department of Agriculture, Rural Electrification Administration, *1939 Report . . .* (Washington, 1940), 47; "U. S. Government Will Finance Locker Plant Installations," *Quick Frozen Foods*, I (Nov. 1938), 34-37 (Dec. 1938), 36-37; Mann, *Refrigerated Food Lockers*; L. B. Mann, *Organizing a Refrigerated Food Locker Associa-*

trained county and home agents that they might help farmers in their communities obtain the maximum advantage from their lockers,[15] while many of the state experiment stations conducted research on freezing preservation and its value to the rural family.[16]

But utility, favorable economic conditions, and encouragement by governmental agencies could not have produced the rapid extension of the locker plant if certain basic technological changes had not taken place. Foremost among these was the perfection of the small, automatic, commercial refrigerating machine. Also important was progress in rural electrification, as was the development of networks of hardsurface country roads, for without easy transportation farmers would have found lockers of little value.[17]

Locker plants tended to be operated in connection with some other enterprise. In 1946, for example, only twenty-nine per cent were independent; the remainder were affiliated mainly with meat and grocery stores, ice and cold-storage concerns, or dairy plants. The trend nevertheless was toward nonaffiliation; between 1941 and 1946 the number of separate units increased perhaps as much as ninety per cent. This probably was a reflection of confidence that the locker plant was here to stay and of a growing recognition that specialized attention was necessary for success.[18]

Ownership in 1946 was concentrated in the hands of individuals, who held forty-nine per cent. Partnerships followed

---

tion (U. S. Department of Agriculture, Farm Credit Administration, Cooperative Research and Service Division, *Miscellaneous Report No. 20*) (Washington, 1939); S. T. Warrington, "Eighteen Points in Planning a Co-op Locker Plant," *News for Farmer Cooperatives*, xi (Oct. 1944), 14-15, 17.

[15] U. S. Department of Agriculture, Extension Service, *Report of Cooperative Extension Work in Agriculture and Home Economics, 1946* (Washington, 1946), 17.

[16] U. S. Department of Agriculture, Agricultural Research Administration, Office of Experiment Stations, *Report of the Agricultural Experiment Stations, 1945* (Washington, 1946), 118-120.

[17] Carver, *loc.cit.*, 7-8.

[18] Mann and Wilkins, *Frozen Food Locker Plants*, i, 7, 9; Warrington, *Frozen Food Locker Plants in the United States, January 1, 1941*, 14.

with twenty-two per cent, corporations with sixteen, and co-operatives with thirteen.[19] Before 1936 more units were owned by corporations than by any other form of business organiza-tion. Since locker plants were increasing in cost and size, it seemed likely that corporate ownership, with the greater re-sources of capital it afforded, would assume something like its old importance.[20]

At first locker concerns provided only storage, but before 1940 a tendency to offer more complete service was apparent. During the forties this became a definite trend, and by 1949 many were moving away from a strictly service-type business and were diverting more and more of their attention to com-mercial processing and merchandising. Many plants would slaughter livestock and poultry, prepare meat for freezing, supervise curing and smoking, and render lard. Some would process fruits and vegetables, though the volume of this trade was comparatively small. Since slaughtering and proc-essing costs were high, a few sought to reduce them by centralizing operations in a single establishment. Merchandis-ing activities were varied. Some sold meats, poultry, fruits, and vegetables. Some distributed commercial frozen foods on a wholesale basis or retailed them directly to their patrons. Some sold home freezers and such supplies as wrappers and containers. Many became miniature cold-storage warehouses and offered bulk storage either at below- or above-freezing temperatures.[21]

---

[19] This estimate was based on the reports of 2,882 plants. Mann and Wilkins, *Frozen Food Locker Plants*, 11. Most, though not all, of the cooperative enterprises were successful in the financial sense. "Frozen Food Lockers," *News for Farmer Cooperatives*, viii (July 1941), 26. For a survey see Mann and Wilkins, *Cooperative Frozen Food Locker Associations*.

[20] Warrington, *Frozen Food Locker Plants in the United States, January 1, 1941*, 18, 20; Mann and Wilkins, *Frozen Food Locker Plants*, i-ii, 12-13.

[21] Mann and Wilkins, *Frozen Food Locker Plants*, ii, 40-41; Mann and Wilkins, *Cooperative Frozen Food Locker Associations*, 15; Mann, "Locker Plants Having Readjustment Pains," *loc.cit.*, 17-18; Mann, "Locker Plants Grow Up," *loc.cit.*, 13; L. B. Mann, "Trends, Prob-lems, and Potential of the Locker Plant Industry," *Quick Frozen Foods and The Locker Plant*, xii (Oct. 1949), 96; J. S. Larson, J. A. Mixon,

The conditions that prevailed during the early years left much to be desired. The physical equipment often was inadequate. Plants might be poorly designed and improperly insulated. Misguided efforts at economy sometimes led to the use of only one compressor where two or three might be required, and only rarely were provided the separate facilities necessary to speed the freezing process. The lockers themselves were often nothing more than wire baskets or crude wooden frames. Techniques of operation frequently were no better than equipment. Commonly meat was packed away in lockers before it had been thoroughly chilled. Packaging was faulty, and sometimes not enough attention was paid to proper sanitation.[22] No doubt the rapid growth of the industry was mainly responsible, but a contributory factor was the failure of many prospective owners to obtain specifications from independent engineers. Many instead made their first contact with salesmen, either of machinery or insulation, who were accused of fitting their recommendations to the amount the prospect was willing to spend rather than to sound principles of design.[23]

Improvements soon were introduced. Proper design and construction became more general, while all-steel lockers and equipment capable of quick freezing were installed. Nor did progress in technique lag behind. Such matters as wrapping the packages and maintaining the proper relative humidity in storage received the attention due them, and by 1939 the American Public Health Association could report that most

---

and E. C. Stokes, *Marketing Frozen Foods—Facilities and Methods* (Washington, 1949), 153; W. E. Guest, "The Storage Locker Business; The Pitfalls and Triumphs in a Lively New Industry," *Refrigerating Engineering*, xxxvi (1938), 237; J. H. Frandsen, "Refrigerated Food Lockers for New England," *Ice and Refrigeration*, xcix (1940), 142; "Freezer Lockers," *Fortune*, xxxvi (Sept. 1947), 194.

[22] S. Bull, "Post-War Problems of the Locker Industry," *Refrigerating Engineering*, xlvi (1943), 310; E. L. Mohr, "Advancement in Locker Plant Equipment," *Refrigerating Engineering*, xxxvi (1938), 239; R. Farquhar, "A History of the Locker Plant," *Quick Frozen Foods and The Locker Plant*, ix (March 1947), 254; G. O. Schlageter, "The Locker Industry Looks Ahead," *Ice and Refrigeration*, cv (1943), 302.

[23] Guest, *loc.cit.*, 237.

of the newly constructed plants were kept remarkably clean.[24]

Responsible for better conditions were several forces. Research by both public and private agencies was fundamental, as was the work of trade associations.[25] Trade journals made available not merely merchandising ideas, but detailed information on technical problems,[26] and by 1946 several colleges were offering courses designed to provide both the engineering and business training necessary.[27] Regulation, imposed in many of the states where lockers were most abundant, no doubt was helpful. In general, statutes required that foods be frozen rapidly, that the temperature of storage rooms not rise above a specified maximum, that only fit foods be frozen, that sanitary practices prevail, and that employees handling food submit periodically to health examinations. Licensing provisions furnished a means of enforcement.[28] Regulation was supported by locker-plant associations in the belief that sound state laws would prevent the industry from being discredited by a few sub-standard operators.[29]

The rate at which locker plants were constructed slowed materially after 1947. A midsummer survey of 1949 showed an increase during the preceding twelve months of only 628, a figure less than fifty per cent of the average for the five years before. This may have been merely a reflection of high building and operating costs, but some feeling existed that the saturation point had been reached. Certainly the disap-

[24] Farquhar, *loc.cit.*, 254; Mohr, *loc.cit.*, 239; American Public Health Association, Committee on Foods, "Public Health Aspects of Frozen Foods," *loc.cit.*, 78; Mann, *Refrigerated Food Lockers*, 7-8; *1944 Guide Book of the Frozen Food Locker Industry* (Des Moines, n. d.), 34; D. K. Tressler and C. P. Evers, *The Freezing Preservation of Foods*, 2nd edn. (New York, 1947), 192. For a diagram of the floor plan of a typical plant see Mann, *Refrigerated Food Lockers*, 11.

[25] *Ice and Refrigeration*, CI (1941), 53.

[26] See *Quick Frozen Foods and The Locker Plant*, Aug. 1938—, and *The Locker Operator*, Aug. 1939—. Titles vary.

[27] Mann, "Locker Plants Grow Up," *loc.cit.*, 13.

[28] W. Carver, "Legislation," H. D. Brown, L. E. Kunkle, and A. R. Winter, eds., *Frozen Foods: Processing and Handling* [Columbus, Ohio, 1946], 127-132; Tressler and Evers, *op.cit.*, 167-168.

[29] Carver, "Legislation," *loc.cit.*, 127; Schlageter, *loc.cit.*, 303; *Ice and Refrigeration*, XCVI (1939), 438.

pearance of food shortages was a factor, as was the keener competition lockers faced from other food processors and distributors as well as from home freezing units.[30]

The principal effect of the locker was to improve rural diet, which during the late winter months lacked variety and appeal. The extent to which this improvement was felt was suggested by the fact that almost three-fourths of the nation's locker patrons were farmers and that perhaps a third of the individuals living on farms benefited.

The refrigerated locker, it appeared, might contribute to the decentralization of the food-processing industry. Already certain activities of the more aggressive concerns pointed in this direction. Some froze meats, fruits, and vegetables for local sale. If they could join with other units to form regional sales organizations, they might be able to distribute their foods in a greater area. Some found it profitable to slaughter livestock and to dress poultry for local markets, hotels, and restaurants. Should these side lines assume significant proportions, the changes effected in our economy might be profound. Since locker plants were located near points of production, received revenue from a variety of services, and should have had low distribution costs, it was conceivable that they might narrow the spread between the price the producer received and that the consumer paid. But established methods of distribution were difficult to change, and before any major alteration could be effected, locker enterprises needed not only sound management and adequate resources of capital, but also provision for the efficient utilization of by-products.[31]

A new device for exploiting the advantages of freezing preservation was the home freezer, which made its appearance late in the thirties. Some of the first models, like the early

[30] Mann, "Locker Plants Having Readjustment Pains," *loc.cit.*, 17; Mann and Wilkins, *Cooperative Frozen Food Locker Associations*, ii.

[31] H. Carlton, *The Frozen Food Industry* (Knoxville, 1941), 16; Mann and Wilkins, *Frozen Food Locker Plants*, 41; Mann, *Refrigerated Food Lockers*, 26-28; Mann, "Trends, Problems, and Potential of the Locker Plant Industry," *loc.cit.*, 96-97; Mann, "Locker Plants Having Readjustment Pains," *loc.cit.*, 17.

household refrigerators, achieved a cooling effect through the agency of chilled brine. Others employed direct-expansion coils exposed within the cabinet. Neither of these systems was entirely satisfactory, for the one was expensive and subject to leaks, while the other frosted too rapidly. But improved freezers soon were evolved which eliminated brine and accomplished their cooling function by means of evaporators built into the walls of the cabinet itself. The modern unit ranged from three to thirty-two cubic feet in capacity, was well constructed and heavily insulated, and often was equipped with a special quick-freezing compartment.[32]

The sale of home freezers gathered momentum up to March 1943, when their manufacture for civilian use was prohibited. During these first years production was in the hands of comparatively small concerns. The giants of the home-appliance industry, such firms as General Electric and Westinghouse, had conducted experiments, but had not built on a large scale.[33] With the end of the war the industry boomed. Two hundred thousand freezers were made in 1946 and twice that many in 1947. Now the manufacturers of household refrigerators entered the field in force along with concerns like International Harvester, whose primary interest lay elsewhere. By 1949 an estimated 1,200,000 units were in use in American homes.[34]

The sales of home freezers expanded rapidly, even though some of the early expectations were not realized. It was not surprising that the freezer was so well received. For one thing, the public had been prepared by its experience with domestic refrigerators, with frozen foods, and with locker plants. For another, the home freezer seemed to some an avenue of escape from the food shortages brought on by the war. Finally,

[32] C. W. DuBois, "Farm Freezer Analysis," *Agricultural Engineering*, xxiv (1943), 344; R. H. Bishop, *Home Food Freezers* (American Society of Refrigerating Engineers, *Refrigerating Engineering Application Data 37*). Published as Section 2 of *Refrigerating Engineering* for January 1947.

[33] "Freezers Are Hot," *Business Week*, Aug. 12, 1944, 44; *Ice and Refrigeration*, cvi (1944), 207-210; B. Sparkes, *Zero Storage in Your Home* (New York, 1944), 67.

[34] Larson, Mixon, and Stokes, *op.cit.*, 160.

many were convinced that the new device offered solid advantages that made it worth having even in times of plenty. To the farm family it meant convenience. It made available in the farmer's own kitchen the essential advantages of the locker plant. Besides, home freezing required less time and labor than canning. The economies possible through use of a freezer were not quite so clear, though it appeared that a rural family which produced a substantial portion of the food it ate could effect savings. To the urban family possession of a freezer afforded the convenience that came from having an ample supply of high-quality edibles immediately at hand. Particularly attractive to the woman who worked outside the home was the opportunity of cooking entire meals in advance and of storing them in a freezer to be withdrawn and warmed for serving as needed. But for a city family to save money it needed access to a supply of low-cost food, either grown in its own garden or purchased on a quantity basis.[35]

The future of the home freezer depended on certain imponderables. Foremost among these was the extent to which consumers were convinced that the advantages were worth the initial and operating costs. With upper-income groups convenience rather than economy would be the determinant, but with middle- and lower-income groups economy would be an important consideration. The opinion was voiced that actual monetary savings must be possible if the home freezer were to achieve widespread acceptance. This might well prove to be true, but if Americans generally came to regard a freezer as a desirable contribution to their standard of living, it was unlikely that expense would prove a serious obstacle. After all, many owners of automobiles would be hard put to justify their possession on strictly economic grounds. Related closely to the problem of economy were the opportunities open to home-freezer owners for quantity buying, particularly of commercial frozen foods. Although many locker plants offered such products to their patrons at a quantity discount, retail distributors did not do so to any considerable extent. If they

[35] *ibid.*, 160, 164-166; E. T. McAllester, "Home Freezers for Farm Families," *Journal of Home Economics*, XLI (1949), 77.

should, the effect on home-freezer acceptance would certainly be favorable.[36]

The impact the freezer might have on the economics of the nation's food supply was still subject to conjecture. Among the first enterprises that would be affected were locker plants. Obviously each cubic foot of freezing storage space in the home represented a foot that might otherwise be rented in a commercial establishment. Both the home freezer and the locker had their advantages; the one offered greater convenience, while the other promised lower cost, less danger of loss in the event of mechanical breakdown, and in many cases relief from the burden of processing.[37] The consensus of opinion was that both would survive. The locker plant might suffer some loss in storage revenue, but might find in the owners of home freezers an expanding market for processing and merchandising services.[38] The patterns of food distribution conceivably might change should a large proportion of American homes be equipped with their own freezing storage. Wholesalers might bypass distributors and sell directly to consumers in quantity lots. This practice, however, did not develop appreciably, partly because wholesalers were reluctant to antagonize their retail outlets.[39] It would have been folly to expect any quick change in this matter, for the channels of distribution in any event were exceedingly resistant to mutation.

[36] Larson, Mixon, and Stokes, *op.cit.*, 160-162, 164-165.

[37] Lenore Sater, head of household equipment research in the Bureau of Human Nutrition and Home Economics, U. S. Department of Agriculture, quoted in McAllester, *loc.cit.*, 78.

[38] Mann and Wilkins, *Frozen Food Locker Plants*, 40; Mann and Wilkins, *Cooperative Frozen Food Locker Associations*, ii; Mann, "Locker Plants Having Readjustment Pains," *loc.cit.*, 17; S. T. Warrington, "Home Units vs. Locker Plants . . . No Knockouts Predicted," *News for Farmer Cooperatives*, XII (May 1945), 14.

[39] Larson, Mixon, and Stokes, *op.cit.*, 161-162.

CHAPTER XIX

THE EXPANDING USE OF REFRIGERATION
1918-1950

O NE of the most pronounced trends in refrigeration was its application to an ever-increasing variety of tasks. The range of its utility was amazing; its assignments varied from such basic matters as assisting in the smelting of iron ore to such purely luxury considerations as the cooling of theaters and beauty salons. The application to uses other than food preservation long antedated the first World War, but only after 1930 did the extension become so dramatically apparent. Several factors were responsible: first, the improvement of machinery, particularly the perfection of small, automatic units; second, new and safe refrigerants; and third, the progress made in the techniques of air conditioning, which in turn was related closely to the development of machines and refrigerants. Air conditioning, it seemed likely, might prove one of the most significant advances in American technology in the twentieth century. Already it had given almost infinite variety to the tasks performed by refrigeration.[1] In the future one of its effects might be to stimulate the industrial development of the South, where high temperatures and humidities had offset the section's advantages in both natural and human resources.[2]

In the metal-working industries the refrigerating machine was a versatile servant. It provided the low temperatures needed in preparing steel for the manufacture of precision tools and gauges and in storing annealed aluminum alloys so

[1] For some indication of the variety of air-conditioning applications see W. B. Henderson, "Air Conditioning and Production Efficiency," American Management Association, *Lighting and Air Conditioning for the Modern Plant* (New York, 1940), 36; American Society of Heating and Ventilating Engineers, *Heating, Ventilating and Air Conditioning Guide, 1946* (New York, 1946), 826-827; J. R. Allen, J. H. Walker, and J. W. James, *Heating and Air Conditioning*, 6th edn. (New York, 1946), 613-616.

[2] "Development of South," *Heating and Ventilating*, XLIII (Oct. 1946), 110, 112.

that they retained their malleability. It made possible the differences in temperature necessary for shrink fits such as valve inserts in automobile cylinder blocks.[3] As a component in air-conditioning installations it helped to provide accurate dimensional controls, to prevent·condensation of moisture on finely machined metal surfaces, bearings for example, and to simulate the climatic conditions required for testing airplane engines and electronic equipment.[4]

Refrigeration in metal working was not confined to the final stages of manufacture, for in the forties it became an important part of blast-furnace operation. Long before the first World War refrigeration had been used to remove moisture from the air blast, but high costs soon demonstrated the practice to be a failure from the economic standpoint. In 1939, however, the Woodward Iron Company revived the technique at its plant at Woodward, Alabama, and found that it brought substantial savings in fuel and significant increase in production. Other companies followed this lead, encouraged by priorities and financial help from the federal government, which saw in the process a more expeditious way of increasing the nation's capacity than the construction of new furnaces. The success of the dry blast in its second trial was due largely to technical improvements associated with air conditioning: the cold-spray dehumidifier and the centrifugal compressor.[5]

[3] R. J. Thompson, "Freon-22," *Refrigerating Engineering*, XLIX (1945), 473; W. A. Phair, "Cold Treatment of Metals," *The Iron Age*, CLI (Feb. 25, 1943), 37-43; H. E. Keeler, "The Use of Refrigeration in National Defense Plants," *Ice and Refrigeration*, CII (1942), 98; "Deepfreeze's Ace," *Business Week*, April 11, 1942, 70.

[4] Henderson, *loc.cit.*, 30-32; W. B. Henderson, "Air Conditioning Goes to War," *Ice and Refrigeration*, CII (1942), 154; "Air Conditioning Aids Precision in Bearing Manufacture," *Scientific American*, CLXXII (1945), 368; *Ice and Refrigeration*, CVIII (May 1945), 20; Thompson, *loc.cit.*, 473.

[5] Henderson, "Air Conditioning and Production Efficiency," *loc.cit.*, 33; Henderson, "Air Conditioning Goes to War," *loc.cit.*, 155; Keeler, *loc.cit.*, 97; L. L. Lewis and R. VD. Dunne, "Why Dry Blast Equipment is Different Now," *Blast Furnace and Steel Plant*, XXIX (1941), 1119-1125; R. VD. Dunne, "Dry Blast and Production for War," *Refrigerating Engineering*, XLIV (1942), 21; "Profits in Air, No. 5," *Heating and Ventilating*, XXXVI (Dec. 1939), 27-28; "Pig-Iron

The chemical industries found refrigeration essential to the accomplishment of many basic processes. It was required in the liquefaction of gases such as oxygen and chlorine and in the solidification of certain liquids. Were it not for refrigeration the mixtures of oils in vegetable shortenings and margarines would have cooled slowly and tended to separate into their constituents as they hardened. Machine-made cold proved particularly helpful in the crystallization of salts from complex solutions. Borax, potassium chloride, and potassium and ammonium nitrate were among the products whose selective separation was facilitated.[6] The resolution of mixtures into their components was another process which benefited. Perhaps the most important example was in the petroleum industry, where low temperatures were used to solidify wax so that it could be removed from lubricating oil.[7] In the regulation of reaction velocities refrigeration was important, particularly as a means of controlling the nitration process used in the manufacture of explosives. After the second World War it was employed to slow down the speed of chemical reaction in the manufacture of "cold rubber," a new synthetic product valuable for its superior resistance to abrasion.[8]

In the food-processing industries refrigeration was applied not only to the storage of raw materials, but to many of the manufacturing operations themselves.[9] Flour mills employed it to provide close control of temperature and humidity, which was necessary if sifting screens were to function properly and

Booster," *Newsweek*, xix (March 2, 1942), 42, 44; "Air-Conditioned War," *Time*, xxxviii (Dec. 29, 1941), 57.

[6] D. H. Killeffer, "Refrigeration in Chemical Industries: A Survey of Applications," *Industrial and Engineering Chemistry*, xxiv (1932), 601-603.

[7] *ibid.*, 603-604; V. Voorhees, "Refrigeration is Indispensable in the Oil Industry," *Refrigerating Engineering*, lvi (1948), 233; Keeler, *loc.cit.*, 97; V. Harrison, "Refrigeration in an Oil Refinery," *Ice and Refrigeration*, xcvi (1939), 211.

[8] Killeffer, *loc.cit.*, 604; D. Chalmers, " 'Cold Rubber'—Its Preparation, Properties, and Applications," *Refrigerating Engineering*, lvii (1949), 641-645, 710.

[9] H. Stiner, "Air Conditioning Reduces Loss in Stored Raw Materials," *Food Industries*, v (1933), 498-500, 528.

uniform quality were to be maintained.[10] Bakeries used machines to keep down the temperature of dough in the mixers and to assure the quick cooling of bread during the warm summer months,[11] while candy plants assigned them the task of creating the conditions necessary for quick, efficient production throughout the year.[12] Refrigeration was almost as useful in manufacturing condensed milk as it was in handling the fresh product itself.[13]

Interest in brewery refrigeration revived with the repeal of the Eighteenth Amendment. Many of the new centrifugal compressors were installed, as well as ammonia equipment in perfected form. Throughout the industry automatic controls brought more effective regulation of temperature and marked economies. But the most important innovation was air conditioning. It provided the optimum environment for the malting process and speeded the work of the coolship, at the same time protecting the wort against air-borne infection. Air conditioning was particularly helpful in brewery fermentation and stock cellars, where it provided even temperatures and checked the mold growth and other contamination that accompanied excessive humidity.[14]

[10] Henderson, "Air Conditioning and Production Efficiency," *loc.cit.*, 30; E. S. Miller, "Air Conditioning a Flour Mill," *Heating, Piping and Air Conditioning*, II (1930), 97-100.

[11] Henderson, "Air Conditioning and Production Efficiency," *loc.cit.*, 28-29; W. J. Hoffmann, *Refrigeration in Breadmaking* (American Society of Refrigerating Engineering, *Refrigerating Engineering Application Data 49*) (Published as Section 2 of *Refrigerating Engineering* for June, 1950), 1-5; W. W. Reece, "Refrigeration as Applied to the Baking Industry," *Ice and Refrigeration*, LXV (1923), 254-255.

[12] Henderson, "Air Conditioning and Production Efficiency," *loc.cit.*, 28-29; W. E. Lowell, "Temperature and Humidity Control in a Candy Plant," *Ice and Refrigeration*, LXXXVII (1934), 153-154.

[13] J. T. Bowen, *Refrigeration in the Handling, Processing, and Storing of Milk and Milk Products* (U. S. Department of Agriculture, *Miscellaneous Publication No. 138*) (Washington, 1932), 49-50.

[14] F. P. Siebel, Sr., "Brewery Refrigeration. History of Progress During the Last Twenty-Five Years," *Ice and Refrigeration*, CI (1941), 55-58; E. H. Vogel, Jr. et al., *The Practical Brewer, A Manual for the Brewing Industry* (St. Louis, 1946), 18-22, 82-83; J. P. Arnold and F. Penman, *History of the Brewing Industry and Brewing Science in America* (Chicago, 1933), 104. The coolship was

A great many other manufacturing activities assigned refrigeration an important place, often in connection with air conditioning. In the tobacco industry, for example, it was used to provide the proper temperature and humidity for the curing process and to preserve the yield of vintage years.[15] In diamond-cutting plants, where excess humidity slowed sawing, it proved valuable in maintaining continuous production and in augmenting capacity.[16] For many years refrigeration was not a factor in textile-mill air conditioning, where the introduction of moisture was the main objective, but with the passage of time textile firms began to install machines to reduce mill temperatures. Since close control of the thermal factor improved the operation of modern high-speed equipment, made working conditions more comfortable, and often could be included in a system without any addition to first costs, it seemed destined to play a greater role in the future.[17] In the synthetic-yarn industry heavy dependence was placed on refrigeration for regulation of temperature and humidity;[18] in printing and lithography the control it afforded was necessary for any real accuracy;[19] and in the photographic industry it was called upon to assist in both the preparation and development of film.[20] The manufacture of pharmaceutical ma-

---

a large, shallow, rectangular basin in which the hot wort was exposed in a thin layer.

[15] H. B. Matzen, "Air Conditioning as Applied to Foods and Industry," *Refrigerating Engineering*, XL (1940), 298.

[16] Henderson, "Air Conditioning Goes to War," *loc.cit.*, 155.

[17] P. L. Davidson, "Refrigeration for Textile Mill Air Conditioning," *Heating and Ventilating*, XLIV (May 1947), 57-62; P. L. Davidson and J. deB. Shepard, "Air Conditioning in Textile Mills," *Refrigerating Engineering*, LVIII (1950), 155.

[18] Henderson, "Air Conditioning and Production Efficiency," *loc.cit.*, 25-26; "Conditioning Essential in Nylon Manufacture," *Refrigerating Engineering*, XL (1940), 21; J. A. Lee, "Controlling Atmospheric Conditions in Rayon Production," *Chemical and Metallurgical Engineering*, XLI (1934), 60.

[19] Henderson, "Air Conditioning and Production Efficiency," *loc.cit.*, 34; American Association of Ice and Refrigeration, Committee on Industrial Refrigeration, "Importance of Mechanical Refrigeration in Various Industries," *Ice and Refrigeration*, LXX (1926), 585.

[20] E. T. Murphy, "The Varied Uses of Refrigeration," *Ice and Refrigeration*, LX (1921), 382; American Association of Ice and Re-

terials benefited from the refrigerating machine. In its air-conditioning function it helped to maintain uniformity of product and to handle such operations as the forming and coating of tablets, the fabrication of gelatin capsules, and the bottling of deliquescent substances.[21] It was essential in the freeze-drying process employed in the preparation of such products as blood plasma, penicillin, and ACTH, the anti-arthritic hormone.[22]

Civil engineers, like manufacturers, sought the aid of temperature control. Freezing still was the best means of checking silt and subterranean water. In the late thirties, for example, a steel test shaft was sunk at the Gilbertsville, Kentucky, dam site. Close to bedrock ground water began to force silt into the shaft. To control this, the surrounding area was frozen by sinking pipes through which chilled brine was circulated.[23] The same technique was used to hold back a silt slide that interfered with the construction of Grand Coulee Dam.[24] Refrigeration proved its worth in another aspect of dam construction, the setting of large concrete masses. A major difficulty was the heat generated by the process of hydration, which if not regulated could cause expansion and contraction and ultimately cracks that might impair the safety of the structure. To eliminate this danger, pipes through which refrigerated water could be pumped were imbedded in the concrete. In the first of the big dams emphasis was on simple heat removal, but in those built later it shifted to the

---

frigeration, Committee on Industrial Refrigeration, *loc.cit.*, 585; A. J. Hess, "Movie Labs Need Refrigeration," *Refrigerating Engineering*, LIII (1947), 507-508.

[21] Henderson, "Air Conditioning and Production Efficiency," *loc.cit.*, 35.

[22] *The New York Times*, May 7, 1950, E13; Oct. 29, 1950, E9; E. W. Flosdorf, "Freeze Drying—Its Application to Penicillin, Blood Plasma, and Orange Juice," *Refrigerating Engineering*, LIV (1947), 226; C. F. Holske, "Refrigeration for Penicillin Manufacture," *Refrigerating Engineering*, LXIX (1945), 21-25.

[23] "Freezing Prevents Seepage at TVA Test Shaft," *Construction Methods and Equipment*, XX (June 1938), 33.

[24] J. A. Newton, "Refrigeration Holds Sliding Silt in Construction of Grand Coulee Dam," *Ice and Refrigeration*, XCII (1937), 77-79.

control of temperature distribution within the concrete in a way designed to prevent the stresses that produced cracks.[25]

The applications of refrigeration were so varied that many defied systematic classification. Scientists discovered controlled cold to be an essential laboratory tool for certain types of research.[26] Aeronautical engineers used refrigerating machinery in such giant installations as the Air Force climatic hangar at Eglin Field, Florida, to test aircraft operation under stratospheric conditions.[27] Surgeons sometimes anesthetized an afflicted limb by packing it in ice. Since this greatly reduced post-operative shock, the practice won special favor when the patient was in poor condition.[28] Experiments were made in cooling even the whole body as a substitute for insulin shock in the treatment of insanity.[29]

The air-conditioning applications discussed thus far were designed primarily to furnish the optimum environment for some product or process. They were indeed important, but formed only a part of the whole picture, for many installations, with refrigeration in control of temperature and humidity, were designed mainly for human comfort. Comfort air conditioning fell into two categories: first, commercial and industrial, and second, domestic.

In the commercial and industrial field comfort conditioning, though usually adopted with profits in mind, was aimed directly at the well-being of either the employee or the customer. For employee-benefit purposes the principal market was the office, either business, professional, or governmental.

[25] C. Rawhouser, "Refrigeration Makes the Big Dams Possible," *Refrigerating Engineering*, LIV (1947), 534; H. N. Royden, "Refrigeration at Hoover Dam," *Ice and Refrigeration*, LXXXVI (1934), 86-87; W. R. Waugh, "Refrigeration in Dam Construction," *Ice and Refrigeration*, CVII (Nov. 1944), 21-23.

[26] J. T. Bowen and W. V. Hukill, "Refrigeration Has Many Uses in the Laboratory," *Refrigerating Engineering*, XL (1940), 287-291.

[27] C. W. Kniffin, "The AAF's Climatic Hangar," *Refrigerating Engineering*, LIV (1947), 128-131, 188, 190.

[28] B. M. Newman, "Shockless Surgery," *Scientific American*, CLXVI (1942), 182-183; A. R. Griffin, *Out of Carnage* (New York, 1945), 81-82.

[29] B. M. Newman, "Refrigeration for Insanity," *Scientific American*, CLXVII (1942), 160-161.

The largest system in this class was that supplied the Pentagon in Washington, where cooling was accomplished by twelve centrifugal machines with a refrigerating capacity equal to the melting of 26,400,000 pounds of ice a day.[30] Some factories and shops were equipped in order to provide a better environment for employees, while certain deep mines had to be air conditioned to make them workable.[31] The economic justification for these installations lay in their effect on worker efficiency and health, since high temperatures and humidities unquestionably had an adverse physiological effect on the human body.[32] In actual practice employers found that air conditioning did increase efficiency, but there was some doubt as to its practical value in reducing the incidence of disease.[33]

Air conditioning designed to stimulate business by providing for the comfort of patrons, though not unknown before 1917,[34] made its real beginning in the nineteen-twenties. The first large users were motion-picture exhibitors, who discovered that it enabled them to operate at a profit throughout the entire year. As early as 1925 it had made the

[30] *Ice and Refrigeration*, CIII (1942), 259.

[31] Henderson, "Air Conditioning Goes to War," *loc.cit.*, 155; Carrier Corporation, *Twenty-Five Years of Air Conditioning, 1915-1940* (n. p. [1940?]), 22.

[32] American Society of Heating and Ventilating Engineers, *op.cit.*, 222-226; "Effects of Different Temperatures on Health and Efficiency," U. S. Department of Labor, Bureau of Labor Statistics, *Monthly Labor Review*, XXXIV (1932), 813-816.

[33] The Philadelphia Electric Company reported a 45.4 per cent decrease in lost time in its air-conditioned offices, while the *Chicago Tribune* experienced a 30 per cent decrease. On the other hand a Metropolitan Life Insurance Company test revealed no measurable effect on absence due to illness, and a U. S. Public Health Service study of air conditioning in a textile plant yielded similar results. H. H. Mather, "The Economic Status of Air Conditioning," *Refrigerating Engineering*, XXXI (1936), 152-153; *Ice and Refrigeration*, XCVII (1939), 135; W. J. McConnell, H. H. Fellows, and M. G. Stephens, "Some Observations on the Health Aspects of Air Conditioning," *Refrigerating Engineering*, XXXV (1938), 23-24; R. H. Britten, J. J. Bloomfield, and J. C. Goddard, *The Health of Workers in a Textile Plant* (U. S. Treasury Department, Public Health Service, *Public Health Bulletin No. 207*) (Washington, 1933), 26.

[34] *Supra*, 182-183.

Balaban and Katz theaters in Chicago more profitable in summer than in winter.[35] About 1926 department stores began to enlist its benefits, and in 1927 a San Antonio office building was conditioned throughout.[36] The turning point came in the early thirties, when the perfection of the small refrigerating machine and the introduction of safe refrigerants made possible new applications. Furthermore, popular demand for air conditioning was undoubtedly stimulated by the Century of Progress Exposition in Chicago, where in the summers of 1933 and 1934 hundreds of thousands first experienced its benefits.[37]

The businesses that sought its advantages varied greatly in character. Retail establishments, from huge department stores to pharmacies and small specialty shops, made installations in the hope of stimulating patronage, particularly in summer. Hotels turned to the new technique in efforts to make their public rooms more attractive and to attain a higher summer guest rate. Restaurants furnished themselves with conditioning apparatus, as did such enterprises as beauty parlors.[38] The Baltimore and Ohio Railroad conditioned a dining car in 1929 and in May 1931 a complete New York-to-Washington train. Soon Pullman cars were equipped, and other roads followed the lead of the Baltimore and Ohio.[39] By 1946 there were over thirteen thousand air-conditioned cars. About thirty per cent of these depended on ice, about

[35] R. E. Cherne, "Developments in Refrigeration as Applied to Air Conditioning," *Ice and Refrigeration*, CI (1941), 29-30; "Weathermakers: Carrier Corp.," *Fortune*, XVII (April 1938), 118; *Ice and Refrigeration*, LXIX (1925), 251.

[36] W. H. Carrier, "Cooling by Conduit Saves Space," *Ice and Refrigeration*, CI (1941), 73.

[37] *Ice and Refrigeration*, LXXXV (1933), 57-60; XCVI (1939), 291.

[38] Chrysler Corporation, *A Complete Report of the Round Table Clinic on Air-Conditioning held at Dayton, Ohio, June 5th, 1941* (Dayton, 1941), 11, 13, 19-22, 27, contains information on gains actually achieved as a result of air conditioning.

[39] E. Harrington, "Air Conditioning for Comfort and Health, Past—Present—Future," *Journal of the Franklin Institute*, CCXV (1933), 660; J. A. Moyer and R. U. Fittz, *Air Conditioning*, 2nd edn. (New York, 1938), 364-365; *Ice and Refrigeration*, LXXXI (1931), 71; LXXXIII (1932), 6.

fifteen per cent on the steam-jet system, and about fifty per cent on compression machines driven either by an internal-combustion engine or by power taken directly or indirectly from the axle.[40] About thirty-five hundred motor busses were equipped by 1946,[41] while as early as 1935 air conditioning was an important factor on ships. The first vessel to have an installation for passenger comfort was probably the *Victoria* of the Lloyd-Triestino Company, which was fitted out in 1930.[42]

Domestic air conditioning made little progress. The chief factor in this branch was the room-cooler, a small, package-type system that could be installed without costly alterations and that in 1949 retailed for about four hundred dollars. Yet between 1946 and 1949 only 235,000 were sold, and many of these certainly went into such non-domestic uses as executive and professional offices.[43] A retarding factor was the cost. Although the initial investment in a room cooler was not prohibitive, operation was relatively expensive. And for conditioning an entire five-room house the cheapest installation in 1950 cost a thousand dollars. The expense might not have proved a serious obstacle had the public been thoroughly sold on the idea. But nothing like general acceptance was won.[44] By no means every householder felt the need of air conditioning, and even those who did and who decided to act often concluded that an inexpensive exhaust fan installed in the attic or an upstairs window was a satisfactory substitute.

[40] American Society of Heating and Ventilating Engineers, *op.cit.*, 869; J. V. Dobbs, "Railway Air Conditioning," *Refrigerating Engineering*, LVII (1949), 458.

[41] American Society of Heating and Ventilating Engineers, *op.cit.*, 871.

[42] Carrier Corporation, *op.cit.*, 28-29.

[43] H. L. Laube et al., " 'Our Embryonic Industry,'—A Discussion of Room Air Conditioners," *Refrigerating Engineering*, LVIII (1950), 137. For a survey of the history of room air-conditioners see R. W. Morgan, "Room Air Conditioners—Past and Present," *Refrigerating Engineering*, LVIII (1950), 34-41.

[44] "That Refrigeration Boom," *Fortune*, XXVIII (Dec. 1943), 255-256; *The New York Times*, July 23, 1950, F6. For a study of the demand for domestic air conditioning in 1938 see "The Fortune Survey," *Fortune*, XVIII (Aug. 1938), 76.

The great expectations held out for air conditioning in the nineteen-thirties did not entirely materialize. Despite the fact that it was applied to almost every type of enclosed space, real sales volume was not achieved. By 1945, however, it was felt that the task of introduction had been completed and that the industry was entering upon its period of most rapid growth.[45]

Refrigeration in all its applications was essential to our military effort in the second World War. Its use in blast furnaces increased pig-iron output and at the same time saved coke. In its absence it would have been impossible to control the tremendous heat generated in atomic piles. As a part of air-conditioning systems it facilitated munitions production. In the manufacture of explosives safety dictated its use, while in the fabrication of such vital equipment as aircraft engines, precision instruments, and photographic film it was indispensable. Without it, the blackout plant would have been impractical.[46]

Another contribution of refrigeration to the solution of logistical problems was its service in the preservation of food. This involved not only the supply of the civilian population but of the armed forces as well. Over sixty per cent of the food served men stationed in the United States was preserved by it. Troops overseas received less, but large quantities were sent abroad in refrigerated ships.[47] The Quartermaster Corps had refrigerated trailers to carry rations into battle areas,[48] while in the Pacific small refrigerated vessels much like tuna clippers were used to transport provisions from central dis-

[45] "That Refrigeration Boom," *loc.cit.*, 249-250, 252; "Weathermakers: Carrier Corp.," *Fortune*, xvii (April 1938), 87; D. W. Russell, "Air Conditioning Prepares for Rapid Expansion," *Industrial Marketing*, xxxi (June 1946), 52, 130.

[46] *Ice and Refrigeration*, cii (1942), 13-15; cxiii (Nov. 1947), 36; Henderson, "Air Conditioning Goes to War," *loc.cit.*, 50-52, 155.

[47] J. A. Hawkins, "Refrigeration—a National Defense Utility," *Ice and Refrigeration*, cii (1942), 23; V. O. Appel, "Freezing and Refrigerated Storage of Poultry and Poultry Products in the United States," *Ice and Refrigeration*, cxv (Sept. 1948), 40.

[48] See inside cover, *Ice and Refrigeration*, cvi (June 1944). In 1949 the Quartermaster Corps undertook to modernize its refrigeration equipment. "War Department is Modernizing Refrigeration Equipment," *Refrigerating Engineering*, lvii (1949), 137.

tributing points to outlying bases, where they were stored in mechanically cooled barges.[49] Thanks to refrigeration, naval personnel on sea duty had food of the highest quality.[50]

Even the combat operation of the Navy benefited. In powder magazines cooled air was essential to reduce the danger of explosions, while without it the conditions in gun turrets, fire-control stations, and plotting rooms would have been almost unbearable.[51]

[49] L. L. Westling, "New Food Ships to Aid Development of Foreign Markets," *Food Industries*, XVII (Aug. 1945), 100; *Ice and Refrigeration*, CVII (Dec. 1944), 47.

[50] *Ice and Refrigeration*, CV (1943), 28.

[51] Henderson, "Air Conditioning Goes to War," *loc.cit.*, 50, 154.

CHAPTER XX

PROSPECT AND RETROSPECT

IN 1950 the future of refrigeration could not be prophesied with certainty,[1] but it was probable that its development would proceed in the direction of trends already established. Refrigerated preservation would be extended to new products. There was good reason, to choose but one example, to believe that freezing techniques would be adapted satisfactorily to milk. Refrigeration no doubt would continue to make contributions to the well-being of producing regions. A striking demonstration of its potentialities in this respect was the work of the Tennessee Valley Authority in applying it to the problems of a large distressed rural area.

Two crucial problems faced in the twentieth century by the valley of the Tennessee—and by much of the South outside the watershed of that river—were soil erosion and the depressed economic and physical condition of the rural population.

Soil erosion in the Tennessee Valley was related intimately to the tremendous acreage of hilly land devoted to the growing of corn. Since the grain was planted in rows and required continuous cultivation, it was a poor hillside crop. Yet year after year steep slopes were seeded, and each year tons and tons of soil were washed away. Sometimes this took place almost imperceptibly by sheet erosion; sometimes disastrous gullying gutted the land. Why did farmers continue to plant corn? To a considerable extent their persistence was to be explained by their dependence on hogs for meat and by the absence of adequate markets for crops more adapted to the rugged terrain.

The poor economic condition of the Valley people was due in part to their inability to find satisfactory sources of cash

[1] For a discussion of the art of invention predicting see S. C. Gilfallen, "The Prediction of Inventions," National Resources Committee, Science Committee, Subcommittee on Technology, *Technological Trends and National Policy, Including the Social Implications of New Inventions* (Washington, 1937), 15-23.

income. They failed to capitalize on the region's potential in meat and dairy products, while their attempts to sell strawberries in glutted markets brought the inevitable low returns. Another reason for economic distress was failure to make full use of home-grown products, a condition which caused an unnecessary cash outlay when commercial substitutes were sought. Nor should waste be neglected as a causative factor, for hogs had to be slaughtered when the weather was cold enough to permit curing before spoilage set in. This not only meant waste in the form of feed for animals that matured before "hog-killing time," but also in the guise of spoiled meat should a miscalculation on the weather be made. The poor physical condition of many farm families was due to a deficient diet, which featured too little fresh fruits, vegetables, meat, and milk and too much fatback and grits.

The amelioration of these problems could be assisted by intelligent application of the techniques of refrigeration. Erosion could be checked if hilly lands were used for pasture or other crops that furnished good cover. Were locker plants and other cooling equipment available, pasture would be profitable economically, for the beef and dairy products yielded by the cattle grazed on it could be preserved.[2] Were freezing facilities available, the cultivation of strawberries, a good cover crop, would be stimulated, for the means to market them advantageously would be at hand. The economic well-being of the farm population could be advanced by refrigeration. Sales of strawberries, cattle, and dairy products would be a ready source of revenue, and home-grown foods could be used more extensively. With adequate refrigeration waste certainly could be checked, for farm animals could be slaughtered not at the dictate of the weather, but when they reached their most profitable weight. Rural dietary habits might also be changed for the better, since no longer would reliance on corn and pork be commanded by necessity.[3]

[2] Beef, it should be remembered, was not particularly palatable when cured.

[3] For discussion of the relationship between refrigeration and the solution of rural problems see C. J. Hurd, "Farm and Community

Basic to any attempt to enlist refrigeration in the cause was rural electrification. In the absence of natural-ice supplies there was no alternative to the refrigerating machine, but in 1933 only three per cent of Tennessee Valley farms had the necessary electric power. Gradually this situation was improved. By 1940 power had been extended to sixteen per cent, by 1945 to twenty-eight per cent, and by 1950 to eighty-two per cent.[4] As soon as significant numbers of farms were supplied, the TVA began its work of investigation, education, and encouragement.

Better refrigeration for the individual farmstead was one of the principal objectives. For the kitchen of the rural home the TVA encouraged the development and sale of efficient lift-top units that retailed at a maximum of eighty dollars.[5] The Authority's rate yardstick had so stimulating an effect that private utilities of the Tennessee Valley which adopted it led the rest of the country in refrigerator sales.[6] For the benefit of small dairy farms an attempt was made in cooperation with the University of Tennessee to determine refrigeration re-

---

Refrigeration in the South," *Refrigerating Engineering*, xxxvi (1938), 295-299; J. P. Ferris, "Refrigeration, Meat and the Soil," *Refrigerating Engineering*, xl (1940), 180, 182; W. R. Woolrich, "The Future of Refrigeration in the South," *Ice and Refrigeration*, xcvii (1939), 16-18; B. H. Luebke and J. R. Kelley, *Some Economic Aspects of the Frozen Fruit Industry in Tennessee* (University of Tennessee, Agricultural Experiment Station, Agricultural Economics and Rural Sociology Department, *Rural Research Series Monograph No. 133*) (Knoxville, 1941), 7-15; E. L. Carpenter and M. Tucker, *Farm and Community Refrigeration* (University of Tennessee Engineering Experiment Station, *Bulletin No. 12*) (Knoxville, 1936), 29; U. S. Department of Agriculture, Rural Electrification Administration, *Profits from Farm Power* (Washington, 1940), 34; and U. S. Department of Agriculture, Rural Electrification Administration, *Electric Power on the Farm.* . . . D. C. Coyle, ed. (Washington, 1936), 41.

[4] Tennessee Valley Authority, *Annual Report* . . . *for the Fiscal Year Ended June 30, 1950* (Washington, 1950), 37, 39.

[5] *Ice and Refrigeration*, lxxxvii (1934), 12; "Model T Appliances," *Business Week*, June 16, 1934, 11; "Small Frigidaire," *Business Week*, June 23, 1934, 8-9.

[6] D. E. Lilienthal, *TVA; Democracy on the March* (New York, 1944), 23.

quirements in the hope of helping manufacturers of machinery better to adapt their products to small-farm needs.[7]

Community refrigeration was encouraged that facilities beyond the means of a single farmer might be available. The simplest of these was the walk-in refrigerator, a cooler in which a dozen or so families could chill and store their meats, fruits, and vegetables. The TVA took the lead in constructing model units, in testing them, and in demonstrating their merits. Not content with making its findings known to manufacturers, it furnished blueprints to all who wished to construct a cooler on their own. By 1941 nineteen were in operation under independent management. Some were conducted as private businesses; others were run on a cooperative basis.[8] But for the growth of interest in locker plants the walk-in refrigerator doubtless would have won greater acceptance. As the potentialities of the locker were recognized, interest shifted in its direction. The TVA studied plants of various sizes with a view to their suitability for rural communities and sought to educate the public to their advantages. By 1950 the Valley was dotted by 250 plants serving nearly 95,000 families.[9] In addition to the encouragement of walk-in coolers and of lockers an effort was made to foster small meat-packing plants. Between 1940 and 1950 about 120 of these establishments, so dependent on refrigeration, made their appearance.[10]

No feature of the Authority's refrigeration program was pressed more persistently than its efforts to promote the freezing preservation of Valley crops. Strawberries were the first to receive attention, partly because Tennessee growers had fallen on evil days. Prices were low, for the entire harvest was thrown on the market within a period of about three weeks.

[7] W. R. Woolrich, R. B. Taylor, and M. Tucker, "Farm and Community Refrigeration in Rural Readjustment," *Refrigerating Engineering*, xxx (1935), 331, 333-334.

[8] Lilienthal, *op.cit.*, 86; H. Finer, *The T.V.A. Lessons for International Application* (Montreal, 1944), 84; Hurd, *loc.cit.*, 295, 297-298; L. N. Baker, "Experiences in the Operation of Rural Community Locker Plants," *Refrigerating Engineering*, xlii (1941), 378; Carpenter and Tucker, *op.cit.*, 21-26.

[9] Baker, *loc.cit.*, 378-379; TVA, *Annual Report, 1950*, 53.

[10] TVA, *Annual Report, 1950*, 53.

The result of depressed prices was a decline in production. In the mid-twenties Tennessee farmers grew about nine and one-half per cent of the nation's total, but ten years later they raised only six and one-half per cent. Freezing preservation, which would permit marketing over a longer period, seemed to be the answer. But there was another even more compelling reason for giving attention first to strawberries: their value as a cover crop. Their tangled roots, their low-lying leaves, and the straw laid over them helped check run-off and hold the soil in place. It was in the interest of erosion control to make strawberry raising profitable. Eventually other fruits and vegetables received attention.[11]

Of the attempts to promote freezing preservation perhaps the most interesting was the development of equipment for quick freezing by immersion of the product in a swiftly flowing stream of refrigerant. The objective in this work was a method well suited to the needs of the Tennessee Valley. High on the list of requisites stood low first cost, ability to adjust economically to large fluctuations in load, ready adaptability to different commodities, and simplicity. The TVA, cooperating with the University of Tennessee and the University of Georgia, made investigations in the agricultural, engineering, and marketing phases of the problem; it conducted laboratory experiments; it set up a pilot plant for testing and demonstration purposes. When results of the tests proved promising, the plant was leased to a farmer's cooperative for commercial operation.[12]

The possibilities of mobile freezing and storage equipment were explored. In 1938 the TVA developed a refrigerated barge that could pick up frozen foods in the producing areas and transport them to any point served by the Mississippi

[11] J. P. Ferris and R. B. Taylor, "Immersion Quick Freezing," *Ice and Refrigeration*, xcvii (1939), 177-179; *Ice and Refrigeration*, cxi (July 1946), 52-53; TVA, *Annual Report, 1950*, 53.

[12] Ferris and Taylor, *loc.cit.*, 177-179; J. G. Woodruff and J. O. Tankersley, "Freezing Fruits and Vegetables by Immersion," *Ice and Refrigeration*, c (1941), 309-310; Lilienthal, *op.cit.*, 112; Tennessee Valley Authority, *Annual Report . . . for the Fiscal Year Ended June 30, 1943* (Washington, 1943), 15.

River system. The barge idea, which was a combination storage and transport facility, was allowed to lie dormant throughout the war, but in 1945 it was revived by the building of another vessel which had the means not only for storage but for preparation and freezing as well.[13] Not long after this the assistance of TVA engineers helped make possible a privately designed and constructed mobile freezer. Consisting of a freezing and holding unit mounted on separate rail cars and of a machinery unit installed on a truck-trailer, it was designed to provide equipment to areas where short seasonal production made operation of a stationary plant unprofitable.[14]

Research in the field of freezing preservation received constant attention. The Authority not only undertook studies itself, but also paid state institutions in the Valley to do likewise. It assisted in coordinating investigation so that duplication of effort might be avoided. For all who wanted help the TVA stood ready to supply technical assistance and economic information.[15]

The efforts to promote freezing preservation brought encouraging results. In 1950 the strawberry-growing industry was clearly reviving, and thirty-eight commercial freezing plants in addition to scores of lockers were operating in the Valley.[16]

The TVA demonstrated that refrigeration could help to solve the problems of the great drainage basin that was its domain. This was not, of course, the only demonstration of the beneficent effects of refrigeration. Often before it had helped solve marketing problems and contributed to the welfare of agricultural regions, while in the thirties and forties problems

[13] Woolrich, "The Future of Refrigeration in the South," *loc.cit.*, 18; Ferris and Taylor, *loc.cit.*, 179-180; D. K. Tressler and C. P. Evers, *The Freezing Preservation of Foods*, 2nd edn. (New York, 1947), 688.

[14] C. F. Ellis, "Mobile Freezer Benefits Single Crop Areas," *Refrigerating Engineering*, LV (1948), 45-46.

[15] Tennessee Valley Authority, *Annual Report . . . for the Fiscal Year Ended June 30, 1946* (Washington, 1946), 57; *Annual Report . . . for the Fiscal Year Ended June 30, 1947* (Washington, 1947), 46-47; *Annual Report, 1950*, 53.

[16] TVA, *Annual Report, 1950*, 53.

like those of the Tennessee Valley were attacked elsewhere with similar weapons and success. But the TVA experience was particularly instructive, for it served as a laboratory and showed what could be done to ameliorate especially difficult conditions when a broad group of techniques were employed imaginatively under coordinated direction. The achievements of the TVA suggested that refrigeration might serve an important function in the regeneration of underdeveloped areas in other parts of the world.

Whatever might be the future of this dynamic technology, it was clear in 1950 that the impact of refrigeration on life in America had been far-reaching. It had helped promote a healthful, well-balanced diet throughout the year not only in city population centers but in rural districts as well. Perhaps greatest in significance was the part it had played in facilitating the urbanization of the United States. But to assert that refrigeration made possible giant cities like New York would be going too far, for the teeming metropolises of the Orient existed without it. It could be affirmed safely that city diet as known in the United States could not be enjoyed in its absence, but even this appraisal might have to be altered in the future should the techniques of dehydration be perfected. As great a boon to producers as to consumers, refrigeration was one of the determining factors in the location of production facilities. Neither the geography nor the economics of fruit and vegetable growing, of dairying, of meat packing could be explained without an understanding of its functions. It had found a place in diverse industrial processes, either directly or by assisting in the creation of a favorable environment, and in many spheres of activity it had made a positive contribution to human comfort. All this it did and much more. By the middle of the twentieth century a new and infinitely complex society had emerged in the United States. To this society the techniques of refrigeration were vital.

A BIBLIOGRAPHICAL NOTE

THE footnotes to this study constitute the best and most detailed guide that I can offer to the sources upon which it is based. But I feel impelled to make certain general bibliographical observations that may help the reader who wishes to pursue the subject further. I wish particularly to comment on bibliographical aids and on the most useful classes of material available.

Among the most valuable aids are the various indexes to periodical literature. *Poole's Index* and *Readers' Guide* are suitable for locating articles in general periodicals, but more specialized tools are necessary to exploit the vast stores of material buried in the files of trade and technical publications. For material published since 1916 that relates to agriculture, *The Agricultural Index* should be consulted. For engineering topics *The Engineering Index*, which under varying titles covers the period since 1884, is indispensable. Another useful guide is *The Industrial Arts Index*, which is devoted not only to engineering periodicals, but to trade journals as well.

Essential to the historian of refrigeration are the several indexes to publications of the United States Department of Agriculture. The first of these from the standpoint of chronology is *A General Index of the Agricultural Reports of the Patent Office, for Twenty-five Years, from 1837 to 1861; and of the Department of Agriculture, for Fifteen Years, from 1862 to 1876*, which the Department published in 1879. For the period to 1902 it is necessary to turn to *List of Publications of the Agriculture Department, 1862-1902, with Analytical Index*, which the Superintendent of Documents brought out in 1904. For the years 1901 to 1940 one is indebted to M. A. Bradley, who between 1932 and 1943 gave us four volumes under the title *Index to Publications of the United States Department of Agriculture*. For more recent material it is necessary to depend on that complete, but difficult-to-use publication of the Department of Agriculture Library, *Bibliography of Agriculture*.

A great number of bibliographies have been prepared that help to shorten the labors of the historian. Some of the most valuable of these are appended to articles in technical journals. Others are to be found in books and manuals, while still others have achieved separate publication. In the latter category an excellent general reference for the period 1915 to 1921 is American Association of Ice and Refrigeration, Committee on Papers and Lectures, *Bibliography of American Literature Relating to Refrigeration, with Synopses of Papers and Reports*, 4 v. (Chicago, 1916-[?]). For broader coverage see the bibliographies in the individual numbers of International Institute of Refrigeration, *International Bulletin of Information on Refrigeration*, which began appearing in August 1910. A guide to early publications by European learned societies which mentions a few American items is J. Bourne, "Contribution to the Bibliography of Mechanical Refrigeration," *Minutes of Proceedings of the Institution of Civil Engineers*, xxxvii (Jan. 20, 1874), 271-282. Bibliographies devoted to the role of refrigeration in the handling of farm products are exceedingly helpful. A well classified and indexed work is L. O. Bercaw, *Refrigeration and Cold Storage: A Selected List of References Covering the Years 1915-1924 and the Early Part of 1925* (U.S. Department of Agriculture Library, *Bibliographical Contributions Number 10*) (Washington, 1925). The coverage is extended to 1931 in L. O. Bercaw and E. M. Colvin, *Bibliography on the Marketing of Agricultural Products* (U.S. Department of Agriculture, *Miscellaneous Publication No. 150*) (Washington, 1932). Not to be overlooked is E. M. Colvin, *Transportation of Agricultural Products in the United States, 1920-June, 1939: A Selected List of References Relating to the Various Phases of Railway, Motor, and Water Carrier Transportation* (U.S. Department of Agriculture, Bureau of Agricultural Economics, *Agricultural Economics Bibliography No. 81*) (Washington, 1939). Another list of references that deals exclusively with the refrigeration of agricultural products is American Society of Refrigerating Engineers, Technical Committee on Agricultural Products Refrigeration, "Bibliography on Refrigeration," *Refrigerating Engineering*, xlii (1941), 247-248. The student wishing an introduction to frozen foods should consult F. S. Erdman, "A Bibliography of Literature on Frozen Foods," *Refrigerating Engineering*, xlviii (1944), 374-380, 414-430. More comprehensive is *Bibliography on Freezing Preservation of Fruits and Vegetables*, compiled by J. A. Berry and H. C. Diehl and published in 1940 by the Bureau of Agricultural Chemistry and Engineering of the Department of Agriculture. Frozen-food lockers are covered through April 1944 by D. W. Graf, *Cold Storage Lockers and Locker Plants: A List of References* (U.S. Department of Agriculture Library, *Library List No. 11*) (Washington, 1944). For a beginning in air conditioning see U.S. Department of Commerce, Bureau of Foreign and Domestic Commerce, *Bibliography of Information on Air Conditioning*, 3rd edn. (Washington, 1934).

Of the many sources of information on the history of refrigeration periodical literature is the most fruitful. This is particularly true of trade and technical journals, which have been rather generally neglected by historians. Scores of these publications are useful, but a few stand out as absolutely indispensable. There is, for example, *Ice Trade Journal* (1877-1904). Without it the full story of the ice industry in the nineteenth century could not be told. Copies of the *Journal* are now very rare; the most complete file is owned by the John Crerar Library in Chicago. More generally available and most helpful of all is *Ice and Refrigeration*, the first issue of which appeared in 1891. It tries to cover most aspects of the field, but its central interest throughout the years has been the manufacturing and merchandising of ice. One of its great services has been the publication of papers presented at the meetings of trade and technical associations. By all odds the best periodical source of technical information is *Refrigerating Engineering*, which as the *A.S.R.E. Journal* and *Transactions of the American Society of Refrigerating Engineers* dates back to 1905. The frozen-food and locker industries are very adequately treated by *Quick Frozen Foods and The Locker Plant*, which began publication in 1938. A journal of superior quality containing much information on the refrigeration of foodstuffs is *Food Industries*. It spans the years since 1928. Periodicals of general scientific scope are helpful for the nineteenth century, but in the twentieth they seldom include much that is not covered better in the technical journals. Particularly useful for the early period are *Journal of the Franklin Institute*, *Scientific American*, and *Popular Science Monthly*. Publications devoted to the examination of business trends sometimes are helpful. High in merit rank the surveys of the various phases of the refrigeration industry that *Fortune* has published from time to time. In the main the learned journals of the social sciences have neglected refrigeration and its history, though occasionally they have published articles dealing with certain of its social implications.

Attention should be directed to the value of the proceedings of trade and technical associations. The most helpful record of the work of a trade organization is the proceedings of the annual meetings of the American Warehousemen's Association. This source, which dates from 1890, is devoted to reports of committees, to papers, and to discussion. The best scientific-association proceedings are those of the International Congresses of Refrigeration, which met at intervals between 1908 and 1936 (another was held in 1951).

Research in the history of refrigeration frequently leads one to public documents, especially those of the federal government. Most important is the printed output of the Department of Agriculture. It is essential to the study of refrigerated transportation and storage, of the frozen-food industry, and of the locker plant. Two random examples of the high quality of Department publications are J. S. Larson, J. A. Mixon, and E. C. Stokes, *Marketing Frozen Foods—Facilities and Methods* (Washington, 1949) and A. W. McKay et al., "Marketing Fruits and Vegetables," U.S. Department of Agriculture, *Agriculture Yearbook, 1925* (Washington, 1926), 623-710. The Census not only contains basic statistics, but on occasion has included historical sketches and interpretive essays. Without the reports of the Federal Trade Commission's investigation of the meat-packing industry the historian would be at a loss, and he would have grave difficulties in tracing the history of the cold-storage industry without the voluminous printed records of Congressional hearings. Of the publications of state organizations the most useful are the accounts of investigations by state agricultural and engineering experiment stations.

Very little has been published in book or pamphlet form on the history of refrigeration. Except for J. A. Ewing, *The Mechanical Production of Cold*, 2nd edn. (Cambridge, 1921) and R. O. Cummings, *The American Ice Harvests; A Historical Study in Technology, 1800-1918* (Berkeley, 1949), the field is practically untouched. There are, however, a number of related secondary works that are very useful. The packing industry is well covered by R. A. Clemen, *The American Livestock and Meat Industry* (New York, 1923) and by L. Corey, *Meat and Man; A Study of Monopoly, Unionism, and Food Policy* (New York, 1950), but better than either of these on refrigerating techniques is *The Packing Industry: A Series of Lectures Given under the Joint Auspices of the School of Commerce and Administration of the University of Chicago and the American Institute of Meat Packers* (Chicago, 1924). Perhaps the best single volume on the fruit and vegetable industry is W. A. Sherman, *Merchandising Fruits and Vegetables; A New Billion Dollar Industry* (Chicago, 1928), while in the cold-storage field the distinction is held by E. A. Duddy, *The Cold-Storage Industry in the United States* (Chicago, 1929). A masterpiece in its class is T. C. Cochran, *The Pabst Brewing Company; The History of an American Business* (New York, 1948). More detailed on some technical aspects, but uncritical and unorganized, is J. P. Arnold and F. Penman, *History of the Brewing Industry and Brewing Science in America* (Chicago, 1933). The milk industry is surveyed by C. L. Roadhouse and J. L. Henderson, *The Market-Milk Industry* (New York, 1941) and an important segment of the nation's fisheries by E. A. Ackerman, *New England's Fishing Industry* (Chicago, 1941). The best secondary account of the private-car problem is L. D. H. Weld, "Private Freight Cars and American Railways" (Faculty of Political Science of Columbia University, eds., *Studies in History, Economics and Public Law*, XXXI, Number 1) (New York, 1908). An early study of considerable merit is H. Carlton, *The Frozen Food Industry* (Knoxville, 1941).

Finally, books and pamphlets which are sources of a primary nature should be mentioned. The principal items that must be grouped under this heading are the numerous manuals and handbooks published for the use of students and engineers. They are of great assistance in piecing together the story of technical development. None of these is more useful than the comprehensive work of D. K. Tressler and C. P. Evers, *The Freezing Preservation of Foods*, 2nd edn. (New York, 1947).

INDEX

ABSORBENT SUBSTANCES, 180n
ACTH, 307
Agricultural colleges, 260-261
Agricultural experiment stations,
222, 260, 294
Air circulation: in cold-storage
warehouses, 25-26, 45, 70, 127,
234, 268; in ice refrigerators, 23-
24, 209, 217; in packing-plant
chill rooms, 55, 143, 244; in refrigerator cars, 48-49, 50-52, 120-
122, 225, 227, 229, 282; in refrigerator ships, 60-61, 125, 231
Air conditioning: definition and explanation of, 179-180; technical
progress in, 180-181, 200-201;
absorption machines developed
for, 193; steam-jet machines used
in, 194; stimulates development
of small refrigerating machines,
200; makes possible new uses for
refrigeration, 302; believed ready
for rapid growth, 312
Air conditioning, applications of:
classification of, 308; early industrial, 180-181; commercial and
industrial primarily for human
comfort, 308-311; food-processing
industries, 304-305; breweries,
305; varied manufacturing industries, 306-307; domestic, 311
Air cooling for human comfort, 181-
183
Aircraft, 194-195
Air purity, in cold-storage warehouses, 234-235
Air transport, of fruits and vegetables, 256
Alaska, 22, 42
Allen, Leicester, 78
American Expeditionary Forces, 187
American Ice Company, N.Y., 109,
117
American Public Health Association,
139, 140, 296-297
American Russian Commercial Company, 22
American Society of Refrigerating
Engineers, 101, 189
American Warehousemen's Associa-
tion, 136, 188, 240
Americans: lead in refrigeration and
its application, 3-4; role in inven-
tion of refrigerating machines,
85; pioneer in publications on re-
frigeration, 101

Ammonia: role in vapor-absorption
machines, 78-79; role in vapor-
compression machines, 84; tables
on properties of, 189; use as refrigerant, 97; use on refrigerator
ships, 125, 231; see also Refrigerating machines, vapor-compression, ammonia
Ancient peoples, methods of cooling
known to, 5-6
Apples: specialized production of,
156; exports of, 162; use of cold
storage for preservation of, 162-
163; investigations in cold storage
of, 164; effect of cold storage on
production of, 165-166; trend toward storage in producing areas,
237-238; shift in rural areas to refrigerated storage of, 258; among
leaders in frozen-fruit field, 261
Apple scald, 257
Arctic Ice Machine Company, 96n
Argentina, 5, 124
Armour, J. Ogden, 147
Armour, Philip D., 51, 158-159
Armour and Company, 122-123,
144; see also Big Five and Big
Four
Armour Car Lines Company, 158-
160
Ascorbic acid, 260
Association of American Railroads,
226, 227, 250
Atchison, Topeka, and Santa Fe
Railway, 159-160
Atlantic cable, 179
Atlantic Ice and Coal Corporation,
117
Atomic piles, 305
Audiffren refrigerating machine,
101, 196
Australia: effect of refrigeration on
development of, 5; need for me-
chanical refrigeration in, 73;
early vapor-compression machines
in, 82-83; early attempt to ship
meat under refrigeration to Eng-
land from, 124
Automatic controls, 196, 200

BACON, FRANCIS, 7
Baker, Ray Stannard, 147, 160
Bakeries, 185, 269, 305
Baltimore and Ohio Railroad, 310
Bananas: brought from Caribbean

tions in, 132-133; economic limitations on speculation in, 133-
134; reasons for adverse criticism
of, 134-137; city regulation of,
137; state regulation of, 137-138,
242; U.S. Food Administration
regulation of, 187-188; proposals
for federal regulation of, 138-139,
239-240; decline in public prejudice against, 139-141, 242-243;
improvements in, 163-165, 256-
258; federal supervision and control of, 241
Coleman, Joseph J., 78, 124
Columbus Iron Works, Columbus,
Ga., 95
Commercial refrigeration, use of ice
for, 28, 54, 115, 219-220
Commercial refrigeration, machines
for, 99-100, 200, 215; *see also*
Markets
Commercial refrigeration, pipe-line
system of, 100, 200
Compressors, improvements in, 97,
190
Compressors, centrifugal, 191, 303,
305
Compressors, multiple-effect, 97
Compressors, multi-stage, 190-191
Compressors, rotary, 197
Compressors, two-stage, 190-191
Condensers: improvements in, 98,
192; location in domestic refrig-
erating machines, 197-198
Confectioneries, 28, 185, 305
Congress: considers proposals for
federal regulation of cold storage,
138-139, 239-240; does not ac-
cept FTC recommendations on
meat-packing industry, 247
Consolidated delivery companies, ice
industry, 116
Cooke, A. H., 204
Coolerator Company, 219
Cooperative marketing associations,
254
Copeland, Edmund J., 195-196
Corn, 261, 314
Countercurrent cooler, 54
Cramer, Stewart W., 180-181
Crane Brothers Manufacturing Com-
pany, Chicago, 93
Cream, frozen, 266
Cream cheese, 266
Creameries, 168-169, 267, 288
Cudahy meat-packing interests, 144;
see also Big Five and Big Four
Cullen, William, 71, 194

Curtain-wall construction of cold-
storage warehouses, 232

DAHL, NICOLAI, 202
Dairy products: strong position of
packers in distribution of, 249;
increase in per capita consump-
tion of, 272
Dairy products, processing and dis-
tribution of, 28, 30, 31-32, 63-64,
69, 166-171, 264-267
Dams, 307-308
Davidson, W. S., 124
Davis, William, 44, 48-49, 52, 62-
63
Davy, Sir Humphry, 73
Defense Plant Corporation, 237
Defrosting problem, in domestic re-
frigerating machines, 198-199
Dehumidification of air, 179-180
Dehydrated foods, 239, 285-286
Dehydration problem: in domestic
refrigerating machines, 198-199;
in cold-storage warehouses, 233-
234
De La Vergne, John C., 94, 182
Department stores, 310
Dew-point control, 180
Diamond-cutting plants, 306
Diehl, H. C., 259-260
Diesel engines, 231
Diet: early American, 8; improved
by French influence, movement
for dietary reform, and growth
of cities, 14; effect of refrigeration on, 177-178, 272; possible
effect of frozen foods on, 286; effect of locker plants on, 289, 298;
in Tennessee Valley, 315
Dimethyl ether of tri-ethylene glycol, 193
Direct connection, of electric motors
to compressors, 98n
Direct-expansion systems, 186, 232-
233
Distilleries, 186
Domestic refrigerating machines,
100-101, 195-199; *see also* Refrigerating machines, domestic
Domestic refrigeration: use of ice
for in Europe and in the U.S., 3;
increasing use of ice for, 27, 53-
54, 114-115; extent of use of ice
for, 219
Double-belt freezer, 204
Double-pipe condensers, 98
Doughs, frozen, 278
Drugs, 185

Index

Dry blast, 184, 303

Dry ice: developed on a commercial basis, 201-202; suggested as refrigerant for refrigerator cars, 225; use in refrigerator cars in frozen-food traffic, 282; use in refrigerator trucks, 230; use in distribution of ice cream, 267

Dutton, John, 74

EARLE, PARKER, 52

Eastman, Timothy C., 60

Economic distress, in Tennessee Valley, 314-315

Eggs: supplies alternate between glut and scarcity, 32-33; use of cold storage in preservation of, 64; quantity stored, 172, 268; temperatures for storage of, 233; improved methods of handling in storage, 267-268; out-of-season production of, 268

Eggs, distribution of: use of refrigeration in, 171-174, 267-269; strong position of packers in, 249

Eggs, frozen, 172-173, 269

Egg-buying program of the U.S. Government, 268-269

Ehrenbaum, 203

Electric power: greater use to drive vapor-compression machines, 97-98; in ice-making plants, 105; replaces steam as primary source of ice-plant power, 207-208; advantages in ice plants, 208; essential to acceptance of domestic refrigerating machines, 197; makes possible small refrigerating machines for commercial and industrial purposes, 200; use to drive refrigerating machinery on ships, 231

Electrification, rural: makes possible storage in producing areas, 238; makes possible use of refrigerating machines on dairy farms, 265; importance to locker plants, 294; basic to use of refrigeration in Tennessee Valley, 316

England: ice shipments to, 18; pioneer attempt to ship meat under refrigeration to, 124; early brine-immersion freezing experiments in, 202; see also Great Britain

Erie Railroad, 30

Ether, 72, 81-84

Europe: methods of cooling known in, 16th and 17th centuries, 6; method of making ice in, 17th century, 6; need for mechanical refrigeration in, 73; development of raw-water ice making in, 104; use of refrigerator cars in, 123-124; cold storage in, 130; early brine-immersion freezing experiments in, 202-203; direct fruit shipments from Pacific coast to, 255

Europeans: role in development and application of refrigeration, 4; role in invention of refrigerating machines, 85; work in reviving absorption machines, 193

Evaporative condensers, 192

Evaporators, improvements in, 98, 192-193

Evans, Oliver, 72

Everybody's, 147

Excavating, 186

Exclusive contracts of private-car lines, 160-169

Expansion valves, in domestic refrigerating machines, 197

Expansion valves, thermostatic, 192-193

Explosives, 187, 312

Express, 32, 66

FACTORIES, 309

Faraday, Michael, 73

Farm freezers, *see* Home freezers

Farmers: use of ice by, 27-28; fail to use ice for domestic refrigeration, 54, 115; use of refrigeration by, 221-223; store eggs carelessly, 171; improved handling of eggs by, 267; greatest users of locker plants and home freezers, 287; advantages of locker plants to, 289-290; constitute three-fourths of nation's locker patrons, 298; advantages of freezers for, 300

Fast-freight lines, 32

Federal Trade Commission: fears packer domination of food distribution, 145; estimates private-car earnings, 147-148; investigates meat-packing industry, 148-149; proposes measures to deal with problem of packer-owned refrigerator cars, 149; orders Coolerator Company to cease unfair claims, 219; reports on monopolistic position of Big Five, 247

Fertilizer, 143-144

Filleting, 271-272

Index

of, 156; use of cold storage for preservation of, 163; investigations in cold storage of, 164; Bureau of Plant Industry studies the transport of, 252-253; among leaders in frozen-fruit field, 261

Peddlers, in ice industry, 116
Peddler cars, 59, 148-149
Penicillin, 307
Pennington, Mary E., 173, 175, 269
Pennsylvania Railroad, 48
Pentagon, Washington, D.C., 309
Perfume, 185
Perkins, Jacob, 72-73
Peru, Illinois, 19-20
Petersen, P. W., 203
Petroleum industry, 185, 304
Pharmaceutical materials, 181, 306-307
Philadelphia, Pa.: receives ice shipments from Boston, 19; ice supply of, 20, 42; appearance of large ice firms in, 21; ice consumption in (1879-80), 53, 109, 114; refrigerating machines installed in breweries in, 90; pipe-line refrigeration in, 100, 200
Photographic industry, 185; 306
Pictet, Raoul P., 85
Pipe-line refrigeration, 100, 200
Piper, Enoch, 62-63
Poetsch, F. H., 186
Plank, 203
Polar Wave Ice and Fuel Company, St. Louis, 116-119
Pontifex, ammonia-absorption machine, 95
Pork: experiments in summer curing of, 25, 35; adaptability to preservation by curing and pickling, 33-34; packed in greater volume than beef and mutton, 34-35; see also Hogs and Hog raising
Potatoes, 163-272
Poultry: use of refrigeration in butchering and marketing of, 32-33, 64-65, 121, 174-175, 269-270; importance of quick freezing in preserving, 206; strong position of big packers in distribution of, 249; greater production of full-drawn or eviscerated, 269-270; generally more palatable when preserved by freezing, 289
Poultry, frozen, 64-65, 174, 269-270
Powell, G. Harold, 152-154, 164
Precision instruments, 181
Precooked foods, frozen, 278-279

Precooling, 153-155, 251-252
Prepackaging, by fresh-produce interests, 264
Printing and lithography, 306
Privately owned refrigerator cars: problem in meat-packing industry, 146-149, 247-248; problem in distribution of groceries, 149, 248-249; problem in fruit and vegetable industry, 158-160, 253
Producing areas: effect of refrigeration on, 30-31, 56-65, 144, 156-162, 165-166; 168, 169-171, 173-174, 176-177, 253, 263, 264, 265-266, 298, 314-320; effect of high-speed transportation on, 66-67; cold storage in, 237-238
Prohibition, 219, 235, 305
Public health: and ice, 110-113; and cold storage, 130-132; and frozen foods, 280-281
Public-utility corporations, 212
Pullman cars, 182, 310-311

Quartermaster Corps, 312-313
Quick freezing: development of techniques of, 202-206; variety in methods of, 204-205; advantages of, 205; not always necessary, 205-206; of fish, 202-204, 271-272; of meats, 245-246; of fruits, 259, 260-261; of vegetables, 259-261; of poultry, 270; equipment installed in locker plants for, 296; TVA develops immersion method of, 318; see also individual products

Railroads: early attitude on furnishing refrigerator cars, 50-51; oppose distribution of meat in refrigerator cars, 57; oppose mechanical refrigerator cars, 123, 225-226; facilitate expansion of fruit and vegetable production, 149-150; furnish refrigerator cars to fruit and vegetable shippers, 159; construct cold-storage warehouses at terminals, 235-236
Railroad cars, 182, 310-311
Rankin, Thomas L., 89
Rankine, William J. M., 73, 78
Refrigerants: development of new, 191-192; for domestic refrigerating machines, 197; new uses for refrigeration made possible by safe, 302; see also individual refrigerants

Index